T0305711

Fundamentals of Additive Manufacturing

Additive manufacturing (AM) is a manufacturing process that has emerged as a viable technology for the production of engineering components. The aspects associated with additive manufacturing, such as less material wastage, ease of manufacturing, less human involvement, fewer tool and fixture requirements, and less post-processing, make the process sustainable for industrial use. Further, this new technology has led to highly optimized product characteristics and functional aspects. This textbook introduces the basics of this new additive manufacturing technology to individuals who will be involved in the grand spectrum of manufacturing finished products.

Fundamentals of Additive Manufacturing Technology: Principles, Technologies, and Applications provides knowledge and insight into various aspects of AM and deals with the basics, categories, materials, tooling, and equipment used. It presents a classified and complete description of the most common and recently developed additive manufacturing methods with applications, solved examples, and review questions. This textbook also emphasizes the fundamentals of the process, its capabilities, typical applications, advantages, and limitations, and also discusses the challenges, needs, and general recommendations for additive manufacturing.

This fundamental textbook is written specifically for undergraduates in manufacturing, mechanical, industrial, and materials engineering disciplines for courses in manufacturing technology taught in engineering colleges and institutions all over the world. It also covers the needs of production and manufacturing engineers and technologists participating in related industries. Additionally, the textbook can be used by students in other disciplines concerned with design and manufacturing, such as automotive, biomedical, and aerospace engineering.

Fundamentals of Additive Manufacturing

Principles, Technologies, and Applications

Helmi Youssef, Hassan El-Hofy,
and Mahmoud Ahmed

CRC Press
Taylor & Francis Group
Boca Raton London New York

CRC Press is an imprint of the
Taylor & Francis Group, an **informa** business

Designed cover image: Helmi Youssef, Hassan El-Hofy and Mahmoud Ahmed

First edition published 2025
by CRC Press
2385 NW Executive Center Drive, Suite 320, Boca Raton FL 33431

and by CRC Press
4 Park Square, Milton Park, Abingdon, Oxon, OX14 4RN

CRC Press is an imprint of Taylor & Francis Group, LLC

ISBN: 978-1-032-58263-4 (hbk)
ISBN: 978-1-032-58780-6 (pbk)
ISBN: 978-1-003-45144-0 (ebk)

DOI: 10.1201/9781003451440

Typeset in Times
by Deanta Global Publishing Services, Chennai, India

Access the Instructor and Student Resources: www.Routledge.com/9781032582634

"If you are not willing to learn,

no one can help you.

If you are determined to learn,

no one can stop you."

Zig Ziglar, 1926–2012

Dedication

to our future blossoms,

H. Youssef: *To Youssef, Nour, Anoreen, Fayrouz, and Yousra*

H. El-Hofy: *To Omar, Youssef, Zaina, Hassan, Hana, Ali, Hala, Sophia, Lilian, and Celine*

M. Ahmed: *To Nada, Jana, Sara, and Ali*

Contents

Preface

Additive manufacturing (AM), unlike other manufacturing processes, being an additive process, has emerged as a viable new technology for the production of engineering components. The aspects associated with AM such as less material wastage, ease of manufacturing, less human involvement, less tool and fixture requirements, and less post-processing, makes the process sustainable for industrial use. Recently, universities have been incorporating additive manufacturing into various curricula of engineering courses. The authors have been involved in setting up programs in their home universities and have written this book because they feel that it covers this new technology in sufficient breadth and depth.

The book in hand, *Fundamentals of Additive Manufacturing: Theories, Techniques, and Applications*, is written specifically for undergraduates in mechanical, industrial, manufacturing, and materials engineering disciplines of the second to fourth levels to cover AM technology related to the courses of manufacturing technology taught in engineering colleges and institutions all over the world. It also covers the needs of production and manufacturing engineers and technologists participating in related industries. Additionally, the book can be used by students in other disciplines concerned with design and manufacturing, such as automotive, biomedical, and aerospace engineering.

The book introduces and defines the AM techniques, provides the concept and features of this new technology. and presents a historical review of AM. The software issues for AM, especially the Standard Triangle Language or STereoLithography (STL) and the concept of "tessellation", are considered. The alternative file formats to the STL format are also discussed and compared. The book presents the classification and concept of additive manufacturing techniques, according to the proposal of the ASTM, F42-Committee, which divides AM into seven categories: vat polymerization, material jetting, material extrusion, powder bed fusion, binder jetting, directed energy deposition, and sheet lamination. The basics and principles of these categories are described and discussed in a simple manner. Then, the suitable technologies for additive manufacturing of engineering materials, such as polymers, composites, metallic materials, and ceramics, are considered in detail.

The topics of design for AM, post-processing, materials for AM processes, and the impact of AM on conventional manufacturing processes and Industry 4.0 are dealt with separately. The book briefly concentrates on the major applications, including the aerospace, automotive, biomedical, electronics, and jewelry industries. The future trends of lattice structures and 4D printing are also included. Finally, challenges and needs are discussed, and recommendations are proposed.

Also, cost analysis and the environmental impact and safety hazards of AM are thoroughly considered. Finally, challenges and needs are discussed, and necessary recommendations are proposed. To enhance full understanding of the topics treated by the book, a number of case studies of additively manufactured products, extracted from different fields of industrial applications, are provided at the end of the book.

The language of additive manufacturing is one of the barriers to entry into the AM techniques. Designers, process developers, computer modelers, laser specialists, motion control technologists, and metallurgists bring their technical terms, definitions, acronyms, and technical slang to the table and do their best to communicate. These terms, definitions, and acronyms can be stringent for newcomers and users of additive manufacturing technology (AMT). In light of this, particular attention is paid to providing acronyms at the beginning of the book and a detailed index at the end of the book for reference.

The book is written in 14 chapters:

1. Introduction to Additive Manufacturing
2. Software Aspects for Additive Manufacturing
3. Basic Additive Manufacturing Techniques
4. Technologies of Additive Manufacturing of Polymers and Composites
5. Technologies of Additive Manufacturing of Metallic Materials
6. Technologies for Additive Manufacturing of Ceramics (AMC)
7. Feedstock Materials for Additive Manufacturing Processes
8. Post-Processing Techniques in AM Processes
9. Design for Additive Manufacturing
10. Impact of Additive Manufacturing on Conventional Manufacturing Processes
11. Manufacturing Cost of Additive Manufacturing
12. Environmental and Health Impacts of Additive Manufacturing
13. Fields of Application of Additive Manufacturing
14. AM Characterization, Challenges and Needs, Future Trends, and Final Recommendations
15. Additive Manufacturing Case Studies

Chapter 1 introduces and defines the additive manufacturing techniques, and provides the concept and features of this new technology. A historical review, features, the concept of rapid tooling technologies, additive manufacturing versus subtractive processes, and the advantages and challenges of additive manufacturing are presented.

Chapter 2 presents the software issues, especially the Standard Triangle Language or STereoLithography (STL) and the concept of "tessellation". The alternative file formats to the STL format are also discussed and compared.

Chapter 3 presents the classification of additive manufacturing techniques according to the ASTM F42-Committee for process categories, which includes vat polymerization, material jetting, material extrusion, powder bed fusion, binder jetting, directed energy deposition, and sheet lamination. The basics and principles of these seven categories of AM are described and discussed in a simple manner along with their related processes, using illustrative diagrams.

Chapter 4 considers in detail the suitable technologies for additive manufacturing of polymers and composites.

Chapter 5 presents additive manufacturing of metals and alloys. These processes correspond to four of the seven categories specified by ASTM F42, namely, powder bed fusion, direct energy deposition, binder jetting, and sheet lamination.

Chapter 6 focuses on different technologies for additive manufacturing of ceramics. A comparison between these technologies is presented.

Chapter 7 covers the materials for additive manufacturing processes including liquid, filament, and powder polymeric materials. It also covers powder metals and alloys, ceramic powders and laminates, and filaments, bonds, and dispersants for powder materials, powder composites, and metal wires and sheets.

Chapter 8 covers post-processing as a final step of the process, where parts receive adjustments that refine the printed part toward its usable, displayable, or marketable final form. Some of these techniques, such as cleaning, de-powdering, support removal, surface finishing, and coloring, are universal, while others, such as curing and sintering, are adopted for specific technologies. Some post-processing techniques may involve thermal treatment to eliminate residual and thermal stresses induced in the printed part to enhance its final properties before printed parts appear in their final usable, displayable, or marketable form.

Chapter 9 covers design aspects for additive manufacturing regarding geometric features and process parameters. It covers design guidelines for selective laser melting and stereolithography. Design guidelines for fused deposition modeling are also covered.

Chapter 10 presents the impact of additive manufacturing on manufacturing processes in general and on each group of the five main conventional processes, namely, casting processes, powder technology processes, metal forming processes, machining processes, and finally, welding processes. For each group, the mutual interactive relations between additive manufacturing and the process are discussed, and examples of specific additive manufacturing–assisted applications are provided.

Chapter 11 provides a comprehensive cost analysis for all the costs incurred in all steps of the additive manufacturing process, namely, pre-processing, processing, and post-processing.

Chapter 12 demonstrates potential environmental implications of additive manufacturing related to key issues including occupational health, waste, and lifecycle impact. It focuses on addressing some pertinent health and safety issues that have emerged from the use of additive manufacturing and associated materials. The exponential use of polymeric materials in the industry has caused an increase in wastage generation and disposal issues, which draws attention to the importance of green composites as well as lifecycle assessment of products and processes, which are dealt with in this chapter.

Chapter 13 briefly concentrates on the major applications, including the aerospace, automotive, biomedical, electronics, and jewelry industries. The future trends of lattice structures and 4D printing are also included.

Chapter 14 presents the challenges, needs, and future trends of additive manufacturing, and corresponding recommendations are proposed.

Chapter 15 presents twelve case studies covering additively manufactured products extracted from different fields of industrial applications. The economic advantages and disadvantages in comparison to conventional technologies are described.

WHY DID WE WRITE THE BOOK?

This book introduces the basics of this new additive manufacturing technology to individuals who will be involved in the grand spectrum of manufacturing finished products.

ADVANTAGES OF THE BOOK

1. It introduces current and new trends in additive manufacturing techniques.
2. It provides a comprehensive knowledge of and insight into various aspects of additive manufacturing technology.
3. It deals with the basics, categories, materials, tooling, and equipment used in additive manufacturing.
4. It presents the industrial applications of this important domain of manufacturing technology.
5. It provides undergraduates in mechanical, industrial, manufacturing, and materials engineering disciplines of the second to fourth levels and readers with the fundamentals and basic knowledge of this new technology.
6. The text material is presented mainly in an up-to-date and comprehensive manner.
7. It covers the fundamentals of the process, its capabilities, typical applications, advantages, and limitations.
8. It discusses the challenges, needs, and general recommendations for additive manufacturing.
9. The book is written in a simple manner that makes the topics easier for the international reader.
10. It includes more descriptive illustrations.
11. It avoids deep mathematical treatments.
12. It covers additive manufacturing materials and technologies.
13. It includes review questions and solved problems.
14. It is supplied with case studies and power point presentations (www.routledge.com/9781032582634).

H. Youssef
H. El-Hofy
M. Ahmed
Alexandria, Egypt
January, 2024

Acknowledgments

It is a pleasure to express our deep gratitude to the Late Professor E. M. Abdel-Rasoul, Mansoura University, Egypt for supplying valuable materials during the preparation of this book. The assistance of Mr. Saied Tieleb of Lord Alexandria Razor Company for his valuable Auto-CAD drawings is highly appreciated. We would like to appreciate the efforts of Dr. Khaled Youssef for his continual assistance in tackling software problems during the preparation of the manuscript. We sincerely appreciate the support, great patience, encouragement, and enthusiasm of our families during the preparation of the manuscript. Thanks, and apologies to many others, whose contributions have been overlooked.

We would like to acknowledge with thanks the dedication and continued help of the editorial board and production staff of CRC Press for their effort in ensuring that the book is accurate and as well designed as possible.

We appreciate very much the permissions from all publishers to reproduce many illustrations and tabulated data from a number of authors, as well as the courtesy of many additive manufacturing companies that provided photographs and drawings of their products to be included in this book. Their generous cooperation is a mark of sincere interest in enhancing the level of engineering education. The credits for all this great help are provided in the captions under the corresponding tables and illustrations.

Finally, CRC Press and the authors would like to thank the reviewers who revised and evaluated the first edition of this book.

About the Authors

Professor Helmi A. A. Youssef, born in August, 1938 in Alexandria, Egypt, acquired his BSc degree with honors in Production Engineering from Alexandria University, Egypt, in 1960. He then completed his scientific education in the Carolo-Wilhelmina Technical University of Braunschweig, Germany, during the period 1961–1967. In June 1964, he acquired his Dipl-Ing degree; then in December 1967, he completed his Dr-Ing degree in the domain of Nontraditional Machining. In 1968, he returned to Alexandria University in the Production Engineering Department as an assistant professor. In 1973, he was promoted to associate professor, and in 1978, to full professor. In the period 1995–1998, Professor Youssef was the chairman of the Production Engineering Department at the Alexandria University. Since 1989, he has been a member of the General Scientific Committee for Promotion of Professors and Associate Professors in the Egyptian Universities.

Based on several research and educational laboratories, which he built, Professor Youssef founded his own scientific school in both Traditional and Nontraditional Machining Technologies. In the early 1970s, he established the first Nontraditional Machining research laboratory in Alexandria University and maybe, in the whole region. Since that time, he has carried out intensive research in his fields of specialization, and supervised many PhD and MSc theses.

Between 1975 and 1998, Professor Youssef was a visiting professor in Arabic universities, such as El-Fateh University, Tripoli, the Technical University, Baghdad, King Saud University, Riyadh, and Beirut Arab University, Beirut. Beside his teaching activities in these universities, he established laboratories and supervised many MSc theses. Moreover, he was a visiting professor at different academic institutions in Egypt and abroad. In 1982, he was a visiting professor at the University of Rostock, Germany, and during the years 1997–1998, he was a visiting professor at the University of Bremen, Germany.

Professor Youssef has organized and participated in many international conferences. He has published many scientific papers in specialized journals. He has authored many books in his fields of specialization, two of which are singly authored. The first is in Arabic, titled *Nontraditional Machining Processes, Theory and Practice*, published in 2005, and another is titled *Machining of Stainless Steels and Superalloys: Traditional and Nontraditional Techniques*, published in 2016. Another two co-authored books are published by Taylor & Francis Group, CRC Press. The first is on *Machining Technology: Machine Tools and Operations*, published in 2008, while the second deals with *Manufacturing Technology: Materials, Processing, and Equipment*, published in 2012. The first book, *Machining Technology*, was published again by Taylor & Francis Group, CRC Press, 2020 in a two-volume set, namely,

Traditional Machining Technology and *Non-Traditional and Advanced Machining Technologies*, while the second, *Manufacturing Technology*, appeared in its second edition in 2023, also published by CRC Press.

Currently, Professor Youssef is an Emeritus Professor in the Production Engineering Department of Alexandria University. His ongoing work involves developing courses and conducting research in the broad domain of Manufacturing Technology. His scientific publications are mainly in the fields of theory of metal cutting, traditional and nontraditional machining, machining of difficult-to-cut materials, and material science and heat treatment technologies.

 **Professor Hassan El-Hofy** received his BSc in Production Engineering in 1976 and his MSc in 1979 from Alexandria University, Egypt. He worked as an assistant lecturer in the same department. In October 1980, he joined Aberdeen University, Scotland, UK, and began his PhD thesis with Professor J. McGeough in electrochemical discharge machining. He won the Overseas Research Student award during the course of his PhD degree, which he duly completed in 1985. He resumed his work as an Assistant Professor at Alexandria University, Egypt, in November 1985. In 1990, he was promoted to the rank of Associate Professor. He was on sabbatical as a Visiting Professor at Al-Fateh University, Tripoli, between 1989 and 1994.

In July 1994, he returned to Alexandria and was promoted to the rank of full Professor in November 1997. Between September 2000 and August 2005, he was a Professor at Qatar University, Qatar. He chaired the ABET accreditation committee for mechanical engineering program toward ABET Substantial Equivalency Recognition, which was granted to the College of Engineering programs (Qatar University, Qatar) in 2005. He received the Qatar Award and a certificate of appreciation for his role in that event.

Professor El-Hofy wrote his first book, entitled *Advanced Machining Processes: Nontraditional and Hybrid Processes*, which was published in 2005. The third edition of his second book, entitled *Fundamentals of Machining Processes— Conventional and Nonconventional Processes*, was published in November 2018 by Taylor & Francis Group, CRC Press. He also co-authored the second edition of his third book, entitled *Machining Technology: Machine Tools and Operations*, published in a two-volume set, namely, *Traditional Machining Technology* and *Non-Traditional and Advanced Machining Technologies*, by Taylor & Francis Group, CRC Press during 2020. In 2011, he released his fourth book, entitled *Manufacturing Technology—Materials, Processes, and Equipment*, which again was published by Taylor & Francis Group, CRC Press. Professor El-Hofy has published over 100 scientific and technical papers and has supervised many graduate students in the area of advanced machining technology. He serves as a consulting editor to many international journals and is a regular participant in many international conferences.

Between August 2007 and July 2010, he was the chairperson of the Department of Production Engineering at Alexandria University, Egypt. In October 2011, he was selected as the Vice Dean for Education and Student's affairs, Faculty of Engineering at Alexandria University, Egypt. Between December 2012 and February 2018, he was the Dean of the Innovative Design Engineering School at Egypt-Japan University of Science and Technology, Alexandria, Egypt. He worked as an acting Vice President of Research from December 2014 to April 2017 at Egypt-Japan University of Science and Technology, Alexandria, Egypt. From February 2018, he became the Professor of Machining Technology at the Department of Industrial and Manufacturing Engineering at Egypt-Japan University of Science and Technology, Egypt. Since December 2022, he has been an Emeritus Professor of Advanced Machining Technology in the Department of Production Engineering at Alexandria University, Egypt.

Professor Mahmoud Hamed Ahmed was born in 1947 in Alexandria, Egypt. He received a BSc degree in Production Engineering from Alexandria University, Egypt in 1970 with first-degree honors. He was assigned as an instructor in the same department, where he obtained an MSc Degree in 1973. Accordingly, he was promoted to the position of Assistant Lecturer. In 1974, he was granted a scholarship from the University of Birmingham, UK, to study for a PhD in the Department of Mechanical Engineering. He pursued his research in the field of "Shearing of Metals" till he obtained his PhD in 1978. During that period, he contributed to the teaching effort in the department as a teaching assistant for some courses.

In 1978, he returned to his homeland and resumed work at Alexandria University, Egypt, as an assistant Professor. He left for the United Arab Emirates on secondment to work for the United Arab Emirates University, UAE, over a period of five years, from 1982 to 1987. He was promoted to the position of an Associate Professor at Alexandria University, Egypt, in 1986. He was seconded again to King Abdul-Aziz University, Saudi Arabia, starting in 1997. He returned home again in 2002 to Alexandria University, Egypt, where he has been working since. Besides the long-term secondments, Professor Ahmed worked as a part-time Visiting Professor for the Arab University of Beirut, Lebanon (1980/81 and 81/82), Qatar University, Qatar (one semester, 1995), and many Egyptian universities, including The Arab Academy for Science and Technology-Alexandria, Egypt (1995–1997, 2002–2007), El-Mansoura University, Egypt (1978–1981), Kima High Institute of Technology-Aswan, Egypt (1979–1982), High Institute of Public Health-Alexandria, Egypt (1980–1982), El-Minia University, Egypt (1978–1982), and El-Menofeyia University, Egypt (1981–1982).

Alongside his career, Professor Ahmed developed and taught numerous graduate and undergraduate courses in the general fields of materials and manufacturing, to name a few: Failure Analysis, Material Selection, Finite Element Analysis,

Fracture Mechanics, Non-destructive Testing, Advanced Manufacturing Processes, Theory of Plasticity, Solid Mechanics, Die Design, Metal Forming, Metal Cutting, Nonconventional Machining, Welding Technology, Engineering Materials, and Manufacturing Technology. Professor Ahmed took part in establishing and developing laboratories in the same fields, including The Material Technology Laboratory (Alexandria University, Egypt), Material Testing, Forming Machines, CNC Machining, Metrology, and Electroplating Laboratories (United Arab Emirates University, UAE), as well as Nonconventional Machining and CNC Machining Laboratories (King Abdul-Aziz University, Saudi Arabia). Over the years, Professor Ahmed has supervised numerous MSc and PhD students, covering the areas of: Electrodischarge Machining, Failure of Welded Joints, Extrusion of Fluted Sections, Plasma Cutting, Ultrasonic Machining, Pulsed Current MIG Welding, Compression of Tubular Sections, Tube Spinning, Characterization of Engineering Materials Using Nodal Analysis, Selection of Nontraditional Machining Processes, Thermo-mechanical Rolling, Functionally Graded Metal Matrix Composites, Residual Stresses in Tube Bending, Multi-Pass Sheet Metal Spinning, Infiltration Casting of Porous Metals, and AM Assisted Manufacturing. Professor Ahmed has a good track record of publishing in numerous national/international conferences and highly ranked journals. He also contributed to the development and improvement of industrial activities in Alexandria, Egypt through consultations related to solving design and manufacturing problems, material and product inspection, failure analysis, plant lay-outs, feasibility and opportunity studies, as well as running crash and training courses in the relevant fields of interest. He is a co-author of a book titled *Manufacturing Technology: Materials, Processes and Equipment*, published by CRC (first edition 2012 and second edition 2024).

List of Acronyms

μSP	Micro-shot peening
3DP	Three-dimensional printing
3DP-BJ	3DP-binder jetting
3MF	3D manufacturing format
2PP	Two-photon photopolymerization
ABS	Acrylonitrile butadiene styrene
ACT	Advanced compaction technology
AFM	Abrasive flow machining
Al$_2$O$_3$	Alumina
ALM	Additive layered manufacturing
AM	Additive manufacturing
AM/SM	Additive/subtractive manufacturing
AMC	Additive manufacturing of ceramics
AMF	Additive manufacturing file
AMT	Additive manufacturing technology
ASA	Acrylonitrile styrene acrylate
ASTM	American society for testing and materials
AT	Additive tooling
ATZ	Alumina-toughened zirconia
BJ	Binder jetting
BJT	Binder jetting technology
BPA	Bisphenol-A
CAD	Computer-aided design
CAM	Computer-aided manufacturing
CAMCs	Carbon matrix composites
CAM-LEM	Computer-aided manufacturing of laminated engineering materials
CBAM	Composite-based additive manufacturing
CCD	Charge-coupled device
CDLP	Continuous digital light processing
CF	Carbon fiber
CFF	Composite filament fabrication
CFRPs	Carbon fiber-reinforced polymers
CIM	Ceramic injection molding
CIP	Cold isostatic pressing
CLIP	Continuous liquid interface production
CM	Conventional manufacturing
CMCs	Ceramic matrix composites
CMNCs	Ceramic matrix nanocomposites
CMP	Chemical mechanical polishing
CNC	Computer numerically controlled
CNT	Carbon nano tube

CODE	Ceramic on demand extrusion
CP	Calcium phosphates/cavitation peening
CRS	Compressive residual stress
CS	Continuous stream
CT	Computed tomography
CVD	Chemical vapor deposition
DDM	Direct digital manufacturing
DED	Directed energy deposition
DfAM	Design for Additive Manufacturing
DFE	Design for environment
DIP	Direct ink printing
DIW	Direct ink writing
DLP	Digital light processing
DM	Desktop manufacturing
DMD	Digital micromirror device
DMLM	Direct metal laser melting
DMLS	Direct metal laser sintering
DOD	Drop-on-demand
DSLS	Direct selective laser sintering
EB	Electron beam
EBAM	Electron beam additive manufacturing
EBF3	Electron beam free form fabrication or
EBM	Electron beam machining/melting
EDM	Electrodischarge machining
EDP	Environmentally degradable plastic
EFF	Extrusion free forming
EISS	Environmental impact scoring systems
ENISE	Ecole nationale d'ingenieurs de saint-etienne
EOS	Electrical overstress
ESD	Electrostatic discharge
EU	European union
FDA	Food and drug administration
FDC	Fused deposition of ceramics
FDM	Fusion deposition modeling
FEA	Finite element analysis
FFF	Fused filament fabrication/freeform fabrication
FGM	Functionally graded materials
FRAM	Fiber reinforced additive manufacturing
GDP	Gross domestic products
GMAW	Gas metal arc welding
GW	Grinding wheel
HA	Hydroxyapatite
HAP	Hydroxyapatite
HDDA	Hexanediol diacrylate
HAZ	Heat-affected zone

HDPE	High density polyethylene
HDT	High heat deflection temperature
HIP	Hot isostatic pressing
HIPS	High-impact polystyrene
HS-LMD	High speed laser metal deposition
HSS	High speed sintering
ICP	Inductively coupled plasma
ICVD	Initiated chemical vapor deposition
IJ3DP	Ink-jet 3D printing
IJP	Inkjet printing
ILT	Institute for laser technology
IPA	Isopropyl alcohol
IPMCs	Ionic polymer–metal composites
LB	Laser beam
LCA	Life cycle assessment (life cycle analysis)
LCAs	Life cycle analyses
LCM	Lithography ceramic manufacturing
LDMD	Laser direct metal deposition
LDPE	HDPE
LENS	Laser engineered net shaping
LM	Layered manufacturing/Laser melting
LMD	Laser metal deposition
LMW	Low molecular weight
LOM	Laminated object manufacturing
LP	Laser polishing
LPBF	Laser powder bed fusion
LPS	liquid-phase sintering
LS	Laser sintering
LSP	Laser shock peening
ME	Material extrusion
MDF	Medium density fiberboard
MIM	Metal injection molding
MIG	Metal inert gas
MIT	Massachusetts institute of technology
MJ	Material jetting
MJF	Multi jet fusion
MIG	Metal inert gas
MLS	Metal laser sintering
MMCs	Metal matrix composites
MRI	Magnetic resonance imaging
NASA	National aeronautics and space administration.
NDT	Non-destructive testing
NIAR	National institute for aviation research
NIR	Near infrared ray
NNS	Near net shaped

NPJ	Nanoparticle jetting
OBJ	Object file format
ODM	On demand manufacturing
OEMs	Original equipment manufacturers.
OJP	Oil jet peening
PA	Aliphatic polyamides (known as nylon)/plasma atomization
PAA	Polyacrylic acid
PAEK	Polyaryletherketones
PAM	Plasma-arc technology
PBF	Powder bed fusion
PC	Polycarbonate
PCB	Printed circuit board
PCPs	Preceramic polymers
PDA	Polydopamine
PDCs	Polymer-derived ceramics
PE	Polyethylene
PEEK	Poly ether ether ketone
PEI	Polyetherimide/polyethylene terephthalate
PEKK	Poly ether ketone ketone
PES	Polyether sulphone
PET	Polyether terephthalate
PETG	Polyethylene terephthalate glycol
PLA	Polylactic acid
PLCs	Programmable logic controllers
PLS	polymer laser sintering
PLY	Polygon file format
PM	Particulate matter
PMMA	Polymethyl methacrylate
PMCs	Polymer matrix composites
PNG	Portable network graphics
PP	Polypropylene
PPE	Personal protective equipment
PPFDA	**Ppoly(1H,1H,2H,2H-perfluorodecyl acrylate)**
PPS	Polypheylene sulphide
PPSF/PPSU	Polyphenylsulfone
PBT	Polybutylene terephthalate
PREP	Plasma rotating electrode method
PS	Polystyrenes
PSD	Particle size distribution
PSL	Plastic sheet lamination
PTFE	Polytetrafluoroethylene
PU	Polyurethane
PVA	Polyvinyl alcohol
PVB	Polyvinyl butyral
PZT	Lead zirconate titanate

RC	Robocasting
RF	Radio frequency
RI	Refractive index
RIC	Rapid investment casting
RM	Rapid manufacturing
ROI	Return on investment
RP	Rapid prototyping
RT	Rapid tooling
SBC	Solid-base curing
SBMF	Sheet-bulk metal forming
S-BJ	Slurry binder jetting
SDL	Selective deposition lamination
SEM	Scanning electron microscopy
SETAC	Society of environmental toxicology and chemistry
SFF	Solid-form fabrication
SGC	Solid ground curing
SHS	Selective heat sintering
SiO2	Silica
S-iSLS	Slurry indirect selective laser sintering
SL	Sheet lamination
SLA	Stereolithography
SLC	Stereolithography of ceramics
SLCOM	Selective lamination composite object manufacturing
SLM	Selective laser melting
SLS	Selective laser sintering
SMP	Shape memory polymer
SP	Shot peening
STEMs	Science, technology, engineering, and mathematics
STL	Standard tessellation language or stereolithography
TCP	Tricalcium phosphate
TIG	Tungsten inert gas
TiO2	Titanium oxide
TPE	Thermoplastic elastomers
TPMS	Triply periodic minimal surface
TPU	Thermoplastic poly urethane
TRSs	Tensile residual stresses
TTCP	Tetracalcium phosphate
UAM	Ultrasonic additive manufacturing
UC	Ultrasonic consolidation
UCAF	Ultrasonic cavitation abrasive finishing
UFP	Ultrafine particulate
UHS	Ultrafast high-temperature sintering
UHTCs	Ultra-high-temperature ceramics
UNEP	United nations environment program
USP	Ultrasonic shot peening

UV	Ultraviolet
VIGA	Vacuum induction melting inert gas atomization
VIM	Vacuum induction melting
VOC	Volatile organic compound
VP	Vat photopolymerization
WAAM	Wire and arc additive manufacture
WJC	Waterjet cutting
WJP	Waterjet peening
XML	Extended markup language
ZTA	Zirconia-toughened alumina

Symbols

Symbols	Description	Unit
α	Angular deviation or angular tolerance	degrees
A	Cost of testing one unit	$/unit
Ai	Area of layer i	mm²
As	Surface area	m²
B	Cost to repair or replace a single unit	$/unit
Cd	Cured depth	mm
CE	Cost of energy consumed by the AM system	$
CE3	Energy cost	$
CI	Machine purchase value	$
CL1	Labor cost in pre-processing	$
CL2	Labor cost in processing	$
CL3	Labor cost in post-processing	$
Cm	Material cost required for the build	$
CM	Machine cost	$
Cm1	Build material cost rate	$/kg
Cm2	Support structure material cost rate	$/kg
Cm3	Material cost in post-processing	$
CM3	Machine cost in post-processing	$
Cm3	Cost of post-processing materials	$
Cmi	Cost rate of the i-th material	$/kg
COH1	Overhead cost in pre-processing	$
COH2	Overhead cost in processing	$
COH3	Overhead cost in post-processing	$
CPC	Processing control cost	$
CT	Inspection cost	$
D	Spot size diameter of the laser beam at the surface	mm
D	Diameter of the particle	mm
Dp	Penetration depth	mm
E	Energy consumed by AM system	kWh
E	Provided energy	kWh
Ec	Minimum energy of polymerization of the monomer	kWh
Ek	Energy consumed by the machine k	kW
ER	Energy cost rate	$/kWh
ERk	Energy cost rate of the machine k in post-processing	$/kWh
H1	Overhead cost rate in pre-processing	$/h
H2	Overhead cost rate in processing	$/h
H3	Overhead cost rate in post-processing	$/h
K_{sf}	The shape factor index	–
M	Machine maintenance cost	$
Mj	Cost of utilizing machine j	$

N	Total number of units	–
n	Number of layers to finish the product/sample number	–
n_t	Normal to the triangle	–
O1	Cost rate of pre-processing operator	$/h
O2	Operator cost rate in processing	$/h
Oq	Operator q cost rate in post-processing	$/h
p	Probability of a nonconforming unit	–
Q	Probability that the sample is accepted	–
S	Scrap value	$
T	Useful life of the AM machine	h
Tb	Built time	h
Tc	Build cycle time	s
Ti	Time to complete layer, defined by the subscript i	s
T_j	Time machine j used in post-processing	h
Tk	Time used by the machine k	h
Tpp	Post-processing time	h
Tprep	Time required to prepare the build	h
Tpro	Time required for doing tasks	h
Tq	Time required for operator q to complete the task in post-processing	h
Tr	Repositioning time between layers	s
Tsb	Time required to set up and build	h
v	Average scanning speed of the laser beam at the surface	mm/s
V	Volume	m³
v1, *v2*, *v3*	Vertices of the triangle	–
W1	Weight of material used for the build	kg
W2	Weight of material used for the support structure	kg
Wc	Cured width	mm
Wi	Weight of the i-th material	kg

1 Introduction to Additive Manufacturing

1.1 BACKGROUND

There are three fundamental fabrication methods. The first way to manufacture something is the well-established "subtractive manufacturing", such as milling or turning, where we begin from raw material and proceed toward the desired product. The next type of manufacturing is "formative manufacturing", such as casting, where the material in solid form is melted into liquid form and the liquid metal is then poured into a specific mold to obtain the object, or forging, in which a block of metal undergoes changes in its dimensions when force is applied. The third type of technology of making things is additive manufacturing (AM), which provides the third supporting pillar of the entire manufacturing technology (Figure 1.1).

The ASTM International Committee F42 on Additive Manufacturing (AM) Technologies (2009) defined AM as the "process of joining materials to make objects from three-dimensional (3D) model data, usually layer by layer, as opposed to subtractive manufacturing methodologies". AM's synonyms include rapid manufacturing (RM), rapid prototyping (RP), layered manufacturing (LM), additive layer manufacturing (ALM), desktop manufacturing (DM), additive tooling (AT), freeform fabrication (FFF), solid-form fabrication (SFF), and direct digital manufacturing (DDM). It will be interesting to watch how the terminology develops in the future. It should be noted that in the literature, most of the terms introduced above are interchangeable. Although each of the names is perfect from the special viewpoint of its creator, many of them cause confusion. Often, this is one reason why newcomers to the field of AM sometimes feel lost.

Conceptually, AM is an approach whereby 3D designs can be built directly from a computer-aided design (CAD) file without part-specific tools or dies. In this freeform layer-wise fabrication, multiple layers are built in the X-Y direction, one on top of the other, generating the Z or third direction and thus producing the 3D design. Once the part is built, it can be used for touch and feel concept models, tested for functional prototypes, or directly used in practice. With new developments in AM, we live in an age on the cusp of industrialized RM taking over as a process to produce many products and making it possible to design and create new ones. The integration of AM participates in allowing people to contribute to the design process from almost any location and will break down the barriers of localized engineering, taking it to global scale. Just as the Internet gave us the ability to spread and access information from any location, CAD and digital designing gave people the ability to make changes

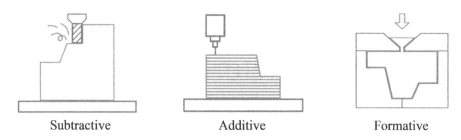

| Subtractive | Additive | Formative |

FIGURE 1.1 Three types of fundamental fabrication processes. (Adapted from Core77 Inco. https://www.core77.com/)

and critique designs from anywhere with very little lead time (Bandyopadhyay and Bose 2016).

Unlike subtractive methods that fabricate products by removing materials from a larger stock or metal sheet, AM creates the final shape by adding materials. It has the ability to make efficient use of raw materials and produce minimal waste while reaching satisfactory geometric accuracy. Using AM, a design in the form of a computerized 3D solid model can be directly transformed into a finished product without the use of additional fixtures and cutting tools. This opens up the possibility of producing parts with complex geometry that are difficult to obtain using material removal (subtractive) processes.

In other words, AM (or RP) is a family of fabrication methods to fabricate engineering prototypes (models, parts, or tools) in the minimum possible lead time based on a CAD model of the item. These methods are generally similar to each other in adding and bonding materials in a layered fashion to form an object. While the sizes of items dealt with in AM range from microscopic size to entire buildings, materials range from paper and plastics to composites, metals, and ceramics. One of the main advantages of AM is its ability to produce complex near net shaped (NNS) parts in difficult-to-machine materials. The traditional method of fabricating a prototype part is molding, forming, or machining, which can require significant lead times (up to several weeks and sometimes longer), depending on part complexity, difficulty in ordering materials, and setting up the production equipment.

The applications of AM include aerospace industries, automotive industries, toys, the oil and gas industry, the marine industry, and prosthetic devices.

1.2 HISTORICAL REVIEW

In this regard, three periods are distinguished, namely:

1. Chuck Starting Period—3D Printing

The development of AM technology was started in the mid-1980s by Charles Chuck, who invented the first form of 3D printing, called stereolithography (SLA).

SLA is a system in which an ultraviolet (UV) light source is focused down into a UV-photocurable liquid polymer bath or vat. Upon contact, the polymer hardens or cures. After a layer of printing is done, the hardened polymer layer moves down on a build plate in the liquid medium, and the next layer of polymer is deposited on the top of the previous one. This process continues until the part is finished, based on the CAD design, and is removed from the liquid medium. In most cases, further curing is needed before the part can be touched. In 1986, Chuck founded the first company, 3D Systems (2014), to develop and manufacture 3D printers. This was the first historical step toward making a RP machine. Chuck was also the first person to find a way to allow a CAD file to communicate with the RP system to build computer-modeled parts. In his method, 3D CAD models had to be sliced; each slice could then be used to build a layer using the 3D printer. The model data, usually in Standard Tessellation Language (STL) format, is first broken down into a series of 2D, very thin cross-sections, which are then fed into an AM machine that adds material layer by layer to fabricate the 3D physical part. After the development of this technology, the patent was approved by the Patent and Trademark Office as the first patent for a RP system (Hull, 1986).

2. Second Period—Development of Other RP Technologies

Simultaneously, other innovators started to develop new types of AM machines that used different methods and materials. Down and his group at the University of Texas started work on a new technology known as selective laser sintering (SLS). SLS operates by having the powdered form of a material spread on a build plate, where a laser selectively sinters the powder in certain areas of the plate. Another layer of powder is distributed on the previous layer, and the process is repeated to complete the 3D part. Dekard and Beaman finished the work on this technology and made the first SLS machine in 1986. They commercialized this technology, creating the first SLS Company, called Nova Automation, which later became DTH Corp., until the company was acquired by 3D Systems in 2001.

Around the same time, Scott and his wife Lisa Crump, two graduates of Washington State University, were developing another AM technology, referred to as fusion deposition modeling (FDM). This technology involves heating a thermoplastic to a semiliquid state, which is deposited onto a substrate, where it builds the part layer by layer. Scott and Lisa founded a company, Stratasys Inc., in 1989, which has continued to grow, and now has many printers that cost from $2000 to $60,000 and has had many patents granted.

Meanwhile, Roy Sanders was developing a new RP method. His company, previously known as Sanders Prototype Inc., now named Solidscape, released its first 3D printer, called the ModelMaker 6Pro, in 1994 (Solidscape 2013). This used an inkjet approach to build a part. This technique acts in the same way as SLA, but instead, hot thermoplastic wax liquid is sprayed onto a plate to build each layer of a part. This technique could be used to make high-resolution wax models for complex investment casting in the jewelry industry. This company had commercial success and was bought by Stratasys, Inc. in 2011.

Once 3D Systems patented their 3D printing technology, SLA companies in other countries started to develop this technology as well. In Japan, two companies, NTT Data CMET and Sony/D-MEC, developed SLA systems in 1988 and 1989, respectively. Also, companies in Europe, such as Electro Optical Systems (EOS) and Quadrax, developed SLA systems in 1990 (Wohlers and Gornet 2011). This technology has sparked interest, and many other companies around the world have started to develop their own systems and come up with new ways to do so.

3. THIRD PERIOD—STEPPING TO AM

At the beginning of this period, most of the technologies were intended to make polymeric objects and were not able to fabricate other materials such as metals, alloys, and ceramics. EOS (H. Langer and H. Steinbichler) launched the first direct metal laser sintering (DMLS) machine in 1995. This process essentially works in the same way as SLS but can sinter metal powders such as Al, Co, Ni, stainless steel, and Ti alloys. In 1997, EOS sold their SLA product line to 3D Systems and took over the global patent rights for laser sintering technology. Since then, they have significantly developed SLS and DMLS and made them the most popular AM processes in the world.

Meanwhile, another type of AM technology that could produce metal parts was being developed in New Mexico: laser engineered net shaping (LENS). The first machine was developed by Sandia National Laboratories, commercialized by Optomec, and sold in 1998 (Optomec 2014). LENS has delivered systems to over 150 customers as of 2012.

A company called Arcam started in 1997, creating a new technology (Arcam History 1997), which was electron beam melting (EBM). EBM works by shooting an electron beam at a powder bed. Once a layer of powder has been melted in selected areas, another layer of powder is laid on top of the previous one, and the process is repeated until the part is complete. Since 2007, more orthopedic implants have been made using EBM technology. EBM is also used in the aerospace industry and in many other applications, which will be presented in detail later.

In conclusion, it can be said that since its inception in the mid-1980s, AM has evolved and blossomed into a plethora of processes, including stereolithography (SLA), fused deposition modeling (FDM), 3D printing (3DP), laminated object manufacturing (LOM), selective laser sintering (SLS), selective laser melting (SLM), laser metal deposition (LMD), inkjet printing, and many others. As AM becomes a promising technology, its impact continues to grow in terms of the total number of fabricated parts, number of new start-up companies, funding opportunities, number of publications and patents, and public awareness.

1.3 FEATURES OF ADDITIVE MANUFACTURING TECHNOLOGY

1.3.1 TECHNOLOGIES OF ADDITIVE MANUFACTURING

As previously explained, AM is a relatively new catchall term for the latest techniques for making things, starting with nothing, building up material layer by layer directly from 3D data. It is an industry standard term according to ASTM F2792 for all techniques that use this process.

There are seven types of AM techniques, all of which are quite different but in most cases achieve what was impossible to produce using traditional methods. The reason they can achieve what was previously impossible is that they build up parts/components/products/things a layer at a time. Each layer is joined to the last by way of a binder, melting, or curing. These technologies, which will be detailed in Chapter 3, can be summarized as:

1. **Material extrusion**—commonly known as FDM or fused deposition modeling, pioneered by Stratasys. This takes a filament of material and pulls it through a heated nozzle to the point of melting. It is then deposited where it is needed.
2. **Binder jetting**—this is the technology that gave 3D printing its name, since it uses inkjet printer heads that selectively bind powder together. Again, the product is built up one layer at a time, with another layer of powder being spread across the top each time. Any powder not touched by the binder is shaken, blown, or brushed off the finished part.
3. **Material jetting**—again, this uses inkjet technology, but this time using material to selectively deposit where needed.
4. **Powder bed fusion**—this is exactly the same as binder jetting, except that instead of a binder being used to fuse the powder granules together, a laser or electron beam melts them together.
5. **Directed energy deposition**—this is similar to FDM but using a metal wire instead of a polymer filament. It can also use a powder fired into place. The powder or wire is then melted and welded onto the previous layer or surface by an electron beam or laser.
6. **Vat photopolymerization**—as the name suggests, this starts with a vat (tank) and a photopolymer (liquid resin that hardens when exposed to a light source). The top layer of liquid is then cured via a light source (DLP) or laser (SLA) and then moves up out of the liquid, or down into the liquid (dependent on the process), for the next layer to be processed.
7. **Sheet lamination**—strictly speaking, this technology is subtractive, but since it builds layer by layer, it slips in as an additive process. It works using a sheet of paper or plastic (or any sheet material), which is cut by a laser or scalpel into shape, and the next sheet layer is cut to size and shape and stuck down on top of it.

Subtractive technology is very easy to define; however, AM is slightly more complex due to the number of different technologies and processes, but it is a relatively straightforward concept.

Finally, the term "3D printing" is often used interchangeably with "additive manufacturing", as mentioned earlier. It comes from binder jetting using 2D printing technology. However, unlike fused deposition modeling (FDM), which was trademarked by Stratasys, the term "3D printing" was never protected. Many experts in the field have tried to differentiate 3D printing from AM by applying their own definitions. For instance, some people say that 3D printing relates to a one-off item manufactured using an additive process, whereas AM leans more toward items produced by a series production method. 3D printing is much easier and quicker to say, and when we have to say it, write it, and type it as often as we do, 3D printing or AM will be just fine. Both 3D printing and AM reflect that the technologies share the theme of sequential-layer material addition and joining through a 3D process.

1.3.2 INTEGRATION OF AM WITH ASSOCIATED TECHNOLOGIES

Currently, the direct fabrication of functional end-use products is becoming the main trend of AM technology, and it has been increasingly implemented to manufacture parts in small or medium quantities. It is important to understand that AM was not developed in isolation from other related technologies. For example, it would not be possible for AM to exist without innovations in areas like 3D graphics and CAD software. Just as 3D CAD is becoming What You See Is What You Get (WYSIWYG), it is the same with AM, and we might just as easily say that What You See Is What You Build (WYSIWYB) (Gibson et al. 2015). Aside from computer technology, there are a number of other technologies that have been developed along with AM that are worthy of note here, since they have served to contribute to further improvement of AM systems. These related technologies are (Gibson et al. 2015):

1. *Lasers:* Many of the earliest AM systems were based on laser technology. The reasons are that lasers provide a high-intensity and highly collimated beam of energy that can be moved very quickly in a controlled manner with the use of directional mirrors. Since AM requires the material in each layer to be solidified or joined in a selective manner, lasers are ideal candidates for use. There are two kinds of laser processing used in AM: curing and heating. With photopolymer resins, the requirement is for laser energy of a specific frequency that will cause the liquid resin to solidify, or "cure". Usually, this laser is in the ultraviolet (UV) range, but other frequencies can be used for heating requirements to provide sufficient thermal energy to cut through a layer of solid material, to cause powder to melt or sheets of material to fuse. For powder processes, for example, the key is to melt the material in a controlled fashion without creating too great a build-up of heat, so that when the laser energy is removed, the molten material rapidly solidifies again.

2. *Printing Technologies:* Inkjet or droplet printing technology has rapidly developed in recent years. Improvements in resolution and reduction in costs has meant that high-resolution printing, often with multiple colors, is available as part of our everyday lives. Initially, colored inks were low in viscosity and fed into the printheads at ambient temperatures. Now, it is possible to generate much higher pressures within the droplet formation chamber, so that materials with much higher viscosity and even molten materials can be printed. This means that droplet deposition can now be

used to print photocurable and molten resins as well as binders for powder systems.

3. *Materials:* Earlier AM technologies were dealing with materials that were already available. However, these original materials were far from ideal for anticipated new applications. For example, the early photocurable resins resulted in AM models that were brittle and warped easily. Powders degraded quickly, and many of the materials used resulted in parts that were quite weak. AM technology is capable of processing a wide variety of functional materials, including structural and conductive metals, ceramics, conductive adhesives, dielectrics, semiconductors, biological materials, and many other materials used to fabricate aerospace, medical, and consumer electronic devices. Materials can even be blended during printing to create new alloys or gradient structures for improved product performance. Parts are now much more accurate, stronger, and longer lasting, and it is even possible to process metals and alloys with some AM technologies. These new materials have resulted in producing even higher-temperature materials and smaller feature sizes.

4. *Programmable Logic Controllers (PLCs):* The input CAD models for AM are large data files generated using standard computer technology. Once they are on the AM machine, however, these files are reduced to a series of process stages that require sensor input and signaling of actuators. Industrial microcontroller systems form the basis of PLCs, which are used to reliably control industrial processes.

5. *Computer Numerically Controlled (CNC):* One of the reasons AM technology was originally developed was because CNC technology was not able to produce satisfactory output within the required time frames. CNC machining was slow, cumbersome, and difficult to operate. AM technology, on the other hand, was quite easy to set up with quick results, but had poor accuracy and limited material capability. As improvements in AM technologies came about, vendors of CNC machining technology realized that there was now growing competition. For geometries that can be machined using a single set-up orientation, CNC machining is often the fastest, most cost-effective method. For parts with complex geometries or parts that require a large proportion of the overall material volume to be machined away as scrap, AM can be used to produce the part more quickly and economically than when using CNC.

1.3.3 Rapid Tooling through Additive Manufacturing

Rapid tooling (RT) is well on its way to becoming the industry standard in many application areas. There is hardly any other area in production that is as sensitive as tool and mold making. Molding tools for presses, punches, injection molding, and die casting machines are the most expensive components in any series production to date. Often, the price of the tool even exceeds the acquisition costs of the machine on which they are used. The reason is that tools are the decisive factor in determining the quality of the products manufactured on them. Until now, the manufacturing processes for production tools have been correspondingly costly. CNC milling machines, eroding machines, and precision grinding machines are also expensive machines that can also only be operated and maintained by specialists. In this sensitive area, AM using RT, which has so far been subject to reservations, is now making inroads. As a manufacturing process for tool and mold making, RT has been introduced in industry for only a short time.

RT can be understood as "fast tool making". The "rapid" in this context usually means "3D printing". Admittedly, modern CNC milling machines with amazing production speeds are also available. Nevertheless, RT denotes manufacturing on a slim timeline as compared with the CNC technology.

RT through AM is intended to produce injection molding tools, testable mold models, deep-drawing tools for thin-film plastic parts, pressing tools for soft and hard sheet materials, and if necessary, also punching tools for thin materials with low tensile strength such as paper, plastic, medium density fiberboard (MDF), and veneer wood.

In addition to the significantly faster and cheaper manufacturing processes for the molded parts, RT in 3D printing offers another unique advantage. The 3D printing process makes it possible to specifically incorporate cooling channels into the tool that are not traditional manufacturing processes in this form. This makes it feasible that with RT, we can incorporate lines for coolant in any cross-sections, radii, and positions. With the previous manufacturing processes for tools, this was only possible by setting with targeted bores. However, a hole dictates its linear orientation and cross section over its entire length. The "printed" cooling channels in RT, however, can be precisely optimized for optimum heat dissipation. In practice, therefore, a trend is gradually gaining ground; inexpensive and thermally optimized tools from RT are gradually replacing channels and cavities made traditionally.

RT is now well on the way to becoming the new standard for processes with a medium technical load, such as injection molding of plastics. The advantages outweigh the disadvantages to such an extent that it would be very surprising if the old processes were still being used in ten years' time. For example, a mold made by RT costs only about 1/100 of what a traditionally manufactured injection mold made of aluminum costs. Therefore, it does not matter that this approach to tool and mold making allows a lower number of cycles through RT. Once the design is complete, a new tool can be produced again within a few hours using RT and costs only a few hundred dollars.

1.3.3.1 Rapid Tooling Technologies

RT generally describes the term "fast tool making". RT through AM or through 3D printing is the exact term for this new area of industrial production. As with all 3D printing, RT primarily requires a computer workstation and a 3D printer. The computer is not only used to develop the model for the prototype. It is also used to precisely design the tool and provide the print files (usually in STL format). In addition to the usual 3D design programs, this may require a variety of other software.

Simulation programs are particularly important for injection molding processes. Temperature distribution in the component and mold, as well as stress distributions in the mold during the production process, can be displayed with the help of these programs. The mold and the production parameters can thus be precisely designed for an optimal result. In plastic injection molding, this applies to the cooling channels, among other things. With their help, the solidification of the injected plastic can be precisely controlled. For example, different surface qualities can be produced during the production process. Lids of housings are often ordered with a high gloss on the visible surface, while the non-visible surface can remain rough. This can only be produced by RT.

RT is not exclusively limited to 3D printing with plastic filaments or resins. AM is now possible with virtually any material. Metal printing in particular has seen tremendous progress in recent years. RT with metal printing produces resilient tools that can partially compensate for the disadvantage of low durability under full load. RT with metal powder produces sintered structures. These can be equipped with any geometries—especially the practical internal cooling channels. However, RT is still a very young technology that will continue to provide innovative breakthroughs in the future.

A suitable additive process for RT is SLA printing. This process uses a synthetic resin that is bombarded with a UV laser. Of all the 3D printing processes, SLA printing offers the smoothest surface. RT by filament printing usually still requires a slight touch-up to smooth the contours. As an alternative to SLA printing, laser-sinter printing (SLS) with plastic powder can also be used for RT. However, the FDM printers in industrial use are compelling because of their very favorable costs for equipment and raw materials. Today, they offer sufficient quality that RT can also be implemented well using them.

1.3.3.2 Steps of RT

RT involves a set of processes aimed at quickly and efficiently producing tools, molds, or dies, primarily using AM technologies. Here are the typical steps involved in RT:

1. **Design Phase:**
   Conceptualization and Designing: This phase involves conceptualizing the tool requirements and creating a digital model using CAD software. Engineers design the tool based on the specifications and requirements of the intended application.
2. **Material Selection:**
   Choosing the appropriate material for the tool is crucial. It depends on factors like the intended application, durability, temperature resistance, and compatibility with the production materials.
3. **AM Process Selection:**
   Selecting the most suitable AM process for creating the tool. Common AM techniques used in RT include selective laser sintering (SLS), SLA, FDM, and DMLS, among others.

4. **Tool Production:**

   Slicing and Printing: The digital model is sliced into layers (if required), and the 3D printer builds the tool layer by layer according to the design. This AM process forms the physical tool.

5. **Post-Processing:**

   After printing, the tool might undergo various post-processing steps:
   - **Surface Smoothing or Polishing:** Smoothing rough surfaces or improving the aesthetic finish.
   - **Heat Treatment:** Enhancing material properties for increased strength or durability.
   - **Machining:** Sometimes, machining processes like milling or drilling are employed to refine specific features or dimensions.

6. **Tool Validation and Testing:**

   Before using the tool in actual production, it undergoes testing and validation to ensure it meets the required specifications and functions effectively. This step helps identify any potential issues or areas for improvement.

7. **Integration into Production:**

   Once validated, the tool is integrated into the manufacturing process for which it was intended. It can be used for producing parts, prototypes, or components as required.

8. **Iterative Improvement:**

   Continuous improvement is often part of the process. Based on feedback from the production process or further design requirements, tools may be refined, redesigned, or reproduced using RT methods.

The specific steps and their sequence might vary depending on the application, the complexity of the tool, the materials used, and the chosen AM technology. RT is beneficial for quick prototyping, small batch production, or situations where traditional tooling methods might be time-consuming or cost-prohibitive.

1.3.3.3 Advantages and Limitations of Rapid Tooling

RT, i.e. tool and mold construction via 3D printing, offers the following advantages over traditional manufacturing processes for mold tools:

- Significantly lower manufacturing costs
- Considerably shorter manufacturing time
- Fewer processing machines required
- Shorter development cycles
- Tools with improved cooling capacity can be produced
- Fewer specialists required

Although there are a lot of advantages to the RT process, RT is not always suited to all projects. One of the drawbacks of RT is precision. RT might not be as precise as conventional tooling. However, this is of less concern nowadays thanks to the technological advancements in the AM and CNC machining technology, which shows great improvement in precision and accuracy. While AM materials have advanced, they might not always match the strength and durability of traditionally manufactured tools.

All that is required for the production of a mold via RT is a 3D printer and CAD software. The 3D printer manufactures additively, i.e. without the production of chips and grinding dust. This makes its material consumption very efficient. If one compares RT with the processing machines that have been required up to now for tool and mold making, the advantage becomes clear: no CNC milling machine, no eroding machine, and no fine grinding machine are required for RT.

1.3.3.4 Rapid Repair

Rapid repair 3D printing processes are particularly interesting for maintenance and repair tasks. Under the keyword "rapid repair", 3D printers are used for the production of spare parts. This is extremely advantageous, especially for self-sufficient, autonomous, or hard-to-reach areas of operation. Already today, submarines, oil drilling platforms, and hard-to-reach expeditions are equipped with powerful 3D printers with which they can produce their own spare parts when needed.

This approach offers several advantages over traditional repair methods:

1. **Speed:** AM allows rapid prototyping and production, which can significantly reduce the time required for repairs compared with conventional manufacturing or repair methods.
2. **Customization:** 3D printing enables the creation of customized parts tailored to the specific repair needs, ensuring a perfect fit and functionality.
3. **Cost-efficiency:** Traditional repair methods often involve extensive machining or fabrication processes, which can be costly. AM minimizes waste and can be more cost-effective, especially for one-off or small batch repairs.
4. **Complexity:** AM techniques can handle complex geometries and intricate designs that may be challenging or impossible to replicate using traditional repair methods.
5. **Material Selection:** Various materials compatible with AM techniques allow flexibility in choosing the most suitable material for the repair, whether it be metals, polymers, ceramics, or composites.
6. **On-demand Production:** AM enables on-demand production of replacement parts, eliminating the need for extensive inventory and reducing lead times.

Common methods for rapid repair using AM include:

- **Direct part replacement:** Printing a new part to replace a damaged or worn-out component.
- **Reinforcement:** Strengthening damaged areas or components by adding material through 3D printing.

- **Reconstruction:** Restoring the original design by using 3D scanning to create a digital model, followed by printing the missing or damaged parts.
- **Hybrid Repair:** Combining AM with traditional repair methods to optimize efficiency and effectiveness.

Industries like aerospace, automotive, healthcare, and manufacturing are increasingly adopting rapid repair solutions using AM due to its numerous advantages in reducing downtime, improving part availability, and enhancing the overall efficiency of maintenance and repair operations.

1.3.4 AM versus Subtractive Manufacturing—Advantages and Challenges

Subtractive manufacturing is an umbrella term for various controlled machining and material removal processes that start with solid blocks, bars, rods of plastic, metal, or other materials that are shaped by removing material through turning, milling, boring, drilling, and grinding. These processes are either performed on general purpose machine tools or more commonly, performed on CNC.

In CNC, a virtual model designed in CAD software serves as input for the fabrication tool. Software simulation is combined with user input to generate toolpaths that guide the cutting tool through the part geometry. These instructions tell the machine how to make necessary cuts, channels, holes, and any other features that require material removal, taking into account the recommended cutting speed and feed rate of the material. CNC tools manufacture parts based on computer-aided manufacturing (CAM) data, with little or no human assistance or interaction.

Subtractive manufacturing processes are ideal for applications that require tight tolerances and geometries that are difficult to mold, cast, or produce with other traditional manufacturing methods. Subtractive manufacturing offers a variety of material and processing methods. Softer materials are much easier to machine to their desired shape but will wear more quickly. Table 1.1 illustrates the CNC machining processes along with some nontraditional machining processes, as well as examples of the machined materials.

In contrast to the subtractive process of removing material from a larger piece, AM or 3D printing processes build objects by adding material one layer at a time, with each successive layer bonding to the preceding layer until the part is complete. This means that in an AM process, materials are usually in the form of powder or wire, and these materials are joined together to produce the part. In subtractive manufacturing, a block of material is machined to cut away certain material to create the final object. This manufacturing principle distinguishes AM and CNC machining and results in different aspects in terms of materials and tools, speed, design freedom, accuracy, and cost.

Just like subtractive CNC tools, AM technologies create parts from CAD models. Preparing models for 3D printing with print preparation or slicer software is mostly automated, making job set-up substantially easier and faster than with CNC tools. Depending on the technology, the 3D printer deposits material, selectively melts and fuses powder, or cures liquid photopolymer materials to create parts based on the

TABLE 1.1

Subtractive Manufacturing Processes

Process	Materials
CNC machining (turning, drilling, boring, milling, reaming)	Hard thermoplastics, thermoset plastics, soft metals, hard metals (industrial machines)
Electrical discharge machining (EDM)	Hard metals
Laser cutting	Thermoplastics, wood, acrylic, fabrics, glass, metals (industrial machines)
Water jet cutting	Plastics, hard and soft metals, stone, glass, granite, composites

CAM data. The 3D printed parts often require some form of cleaning and finishing to achieve their final properties and appearance before they're ready to use. The most common materials used in AM are plastics, metals, and ceramic materials. Desktop and bench-top 3D printers offer an affordable solution to create parts from these materials.

Significant progress has been made since the innovation of AM, and there is an expectation that this technology can provide various benefits to society, mainly reduced raw material usage and energy consumption. AM has great potential to provide unprecedented control over the shape, composition, and function of fabricated products, as well as a high degree of personalization for individuals. In terms of economy and environmental sustainability, AM offers multiple advantages over conventional manufacturing technologies, including reduced material waste and energy consumption, shortened time to market, just-in-time production, and fabrication of structures not possible by traditional means. From the education and training perspective, AM holds the potential to promote science, technology, engineering, and mathematics (STEM) education because it can engage a broad population of both students and adults in both formal and informal domains.

The sustainable aspects of AM, such as less material wastage, less post-processing, and much less cost even for manufacturing complex parts makes AM a technology of the future. The other sustainable aspects include the potential of AM to reuse plastics, recycle, and reduce emissions. The technology is also capable of producing designs with complex and optimized geometries, which helps in developing parts with light weight and better strength-to-weight ratios. Therefore, the use of AM helps to produce designs that are sustainable (Jandyal et al. 2022).

The subtractive processes include both traditional and nontraditional machining processes (TMPs, and NTMPs). Compared with subtractive processes, this building-up process of AM layer by layer from geometry described in a 3D design model can provide the following perceived advantages:

- AM increases design flexibility, and any changes can be performed without restriction.

- AM is capable of building complex three-dimensional geometrical shapes. Moreover, it is possible to build a single part with varying mechanical properties. This opens up opportunities for design innovation.
- AM machines possess high production flexibility, as they do not require costly setups and hence, are economical in small batch production.
- AM improves end-product performance, reduces time to market, and extends product life.
- It is automated based on CAD models.
- Subtractive processes require auxiliary resources such as jigs, fixtures, cutting tools, and coolants in addition to the main machine tool itself. AM does not require these additional resources.
- AM requires minimal or no human intervention and set-up time.
- AM produces accurate prototypes in a short time at a minimum cost as compared with subtractive processes.
- Unlike subtractive processes, where a large amount of material needs to be removed, AM uses raw materials efficiently by building parts layer by layer. Leftover materials can often be reused with minimum losses. Moreover, material is saved due to NNS.
- Compared with subtractive processes, AM is more efficient in terms of reduced environmental impact for manufacturing sustainability. It does not require the use of coolant and other process inputs, and thus produces less pollution of the terrestrial, aquatic, and atmospheric systems, and requires less landfill.

Finally, let's consider the "rapid" character of this technology. The speed advantage is not just in terms of the time it takes to build parts. The speeding up of the whole product development process relies greatly on the fact that we are using computers throughout. Regardless of the complexity of parts to be built, building within an AM machine is generally performed in a single step. Most subtractive processes would require multiple and iterative stages to be carried out. As more features in a design are included, the number of these stages may increase dramatically. In the case of conventional technologies, if a skilled craftsman is requested to build a prototype, he may find that he must manufacture the part in a number of stages. This may be because he must employ a variety of steps, ranging from hand carving, through molding and forming techniques, to CNC machining. Hand carving and similar operations are tedious, difficult, and prone to error. Molding technology can be messy and obviously requires the building of one or more molds. CNC machining requires careful planning and a sequential approach that may also require construction of fixtures before the part itself can be made. AM can be used to remove or at least simplify many of these multistage processes. Workshops that adopt AM technology can be much cleaner, more streamlined, and more versatile (Gibson et al. 2015).

These previously mentioned advantages are driving interest in AM technology for applications throughout the product lifecycle, from new product development to volume manufacturing to product repair. Consequently, leading industry authorities have declared that AM is the first manufacturing "revolution" of the 21st century.

However, AM technology still cannot fully compete with subtractive technology, especially in the case of mass production, because of the following drawbacks (Stein 2012):

- Sometimes, supporting materials are needed.
- Post-processing is sometimes needed for optimum surface quality and for the removal of supporting materials.
- AM processes often use liquid polymers, or a powder, as starting materials to build object layers. These materials make AM unable to produce large-sized objects due to lack of material strength. Large-sized objects also are often impractical due to the extended processing time needed to complete the object.
- AM equipment is an expensive investment compared with subtractive processes. Simple 3D printers average approximately $5000 and can go as high as $50,000 for large-sized models, not including the cost of accessories and resins or other operational materials.
- The maintenance of some AM machines being more complex than others, the maintenance requirements will differ. Some companies will add cost to their machines to ensure that they are better supported.
- Material restrictions: some machines can only process one or two materials, while others can process more, including composites.

Researchers have been working on improving AM technologies to overcome the previously mentioned drawbacks. Finally, it should be noted, however, that modern CNC machine tools also provide the capability of producing complex shapes quickly and are a viable option for AM.

1.4 REVIEW QUESTIONS

1. Explain what is meant by near net shaping.
2. Who was the main founder of AM? List only five of AM's synonyms, and give their acronyms.
3. Mention the advantages of AM techniques.
4. Give an example for an AM method based on liquid material, powders, and solid materials.
5. What is the abbreviation of STL file format?
6. When did Arcam start to create the EBM additive technology for metallic products?
7. What are the main technologies that have been developed and integrated along with AM techniques? Write short notes to describe these technologies, illustrating how they did so.
8. Enumerate advantages of AM technologies over subtractive manufacturing techniques.
9. Why does AM not fully compete with subtractive technology? Enumerate the disadvantages of AM.

BIBLIOGRAPHY

3D Systems. (2014) *The Journey of a Lifetime*. 3D Systems, Inc. www.3dsystems.com

Arcam. Arcam History. (1997) Arcam AB CAD to Metal. http://www.arcam.com/company/about-arcam/history/

ASTM. (2009) *International Committee F42 on Additive Manufacturing Technologies (ASTM F2792–10)*. West Conshohocken, PA: Standard Terminology for Additive Manufacturing Technologies.

Bandyopadhyay, Amit and Bose, Susmita (2016) *Additive Manufacturing*. Boca Raton, London, New York: CRC Press.

Gibson, Ian, Rosen, David, and Stucker, Brent. (2015) *Additive Manufacturing Technologies*. 2nd ed. New York: Springer.

Hull, C. W. (1986) Apparatus for Production of Three-dimensional Objects by Stereolithography. *Uvp. Assignee*. US Patent 4575330 A.

Jandyal, Anketa, Chaturvedi, Ikshita, Wazir, Ishika, Raina, Ankush, and Haq, Mir Irfan Ul. (2022) 3D Printing – A Review of Processes, Materials and Applications in Industry 4.0. *Sustainable Operations and Computers*, 3, 33–42. https://doi.org/10.1016/j.susoc.2021.09.004

Optomec. Company Milestones Optomec. (2014) Production Grade 3D Printers with a Material Difference. http://www.optomec.com/company/milestones/

Solidscape about Us. Solidscape. (2013) A Stratasys Company. Solid-scape.com

Stein, A. (2012) Disadvantages of 3D Printers. *eHow TECH*. http://www.ehow.com/facts_7652991_disadvantages-3d-printers.html. Accessed 01 Aug 2012

Wohlers, T. and Gornet, T. (2011) History of Additive Manufacturing. State of the Industry. Wohlers Report. Fort Collins, CO: Wohlers Associates.

2 Software Aspects for Additive Manufacturing

2.1 ADDITIVE MANUFACTURING PRINTING PROCESS

As mentioned in Chapter 1, additive manufacturing (AM) is basically an additive manufacturing process, wherein the parts are developed additively layer by layer. In AM, we start with the fundamental design of the part we want to model. This design is created on a computer software that is attachable to 3D printers. This software then generates a special type of file to be sent to the printer. The 3D printer reads that file and creates the product by adjoining one layer over the other. 3D printers read the parts as a single two-dimensional layer at a time rather than a whole single part. The working of 3D printers, as shown in Figure 2.1 (Jandyal et al. 2022), is based on the fact that they are designed to read Standard Tessellation Language (STL) file type.

The AM production process begins with a 3D model that is designed using computer-aided design (CAD) software (or 3D scanned from a physical object).The pre-processing stage considers the technical specifications and boundary conditions, and the last stage is the final CAD model of the 3D print.

The second production stage is called processing. As a 3D printed object is fabricated layer by layer, each consecutive layer has to be supported by either the platform, the preceding layer, or extra support elements. After the correct and optimal orientation of the model and the supports has been designed, the STL model, including the supports, is sliced into layers with a plane parallel to the platform surface, namely, the x-y plane. Each layer is then built consecutively in the z direction. The layer thickness depends on the printer, AM technology, and quality requirements. The sliced model is subsequently sent to the printer.

After the printing is finished, the model is removed from the platform, and other technical processing procedures are used to refine the printed object in a stage called post-processing. A typical AM printing process is shown in Figure 2.2.

2.2 STANDARD TRIANGLE LANGUAGE (STL FILE FORMAT)

The Standard Triangle Language or STereoLithography (STL) is the industry standard file type for 3D printing. It uses a series of triangles to represent the surfaces of a solid model. All modern CAD software allows its native file format to be exported into STL. The 3D model is then converted into machine language (G-code) through a process called "slicing" and is ready to print.

The STL file format uses a series of linked triangles to recreate the surface geometry of a solid model. When higher resolution is required, more triangles should be

DOI: 10.1201/9781003451440-2

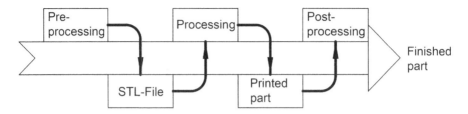

FIGURE 2.1 Basic process of 3D printers to create 3D object. (Adapted from Jandyal 2022).

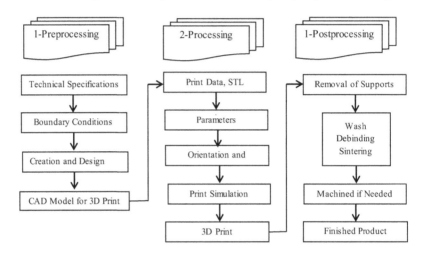

FIGURE 2.2 AM printing process.

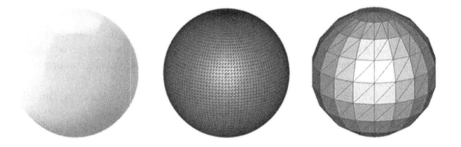

FIGURE 2.3 The perfect spherical surface on the left CAD model (a) is approximated by tessellations. The figure on the right (c) uses big triangles, resulting in a coarse model. The figure in the center (b) uses smaller triangles and achieves a smoother approximation. (From https://i.materialise.com/)

used to approximate the surfaces of the 3D model better; however, the size of the STL file will increase. If too low resolution is used, the model will have visible triangles on its surface when it is printed (Figure 2.3). Increasing the resolution above a certain point is also not recommended, as it brings no additional benefit, and the size of the file will be unnecessarily increased, making it more difficult to handle and process.

An STL file stores information about 3D models. This format describes only the surface geometry of a three-dimensional object without any representation of color, texture, or other common model attributes. The STL file format is the most commonly used file format for 3D printing. When used in conjunction with a 3D slicer, it allows a computer to communicate with 3D printer hardware. Since its humble beginnings, the STL file format has been adopted and supported by many other CAD software packages, and today, it is widely used for rapid prototyping, 3D printing, and computer-aided manufacturing (CAM).

2.2.1 How Does the STL File Format Store a 3D Model?

The main purpose of the STL file format is to encode the surface geometry of a 3D object (CAD model). It encodes this information using a simple concept called "tessellation" (STL model) (Figure 2.4).

1. *Tessellation*

Tessellation is the process of tiling a surface with one or more geometric shapes such that there are no overlaps or gaps. A tiled floor or wall is a good real-life example of tessellation. Tessellation can involve simple geometric shapes or very complicated shapes.

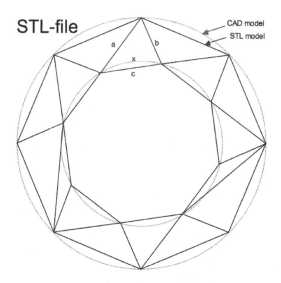

FIGURE 2.4 Difference between a CAD model and its implementation as an STL file. (From Laurens Van Lieshout/Wikipedia.)

2. *Exploiting tessellation to encode surface geometry*

Back in 1986, Chuck Hull had just invented the first stereolithographic 3D printer (Patent US 4575330 A), and the Albert Consulting Group for 3D Systems were trying to figure out a way to transfer information about 3D CAD models to the 3D printer. They realized that they could use tessellations of the 3D model's surface to encode this information. The basic idea was to tessellate the two-dimensional outer surface of 3D models using tiny triangles (also called "facets") and store information about the facets in a file.

Consider a few examples to understand how this works. For example, consider a simple 3D cube; this can be covered by 12 triangles, as shown in Figure 2.5. A 3D model of a sphere can be covered by many small triangles, also shown in the same figure. Another example of a very complicated 3D shape (tessellation of a 3D pig), which has also been tessellated with triangles, is shown in Figure 2.6.The Albert Consulting Group for 3D Systems realized that if they could store the information

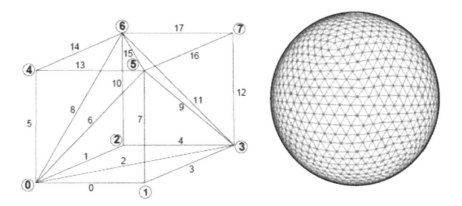

FIGURE 2.5   Tessellations of a cube and a sphere.

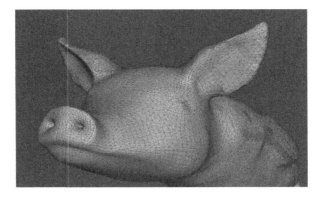

FIGURE 2.6   Tessellation of a 3D pig. (From https://i.materialise.com/)

about these tiny triangles in a file, then this file could completely describe the surface of an arbitrary 3D model. This formed the basic idea behind the STL file format.

2.2.2 How Does an STL File Store Information on Triangular Facets?

The STL file format provides two different ways of storing information about the triangular facets that tile the object surface. These are called *ASCII encoding* and *binary encoding*. Both usually have a filename extension of ".stl". In both formats, the following information about each triangle is stored:

1. The coordinates of the vertices.
2. The components of the unit normal vector to the triangle. The normal vector should point outward with respect to the 3D model.

2.2.2.1 The ASCII STL File Format

The ASCII STL file starts with the mandatory line:

```
solid<name>
```

where <*name*> is the name of the 3D model. Name can be left blank, but there must be a space after the word "solid" in that case.

The file continues with information about the covering triangles. Information about the vertices and the normal vector is represented as follows:

```
facet normal nₓnᵧn_z
  outer loop
    vertex v1ₓ v1ᵧ v1_z
    vertex v2ₓ v2ᵧ v2_z
    vertex v3ₓ v3ᵧ v3_z
  endloop
endfacet
```

Here, n is the normal to the triangle, and v1, v2, and v3 are the vertices of the triangle. Coordinate values are represented as a floating point number with sign-mantissa-e-sign-exponent format, e.g. "3.245000e-002".

The file ends with the mandatory line:
```
endsolid<name>
```

2.2.2.2 The Binary STL File Format

If the tessellation involves many small triangles, the ASCII STL file can become huge. This is why a more compact binary version exists.

The binary STL file starts with an 80-character header. This is generally ignored by most STL file readers, with some notable exceptions. After the header, the total number of triangles (n_t) is indicated using a 4-byte unsigned integer.

```
UINT8[80] - Header
UINT32 - Number of triangles
```

The information about triangles follows subsequently. The file simply ends after the last triangle. Each triangle is represented by twelve 32-bit floating point numbers. Just like the ASCII STL file, three numbers are for the 3D Cartesian coordinates of the normal to the triangle. The remaining nine numbers are for the coordinates of the vertices (three each). Here's how this looks:

```
for each triangle
REAL32[3] - Normal vector
REAL32[3] - Vertex 1
REAL32[3] - Vertex 2
REAL32[3] - Vertex 3
UINT16 - Attribute byte count
end
```

Note that after each triangle, there is a 2-byte sequence called the "attribute byte count". In most cases, this is set to zero and acts a spacer between two triangles. But, some software also uses these 2 bytes to encode additional information about the triangle.

2.2.2.3   Which Format Should Be Selected, Binary or ASCII Format?

Finally, we have a choice of exporting the STL file in binary or ASCII format. The binary format is always recommended for 3D printing, since it results in smaller file sizes. However, if it is required to manually inspect the STL file for debugging, then ASCII is preferable because it is easier to read.

2.2.3   Rules for the STL File Format

The STL specification has some special rules for tessellation and for storing information:

1. **Vertex rule:** Each triangle must share two vertices with its neighboring triangles (Figure 2.7).
2. **Orientation rule:** The orientation of each facet is specified in two ways: by the direction of the normal vector and by the ordering of the vertices (Figure 2.8).

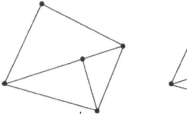

FIGURE 2.7   Vertex rule for STL files: the figure on the left is an invalid tessellation, while the figure on the right is acceptable.

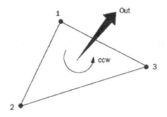

FIGURE 2.8 An STL file stores the coordinates of the vertices and the components of the unit normal vector to the facets.

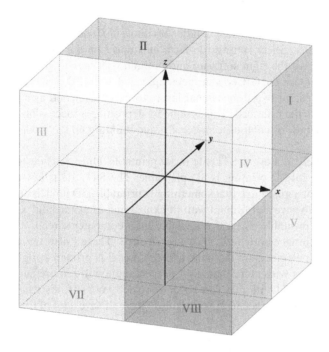

FIGURE 2.9 Octant I is the all-positive octant.

3. *The all-positive octant rule:* This says that the coordinates of the triangle vertices must all be positive (Figure 2.9). This implies that the 3D object lives in the all-positive octant of the 3D Cartesian coordinate system (hence the name). The rationale behind this rule is to save space. If the 3D object were allowed to live anywhere in the coordinate space, it would be necessary to deal with negative coordinates. To store negative coordinates, one needs to use *signed floating point numbers*, which require one additional bit to store the sign (+/–). By ensuring that all coordinates are positive, this rule makes sure that we are able to use *unsigned numbers* for the coordinates and save a bit for every coordinate value we store.

4. *The triangle sorting rule:* This recommends that the triangles appear in ascending z-value order. This helps slicers to slice the 3D models faster. However, this rule is not strictly enforced; it is only recommended.

2.2.4   3D Printing the STL File

For 3D printing, the STL file has to be opened in a dedicated slicer. A slicer is a piece of 3D printing software that converts digital 3D models into printing instructions for the 3D printer to create an object. The slicer chops up the STL file into hundreds (sometimes thousands) of flat horizontal layers based on the chosen settings and calculates how much material the printer will need to build the object and how long it will take to do it.

All of this information is then bundled up into a G-Code file, the native language of the 3D printer. Slicer settings do have an impact on the print quality, so it is important to have the right software and optimum settings to get the best-quality print possible. Once the G-Code has been uploaded to the 3D printer, the next stage is for those separate two-dimensional layers to be reassembled as a three-dimensional object on the print bed. This is done by depositing a succession of thin layers of plastics, metals, or composite materials, and building up the model one layer at a time.

Unfortunately, not every STL file is 3D printable. Only a 3D design that's specifically made for 3D printing is 3D printable. The STL file is just the container for the data, not a guarantee that something is printable. 3D models suitable for 3D printing need to have a minimum wall thickness and a "watertight" surface geometry to be 3D printable. Even if it's visible on a computer screen, it's impossible to print something with a wall thickness of zero. There's also the consideration of overhanging elements on the model. If the model is printed upright, then overhanging elements with more than a 45-degree angle will require supports. When downloading an STL file that you haven't created yourself, it's worth taking the time to verify that it is indeed 3D printable. This will save a lot of time and wasted filament.

2.2.5   Optimizing an STL File for Best 3D Printing Performance

The STL file format approximates the surface of a CAD model with triangles. The approximation is never perfect, and the facets introduce coarseness to the model. The 3D printer will print the object with the same coarseness as specified by the STL file. Of course, by making the triangles smaller and smaller, the approximation can be made better and better, resulting in good-quality prints. However, as the size of the triangle is decreased, the number of triangles needed to cover the surface also increases. This leads to a gigantic STL file, which 3D printers cannot handle. It's also difficult to share or upload huge files like that. It is, therefore, very important to find the right balance between file size and print quality. Most CAD software offers a couple of settings when exporting STL files. These settings control the size of the facets and hence, print quality and file size.

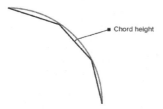

FIGURE 2.10   A visual illustration of chord height (tolerance). The chord height is the height between the STL mesh and the actual surface. (From http://3dhubs.com)

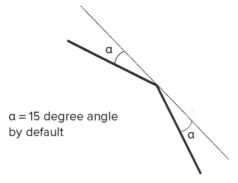

FIGURE 2.11   Angular tolerance is the angle α = 15° by default. Angular tolerance is the angle between the normals of adjacent triangles. (From http://3dhubs.com)

The most important settings and their optimum values:

1. ***Chord height or tolerance:*** Most CAD software will let you choose a parameter called chord height or tolerance (Figure 2.10). The chord height is the maximum distance from the surface of the original design and the STL mesh. If you choose the right tolerance, your prints will look smooth and not pixelated. It's quite obvious that the smaller the chord height, the more accurately the facets represent the actual surface of the model. It is recommended to set the tolerance between 10 and 1 μm. This usually results in good-quality prints.
2. ***Angular deviation or angular tolerance:*** Angular tolerance limits the angle between the normal of adjacent triangles. The default angle is usually set at 15 degrees (Figure 2.11). Decreasing the tolerance (which can range from to 0 to 1) improves print resolution.

2.3   ALTERNATIVES TO THE STL FILE FORMAT

Besides STL, there are over 30 file formats for 3D printing. The most important file formats are: .AMF, .OBJ, .3DM, .3DS, .SKP, and .STEP. Another option is the

Polygon file format (PLY), which was originally used for storing 3D scanned objects (Table 2.1).

These file formats vary in their capabilities, their level of detail, and the information they can store. The choice of format often depends on the specific requirements of the 3D printing process, software compatibility, and the complexity of the object being printed.

More recently, there have been efforts to launch a new file type by the 3MF Consortium, which is proposing a new 3D printing file format called 3MF. The 3D Manufacturing Format (3MF) should make it possible to transfer color, material, and other information, which is currently not possible with STL. It is claimed that 3MF will streamline and improve the 3D printing process. The most ubiquitous of the alternative file formats to the STL format are OBJ, AMF, and 3MF. So, what exactly is the reason for their ubiquity? This will be briefly discussed.

TABLE 2.1

Various File Formats Commonly Used in Additive Manufacturing (AM) with Their Descriptions and Applicability

File Format	Description	Applicability
STL (Stereolithography)	Basic triangular mesh representation of the surface geometry	Widely supported, but lacks color and texture information, used for printing models
OBJ (Wavefront OBJ)	Supports geometry, texture, and material information	Commonly used for 3D model interchange between software applications
3MF (3D Manufacturing Format)	Contains data about the object's geometry, materials, colors, and textures	Designed to be a comprehensive 3D printing format
AMF (Additive Manufacturing File Format)	Encodes rich information such as material properties, color, texture, and multiple objects within a single file	Designed specifically for additive manufacturing, offers more comprehensive data than STL
STEP (Standard for the Exchange of Product Data)	Encodes geometric and technical data in a neutral format	Used for CAD modeling and widely accepted in manufacturing
IGES (Initial Graphics Exchange Specification)	Encodes 2D/3D data for CAD data exchange	Commonly used for transferring surface geometry between systems
PLY (Polygon File Format)	Stores three-dimensional data as a collection of vertices, faces, and other elements	Used for both data exchange and printing in 3D
X3D (Extensible 3D)	Open standard for 3D graphics, often used for web-based applications	Supports various features for visualization and can be used in 3D printing

1. *The OBJ file format:*

Developed by WaveFront Technologies, the OBJ file format was originally used in graphics design. With the development of multi-color and multi-material printing, the file format was later adopted by the 3D printing industry.

The OBJ file format can store geometry and color (all3dp.com). In terms of popularity, OBJ is second to STL. However, unlike STL, which only stores geometrical data, OBJ can store geometry, color, texture, and material data. Color data is stored in a separate companion MTL (material template).

Another key characteristic of the OBJ 3D printing file format is that it enables choosing the way the geometry of the model is encoded. It can create tessellations using various shapes such as polygons and quadrilaterals, not just triangles. It is possible also to use more advanced and precise methods, such as free-form curves and surfaces. These allow OBJ files to store models with far greater accuracy.

Advantages

- Stores data accurately
- Stores geometry, color, texture, and material data

Disadvantages

- Not as popular as STL, hence it has limited compatibility
- Contains a large amount of complex data. This makes sharing or editing time-consuming
- Color and texture data are stored in a separate file

2. *The Additive Manufacturing File Format (AMF format):*

While effective, there are numerous difficulties surrounding the STL format. As AM technologies move forward to include multiple materials, colors, lattice structures, and textured surfaces, it is likely that an alternative format will be required. The ASTM Committee F42 on Additive Manufacturing Technologies released the ASTM 2915-12, AMF Standard Specification for AMF Format (2020). This file format is still very much under development but has already been implemented in some commercial software. Considerably more complex than the STL format, AMF aims to embrace a whole host of new part descriptions that have hindered the development of current AM technologies. The basic advantage of AMF over STL is the capability to include the following additional aspects. More details can be found in https://blog .ansi.org/?p=7189.

- *Curved triangles:* In order to specify the surface accurately with a lower number of triangles, curved triangles can be used. By specifying the triangles in this way, many fewer triangles need to be used for a typical CAD model. This addresses problems associated with large STL files resulting from complex geometry models for high-resolution systems, leading to higher resolution and smaller file size in the case of AMF format.

- *Color specification:* Supports colors for the technologies that can utilize colors.
- *Texture maps:* Supports different surface textures for the technologies that can produce textured parts. It should be noted that this is an image texturing process similar to computer graphics.
- *Material specification:* To make parts with different materials. This will be used with AM technologies that can use multiple materials for producing parts. Currently, the machines from Stratasys and some other extrusion-based systems have the capacity to build multiple material parts.
- *Material variants:* AMF operators can be used to modify the basic structure of the part to be fabricated. Some AM technologies would be capable of making parts from materials that gradually blend with others.
- *Constellations:* These represent multiple groups of parts manufactured at one time. Constellations can also be used to specify the locations and orientations of the parts when they are manufactured.
- *Additional metadata:* Metadata can contain any kind of information that the designer would like to include with the file.
- *Formulae:* Any equations representing the surface geometry or equations that govern the construction of support structures of AM parts can be included.

Advantages

- Can store all possible data and metadata about a model
- Scale can be specified in various units
- Very low possibility of error
- Small file size

Disadvantages

- Has limited support
- Slow adoption

3. The 3D Manufacturing Format (3MF Consortium)

After analyzing the shortcomings and slow adoption of the AMF file format, some of the biggest names in 3D printing, including Autodesk, 3D Systems, Stratasys, HP, and Microsoft, came together to form a body known as the 3MF Consortium. This body developed the 3MF 3D printing file format, which is very similar to, but much more widely accepted than, the AMF format. 3MF has all the technical properties of AMF. It used curved triangular tessellations to encode geometry. It can also store color, texture, material, and orientation data, and is highly accurate. 3MF files are mostly error-free and are considered to be ready to print, something that is very much appreciated in 3D printing.

Advantages

- Relatively popular and compatible with dozens of companies
- Stores geometry data accurately
- Can store all data related to a model
- Stores all data, metadata, and properties is a single archive

Disadvantage

- May become proprietary

2.4 STL FILE FORMAT VERSUS ALTERNATIVES

Finally, one important question arises, namely, when is an STL file recommended or not recommended to be used in AM?

1. *Recommended:* If it is required to print with a single color or material, which is most often the case, then STL is better than OBJ, since it is simpler, leading to smaller file sizes and faster processing. Moreover, the STL file format is universal and supported by nearly all 3D printers. This cannot be said for the OBJ format, even though it enjoys reasonable adoption. The AMF and 3MF formats are not currently widely supported.

   The STL format is so popular because it is so simple; that is why it is used by hobby makers, professionals, engineers, and 3D printing services. Due to the radical reduction of data volume, STL files are small and can usually be sent by email without problems. They do not pose any problems during upload and download either. In addition, most software programs, 3D printers, and other machines can handle this type of file without any difficulty.

2. *Not recommended:* As seen earlier, the STL file format cannot store additional information such as color, material, and so on of the facets or triangles. It only stores information about the vertices and the normal vector. This means that if it is required to use multiple colors or multiple materials for the prints, the STL file format is not the right choice. The OBJ format is a popular format enjoying good support which has a way of specifying color, material, and so on. Moreover, there are some glaring disadvantages to using STL as well. As the fidelity of printing processes embraces micron-scale resolution, the number of triangles required to describe smooth curved surfaces can result in massive file sizes. It's also impossible to include metadata (such as authorship and copyright information) in a STL file.

Conclusion: If your 3D printing needs are simple, then perhaps there is no reason to move away from the STL file format. However, for more advanced prints using multiple materials and colors, it is perhaps advisable to try the OBJ or other available formats.

2.5   REVIEW QUESTIONS

1. What are the most important file formats that are used beside STL file format?
2. What are the advantages and disadvantages of the following file formats as used in AM techniques?
   OBJ, AMF, 3MF.
3. In which cases do you recommend the STL file format as used in AM techniques?
4. What is the relationship between generative manufacturing, AM, and layer-based manufacturing?
5. What is the principle of AM?
6. Why do AM parts show stair-stepping effects?
7. What are the two main steps of every AM process?
8. What is the typical layer thickness of AM parts made of plastics?

BIBLIOGRAPHY

AMF, F2915–12 (2020) File Format, Version 1.2.

Hull, C. W. (1986) Apparatus for Production of Three-dimensional Objects by Stereolithography. *Uvp. Assignee.* US Patent 4575330 A.

http://3dhubs.com

https://blog.ansi.org/?p=7189

https://i.materialise.com/en

Jandyal, Anketa, Chaturvedi, Ikshita, Wazir, Ishika, Raina, Ankush, and Haq, Mir Irfan Ul. (2022) 3D Printing – A Review of Processes, Materials and Applications in Industry 4.0. *Sustainable Operations and Computers*, 3, 33–42. https://doi.org/10.1016/j.susoc.2021.09.004

Laurens van Lieshout / Wikipedia.

3 Basic Additive Manufacturing Techniques

3.1 TECHNIQUES OF ADDITIVE CATEGORIZED ACCORDING TO THE ASTM

There are actually lots of many individual additive manufacturing (AM) processes, which vary in their method of layer manufacturing. Individual processes will differ depending on the material and machine technology used. The American Society for Testing and Materials (ASTM) "ASTM F42—Additive Manufacturing" group formulated a set of standards that classify the range of AM processes into seven categories (ASTM 2009). These are vat polymerization, material jetting (MJ), material extrusion (ME), powder bed fusion (PBF), binder jetting (BJ), directed energy deposition (DED), and sheet lamination (SL). Each category incorporates a number of processes, including stereolithography (SLA), fused deposition modeling (FDM), 3D printing (3DP), laminated object manufacturing (LOM), selective laser sintering (SLS), selective laser melting (SLM), laser metal deposition (LMD), inkjet printing (IJP), and others (Figure 3.1). This classification system is adopted in this chapter.

These individual processes will differ depending on the material and machine technology used. Actually, there are different ways to categorize AM processes. For example, one approach is to classify them into three categories based on whether the starting materials exist as a liquid, powder, or solid. They also may be categorized based on the types of part materials, such as polymers, metals, ceramics, composites, and biological materials. An important classification of AM processes may also be based on the physical process for adhesion.

3.1.1 VAT PHOTOPOLYMERIZATION

Vat photopolymerization uses a vat of liquid photopolymer resin, out of which the model is constructed layer by layer. An ultraviolet (UV) light is used to cure or harden the resin where required, while a platform moves the object being made downward after each new layer is cured. As the process uses liquid to form objects, there is no structural support of the material during the build phase, unlike powder-based methods, where support is given from the unbound material. In this case, support structures will often need to be added. Polymers and ceramics can be processed.

DOI: 10.1201/9781003451440-3

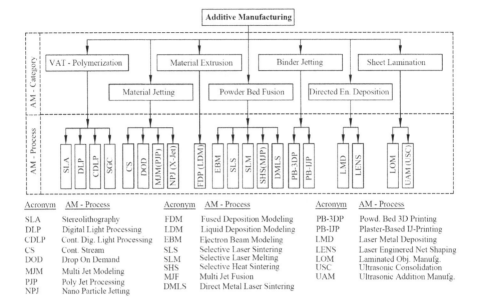

FIGURE 3.1 Classification of AM according to its categories and processes.

The three most common types of this technology are stereolithography (SLA), digital light processing (DLP), and continuous digital light processing (CDLP). The solid ground curing (SGC) process has not been used since 1999.

Advantages of Vat Photopolymerization:

- High level of accuracy and good finish
- Relatively quick process
- Large build areas

Disadvantages of Vat Photo Polymerization:

- Relatively expensive
- Lengthy post-processing time and removal from resin
- Limited to photo-resin materials
- Can still be affected by UV light after printing
- May require support structures and post-curing for parts to be strong enough for structural use

1. Stereolithography (SLA)

The US Patent 4575330 for stereolithography (SLA) was awarded on March 11, 1986, based on the work of inventor Charles Hull. It was the first material addition technology, introduced by 3DSystems, Inc. SLA uses photosensitive monomer resins that react to the radiation of lasers in the ultraviolet (UV) range. Upon irradiation, these materials undergo a chemical reaction to become solid. This reaction is called photopolymerization, and is typically complex, involving many chemical details. Charles Hull discovered that solid polymer patterns could be produced. By curing one layer over a previous layer, he could fabricate a solid 3D part. This was the beginning of stereolithography (SLA) technology.

Figure 3.2 is a schematic illustration of the setup of the SLA process. It consists of a platform that can be moved vertically inside a vessel (vat) containing the photosensitive polymer, and a laser whose beam can be controlled in the x-y direction. At the beginning, the platform is positioned vertically near the surface of the liquid photopolymer, and a laser beam is directed through a curing path that comprises an area corresponding to the base (bottom layer) of the part. This and subsequent curing paths are defined by the previously prepared stereolithography–computer-aided design (STL-CAD) file. By controlling the movements of the beam and the platform through a servo-control system, a variety of parts can be formed by this process. Total cycle times range from a few hours to a day.

The action of the laser is to harden (cure) the photosensitive polymer where the beam strikes the liquid, forming a solid layer of plastic that adheres to the platform. When the initial layer is completed, the platform is lowered by a distance equal to the layer thickness, and a second layer is formed on the top of the first by the laser, and so on. Before each new layer is cured, a wiper blade is passed over the viscous liquid resin to ensure that its level is the same throughout the surface. Each layer ranges typically from 0.075 to 0.50 mm. Thinner layers provide better resolution and allow more intricate shapes to be produced. However, processing time is greater. The starting materials are liquid monomers, typically acrylic, although epoxy has also been used for SLA. Polymerization occurs upon exposure to UV light produced by helium-cadmium or argon ion lasers. The scan speed of SLA lasers typically ranges between 500 and 2500 mm/s.

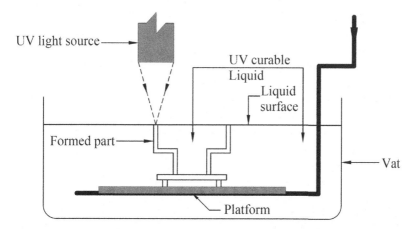

FIGURE 3.2 Setup of stereolithography (SLA). (Courtesy of A.S. Alpert, 3D Systems)

Time Estimation of SLA Process:

The time required to build a part by the SLA process ranges from 1 hour for small parts of simple geometry up to several dozen hours for complex parts. Other factors that affect cycle time are scan speed and layer thickness. The part build time T_c can be estimated by determining the time to complete each layer T_i (Equation 3.1) and then summing the times for all layers to get T_c (Equation (3.2)).

$$T_i = \frac{A_i}{vD} + T_r \tag{3.1}$$

Where Ti = time to complete layer, defined by the subscript, s

  A_i = area of layer i, mm²
  v = average scanning speed of the laser beam at the surface, mm/s
  D = spot size diameter of the laser beam at the surface, mm
  Tr = repositioning time between layers, s

In the case of SLA, the repositioning time involves lowering the worktable in preparation for the next layer to be fabricated. Other AM techniques require analogous repositioning steps between layers. The average scanning speed v must include any effects of interruptions in the scanning path (e.g., because of gaps between areas of the part in a given layer). Once the T_i values have been determined for all layers, then the build cycle time T_c can be determined accordingly:

$$T_e = \sum_{i=1}^{n} T_i \tag{3.2}$$

Where n = number of layers to finish the product

Although Eqs (3.1) and (3.2) have been developed for stereolithography, similar formulas can be developed for the other AM material addition technologies, because they all use the same layer-by-layer fabrication method.

After all layers have been formed, the cured part should be "baked" in a fluorescent oven until completely solidified. Excess polymer is removed with alcohol. Light sanding is sometimes used to improve smoothness and appearance.

In SLA, support structures are sometimes required to attach the part to the build platform. Depending on its design and orientation, the part may contain overhanging features that should be supported. Once the part is completed, the object is elevated from the liquid, and the support structures are cut off and removed manually.

In the SLA technique, progress is being made toward improvements in:

• Accuracy and dimensional stability of prototypes produced
• Less expensive liquid modeling materials used
• Strength so that the produced prototypes can be truly used as models in the traditional sense

Advantages and Limitations of SLA:

Advantages:

SLA is particularly suitable in the manufacturing industry for the following reasons:

- High level of accuracy and good finish
- Relatively quick process
- Typically large build areas: Object 1000 × 800 × 500 and maximum model weight 200 kg
- The process is fully automated
- Semitransparent polymers can be processed for optical reasons

Limitations: The main limitations of SLA are:

- Relatively expensive. The photopolymer alone costs $300 to $500 per gallon, not to mention the machine itself, which may range from $100,000 to $500,000
- Working with liquids can be messy
- Lengthy post-processing time and removal from resin
- Materials used in SLA are relatively limited compared with other AM processes
- Often requires support structures and post-curing for parts to be strong enough for structural use

Main Applications of Parts Produced by SLA:

The automotive, aerospace, electronics, and medical industries are among the industrial applications that use SLA as a rapid and inexpensive method of producing prototypes, aiming at reducing product development cycle times. One major application of SLA is in the area of making molds and dies for casting and injection molding (Kalpakjian and Schmid 2003).

2. Digital Light Processing (DLP)

There are two main vat polymerization technologies, SLA and DLP. Both of these technologies are similar but at the same time different. This variation of the photopolymerization process works with a commercial DLP projector as UV light source. It projects the complete contour of a cross section of the actual layer and initiates the solidification simultaneously.

The main components of a DLP 3D printer are the following: the digital light projector screen, the digital micromirror device (DMD), the vat (resin tank), the build plate, and the elevator for the build plate (Figure 3.3).

1. The digital light projector is the light source of a DLP 3D printer. The DMD is a component, made of thousands of micromirrors used for navigating the light beam projected by the DLP. The projector is mounted into the lower part of the machine body (under the vat).

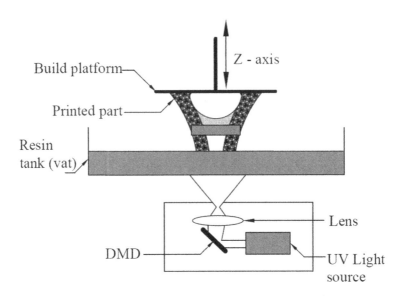

FIGURE 3.3 Schematic of a DLP 3D printer. (Courtesy of Wevolver)

2. The vat is basically a tank for the resin. However, the vat needs to have a transparent bottom so that the light projected by the DLP reaches the resin and cures it.
3. The build platform is simply the surface the printed objects stick to during printing. The z axis is the direction for slowly lifting the build platform during the printing process.

To make sure that the 3D models are well prepared for 3D printing, the user makes use of so-called "slicer software", which is provided by either a third party or the printer manufacturer. The main purpose of slicer software is to give the user the ability to set all the parameters for the printing job, and then prepare a file that can tell the printer what to print. For example, the users define the print speed, layer height, and support material positioning in the slicer. After that, the slicer virtually slices the 3D model into hundreds of layers. After the slicing is done, a portable network graphics (PNG) stack of images is produced, which can be flipped through like a picture book, one layer at a time, until the last layer of the model is reached. Assuming that the default layer thickness is 100 μm, if a part is 100 mm tall, the PNG number is 100 mm/100 μm. Each image is displayed on the UV projector, which cures the UV photopolymer 100 μm at a time.

The first step of the printing process is uploading a 3D model to the printer. When that's done, the resin needs to be poured into the vat. That is then followed by lowering the build platform into the resin tank and the resin. The build platform is lowered into the resin to the point when only a tiny bit of space is left between the vat bottom and the build plate. The tiny bit of space left is specified by the layer height of the future part. If the desired layer height for the part is 50 μm, then the space left between the two is set to 50 μm.

When that is all set and done, the DLP starts its work. It flashes an image of the individual layer. The projected light making the image of the layer is then guided to the transparent bottom of the vat in the pattern of the layer by the DMD. When the image of the layer reaches the vat bottom, the resin is cured into a solid forming the first layer. In order to create the space in the vat for the next layer to be cured, the build platform moves up one layer in height. To be more precise, the platform usually needs to move up more than one layer thickness to allow resin to flow back under the build head, which is especially true in the case of higher-viscosity resins. Then, once again, the digital light projector flashes an image of the layer to the vat bottom, causing yet another layer to cure into a solid. That process is repeated until the entire part is finished.

Due to the small reservoir, the process is designed for small parts. A wide variety of photosensitive plastic materials are available, including biocompatible grades that can be used to make hearing aid housings or masters for dental prostheses.

Comparison between DLP and SLA:
The core difference between SLA and DLP is the light source used for the solidification of the resin. DLP uses a digital light projector screen, which flashes the image of the layer and therefore cures the resin in the form of the layer. However, SLA 3D printers cure resin in a very different way. Instead, they use a laser from the vat top that cures the resin into layers by "drawing" the layer's pattern on the bottom of the resin tank. This leads to the first benefit of DLP when compared with SLA. Because the digital light projector of a DLP 3D printer flashes the entire image of a layer at once and therefore cures the layer, layers are made fast. In other words, the laser in the SLA printer has to go "point by point" to cure a single layer, while DLP does it more quickly with the single flash to cure the layer's image at once.

Traditionally, DLP and SLA usually were not used for the manufacturing of parts that are under load, but more for parts that prioritize aesthetics and dimensional accuracy. Alongside SLA, DLP is considered to be one of the most accurate 3D printing technologies.

Resins are tricky to handle, and their shelf life is also limited. Most of the time, resins have a shelf life of about 1 year. Great dimensional accuracy and the fast print speed make DLP a rather desirable choice when it comes to the manufacturing of parts. It's important to keep in mind the fact that DLP is unable to produce parts with great strength; it's much more suited to making accurate and beautiful parts that are not intended to deal with load.

Advantages and Disadvantages of DLP:
Advantages of DLP

- Very intricate designs, more accurate than FDM or SLS
- Fast, almost always faster than SLA printing
- Lower running costs than SLA, as usually a shallower resin vat is used, reducing waste

Disadvantages of DLP

- As with SLA, parts cannot be left out in the sun or they will degrade
- Parts overall have worse mechanical properties than FDM; they break or crack more easily and are at risk of deteriorating over time
- More expensive to run than FDM; resins are far more expensive than filaments, and the regular replacements of resin tanks and occasionally print platforms also add up

Main Applications of Parts Produced by DLP:
DLP 3D printing is an interesting and complex technology that is currently undergoing rapid developments. Because of its speed, precision, and material versatility, it holds the promise of enabling mass production. That leads to the expectation of more and more parts printed by DLP, and in the near future, many new applications will appear. The most common applications of DLP include the dental, medical, and jewelry industries.

3. Continuous Digital Light Processing (CDLP)

CDLP or continuous liquid interface production (CLIP) technology is an innovation based on DLP technology. In particular, unlike SLA and DLP technology, CDLP employs digital projection with light-emitting diodes (LEDs) and an oxygen-permeable window instead of a normal glass window. This oxygen-permeable window forms a so-called dead zone as thick as a human hair, which allows the liquid resin to flow between the interface of the printed part and the window. This uncured resin flow remarkably increases the resolution of the printed part as well as decreasing the risk of printing failure due to the peeling force. Moreover, as opposed to the layer-by-layer method, CDLP machines are designed with continuous movement of the build platform, thus allowing undisrupted prototype printing at speeds of several hundred millimeters per hour. A typical CDLP machine and its components are illustrated in Figure 3.4a. In contrast, Figure 3.4b illustrates the layer-by-layer movement of the build platform in the case of DLP.

Thanks to the continuous printing process, the problem regarding layer connection is eliminated, and the visible staircase effect is minimized. Therefore, the fabricated parts have isotropic mechanical properties, although they may appear anisotropic (Janusziewicz and coworkers 2016).

4. Solid Ground Curing (SGC)

The solid ground curing process is also called solid-base curing (SBC).This process has not been used since 1999. The disappearance of SGC is mainly due to the fact that the production system was too complicated and cumbersome, and required high maintenance and skilled supervision. It suffered from high initial and operating costs. Like SLA, DLP, and CDLP, SGC works by curing a photosensitive polymer layer by layer to create a solid model based on CAD geometric data. Instead of using

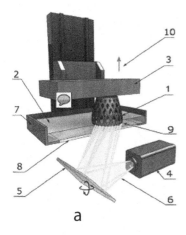
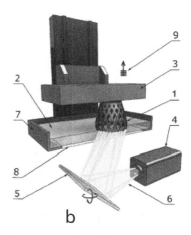

1-printed part, 2-liquid resin,
3-building platform, 4-light source,
5-digital projector, 6-light beam,
7-resin tank, 8-oxygenpermeable window,
9-dead zone, and 10-continuous elevation

1-printed part, 2-liquid resin
3-building platform, 4-light source
5-digital projector, 6-light beam,
7-resin tank, 8-window, and
9-layer by layer elevation

FIGURE 3.4 Differences between DLP and CDLP. (From Pagac, M. et al., *Polymers*, 13, 598, 2021.)

a scanning laser beam to accomplish the curing of a given layer, the entire layer is exposed to an UV light source through a mask that is positioned above the surface of the liquid polymer. The hardening (curing) process takes 2 to 3 seconds, while 2 minutes are needed to finish each layer.

A schematic diagram for the SGC process is shown in Figure 3.5. A CAD geometric model of the part has been sliced into layers. For each layer, the step-by-step procedure in SGC is performed in the following steps:

- A mask is created on a glass plate by electrostatically charging a negative image of the layer.
- A thin layer of photosensitive polymer is first sprayed onto the work platform.
- The mask plate is then placed on top of the platform and exposed to a high-powered (e.g. 2000 W) UV lamp. The portions of the liquid polymer layer that are unprotected by the mask are solidified in about 2 seconds.
- The mask is removed, cleaned, and made ready for a subsequent layer as in step 1. Simultaneously, the liquid polymer remaining on the surface is removed.
- The area of the layer is filled in with hot wax. When hardened, the wax acts to support overhanging sections of the part.
- When the wax has cooled and solidified, the polymer–wax surface is milled to form a flat layer of specified thickness, ready to receive the next layer.

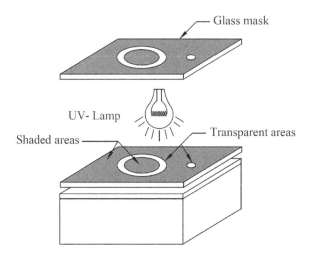

FIGURE 3.5 SGC process—a schematic. (Adapted from Groover, M.P., *Fundamentals of Modern Manufacturing-Materials, Processes, and Systems.* 4th ed., John Wiley & Sons, Inc., Hoboken, New Jersey, 2010.)

The process is repeated for the next slice until the object is completely finished. The wax provides support for fragile and overhanging features of the part during fabrication but can be melted away later to leave the free-standing part. No post-curing of the completed prototype model is required, as in stereolithography.

Advantages and Limitations of SGC Process:
Advantages:
SGC has the advantage of high production rate, both because the entire slices are produced at once, and because two glass screens are used concurrently. That is, while one screen is being used to expose the polymer, the next screen is already being prepared, and it is ready as soon as the milling operation is completed (Kalpakjian and Schmid 2003). The wax support may be immediately removed, or it may remain in place for protection during shipping of the part. The use of the wax support means that any complicated and large shapes can be processed. Usually, no finishing operations are required. The SGC process attains high accuracy and finish. Layers are milled to enhance the accuracy.

Limitations:
This process is rather complicated, and the equipment requires high maintenance and skilled supervision. As with SLA, the range of materials that can be processed is limited. Moreover, the process is noisy and messy, with high equipment costs, and needs a large space.

3.1.2 MATERIAL JETTING

Material jetting (MJ) is an AM process that operates in a similar fashion to 2D inkjet printers. The principle of printing as a three-dimensional building method was first demonstrated as a patent in the 1980s, and the first commercially successful technology was developed by "Sanders Prototype" (now Solidscape Inc., one of the leading manufacturers of high-precision 3D printers), introduced in 1994, which printed a basic wax material that was heated to liquid state. All the next members of the first generation of MJ machines relied on heated waxy thermoplastics as their build material; they are therefore most appropriate for concept modeling and investment casting patterns. More recently, the focus of development has been on the deposition of acrylate photopolymer, wherein droplets of liquid monomer are formed and then exposed to UV light to promote polymerization. MJ machines are based on jetting the material in the form of liquid droplets through nozzles in a printhead, which moves horizontally onto a build platform to form a thin layer. When the layer solidifies, the platform is moved down, a second layer is deposited, and so on, building the part layer by layer. Droplets that are not used are recycled back into the printing system. The material layers are then cured or hardened using UV light. Support structures are always required in MJ and need post-processing to be removed. Figure 3.6 represents the main parts of an MJ machine.

Materials Processed by MJ:
While industry players have so far introduced printing machines that use waxy polymers and acrylic photopolymers exclusively, research groups around the world have experimented with the potential for printing machines that could build in other materials, including polymers, ceramics, and metals.

To facilitate jetting, materials that are solid at room temperature must be heated so that they liquefy. For high-viscosity fluids, the viscosity of the fluid must be lowered to enable jetting. The most common practices are to use heat or solvents or other low-viscosity components in the fluid. The limitation on viscosity quickly becomes the most problematic aspect for droplet formation in MJ.

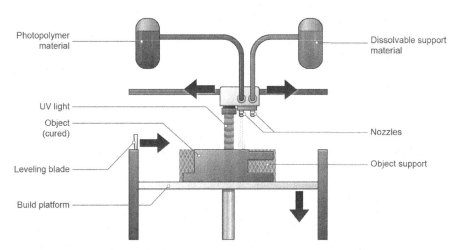

FIGURE 3.6 Material jetting machine by Dassault Systems 2018. (From https://make.3dexperience.3ds.com/)

Advantages of Material Jetting:

MJ already has a strong foothold in terms of becoming a successful AM technology. The process can offer many advantages, including:

- High speed and scalability: by using print heads with hundreds or thousands of nozzles, it is possible to deposit a lot of material quickly and over a considerable area. Scalability means that printing speed can be increased by adding another printhead to a machine
- Very big, intricate parts can be printed with great accuracy. The typical build size is approximately $380 \times 250 \times 200$ mm, while large industrial systems can be as big as $1000 \times 800 \times 500$ mm
- Ease of building parts that are fully transparent or in multiple materials, and multiple colors. There are MJ machines that print in high resolution 24-bit color
- Parts created with MJ have homogeneous mechanical and thermal properties
- MJ products have very high dimensional accuracy and excellent surface finish. MJ systems have a dimensional accuracy of $\pm0.1\%$ with a typical lower limit of ±0.1 mm (sometimes as low as ±0.02 mm)
- The level of post-processing required to enhance the properties is limited compared with other AM processes, and the functional and aesthetic qualities of a part are largely determined during the printing stage
- The process benefits from highly accurate deposition of droplets and therefore, low waste
- MJ offers the option of printing parts in either a glossy or a matte setting. The glossy setting should be used when a smooth, shiny surface is desired
- Printing machines are much lower in cost than other AM machines, particularly the ones that use lasers

Limitations of Material Jetting:

A few disadvantages of MJ will provide a more balanced evaluation of the process capabilities.

- The choice of materials is limited; only waxes and photopolymers are commercially available
- Part accuracy, particularly for large parts, is generally not as good as with some other processes, notably vat photopolymerization and ME
- Warping can occur during build-up, but it is not as common as in other 3D printing technologies due to the relatively lower working temperature
- If the material is not in liquid form or suitable viscosity to begin with, additional techniques may be necessary, such as suspending particles in a carrier liquid, dissolving materials in a solvent, melting a solid polymer, mixing a formulation of monomer or prepolymer with a polymerization initiator, or addition of surfactants to attain acceptable characteristics

- Much attention must be given to monitoring and maintaining nozzle performance during operation to avoid clogging. Purging, cleaning, and wiping cycles are periodically necessary
- The most often used MJ materials are photosensitive, and their mechanical properties degrade over time
- Support material is often required
- Slow build process

1. Continuous Stream (CS) or Drop on Demand (DOD) Stream

There are two distinctive techniques for creating and expelling droplets from the printhead through the nozzle. These are mainly the continuous stream (CS) or the Drop on Demand (DOD) stream. The distinction refers to the form in which the liquid exits the nozzle, either as a continuous column of liquid or as discrete droplets. In CS mode, a steady pressure is applied to the fluid reservoir, causing a pressurized column of fluid to be ejected from the nozzle. After departing the nozzle, this stream breaks into droplets due to Rayleigh instability. The breakup can be made more consistent by vibrating, perturbing, or modulating the jet at a fixed frequency close to the spontaneous droplet formation rate, in which case the droplet formation process is synchronized with the forced vibration, and ink droplets of uniform mass are ejected. Because droplets are produced at constant intervals, their deposition must be controlled after they separate from the jet. To achieve this, they are introduced to a charging field and thus attain an electrostatic charge. These charged particles then pass through a deflection field, which directs the particles to their desired destinations—either a location on the substrate or a container of material to be recycled or disposed. An advantage of CS deposition is the high throughput rate; it has therefore seen widespread use in applications such as food and pharmaceutical labeling. A schematic of CS is shown in Figure 3.7a.

In DOD mode, individual droplets are produced directly from the nozzle. Droplets are formed only when individual pressure pulses in the nozzle cause the fluid to be expelled; these pressure pulses are created at specific times by thermal, electrostatic, piezoelectric, acoustic, or other actuators. DOD methods can deposit droplets of 25–120 μm at a rate of 0–2000 drops per second. In the current DOD printing industry, thermal (bubble-jet) and piezoelectric actuator technologies dominate. Thermal actuators rely on a resistor to heat the liquid within a reservoir until a bubble expands in it, forcing a droplet out of the nozzle. Piezoelectric actuators rely upon the deformation of a piezoelectric element to reduce the volume of the liquid reservoir, which causes a droplet to be ejected. Piezoelectric DOD is more widely applicable than thermal because it does not rely on the formation of a vapor bubble or on heating, which can damage sensitive materials. The DOD technique is presented in Figure 3.7b. The waveforms employed in piezoelectrically driven DOD systems (shown in the figure) can vary from simple positive square waves to complex negative–positive–negative waves in which the amplitude, duration, and other parameters are carefully modulated to create the droplets as desired.

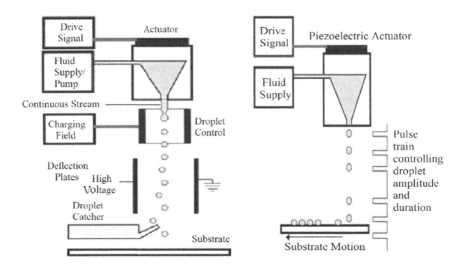

FIGURE 3.7 Schematic of the (a) continuous stream (CS) printing, and (b) Drop on Demand (DOD) printing techniques. (With kind permission from Springer Science+Business Media: *Additive Manufacturing Technologies*, 2021, Gibson, I., Rosen, D., Stucker, B., and Khorasani, M.)

The three main companies involved in the development of the MJ printing industry are still the main players offering printing-based machines. These are Solidscape, 3D Systems, and Stratasys. Each of these machines employs two single jets, one to deposit a thermoplastic part material and one to deposit a waxy support material, to form layers of about 12.5 μm thick. It should be noted that these machines also fly-cut layers after deposition to ensure that the layer is flat for the subsequent layer. Because of the slow and accurate build style as well as the waxy materials, these machines are often used to fabricate investment castings for the jewelry and dentistry industries.

3D Systems and Stratasys offer machines using the ability to print and cure acrylic photopolymers. Stratasys markets a series of printers that print a number of different acrylic-based photopolymer materials in 15-μm layers from heads containing 1536 individual nozzles, resulting in rapid, line-wise deposition efficiency, as opposed to the slower, point-wise approach used by Solidscape. Each photopolymer layer is cured by UV light immediately as it is printed, producing fully cured models without post-curing. Support structures are built in a gel-like material, which is removed by hand and water jetting. By automatically adjusting build styles, the machine can print up to 25 different effective materials by varying the relative composition of the two photopolymers. Machines are emerging that print increasing numbers of materials. Figure 3.8 presents typical applications involving multi-material full-color products of MJ.

5. PolyJet or Multi-Jet Modeling (PJ/MJM)

In 2003, 3D Systems launched a competing technology identified as multi-jet modeling (MJM), where the number of jetting nozzles is increased, changing the material

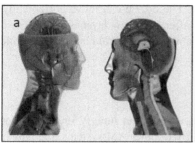

FIGURE 3.8 Typical MJ products: (a) full-color printing of educational medical models, (b) model for industrial part, (c) bicycle helmet and shoe. (Parts (a) and (c) courtesy of Stratasys; part (b) courtesy of Solidscape.)

solidification strategy. Recently, machines jetting photopolymers using print heads with over 1500 nozzles have been developed.

PolyJet (PJ) or MJM 3D printing technology was first patented by the Objet Company, now a Stratasys brand. The photopolymer materials are jetted in ultra-thin layers onto a build tray in a similar fashion to inkjet document printing. Each photopolymer layer is cured by UV light immediately after being jetted. The repetition of jetting and curing steps, layer after layer, produces fully cured models that can be handled and used immediately. The gel-like support material, which is specially designed to support complex geometries, can easily be removed by hand or by using a water jet.

6. Nanoparticle Jetting (NPJ)

This MJ technology, patented by XJet, uses a liquid that contains building nanoparticles or support nanoparticles, which is loaded into the printer as a cartridge and jetted onto the build tray in extremely thin layers of droplets. High temperatures inside the build envelope cause the liquid to evaporate, leaving behind parts made from the building material. This technique is suitable for metals and ceramics.

The key to the NPJ technique begins with its unique liquid dispersion methodology. Liquid suspensions containing solid nanoparticles of selected support and construction materials are injected into the manufacturing tray to additively manufacture detailed parts. The liquid suspensions are delivered and installed in sealed cartridges without problems. The precision of the inkjet printheads and the use of ultra-thin layers will create a super-sharp z resolution, enabling extremely neat parts. This is crucial to achieve excellent shape and dimensional tolerance. It is expected that in the coming years, more actors will continue to contribute to the evolution of this technology.

3.1.3 Material Extrusion

ME is a process in which material in the form of a continuous filament of thermoplastic or composite is drawn through a nozzle, where it is heated and is then deposited layer by layer to construct 3D parts. The nozzle can move horizontally, and a platform moves up and down vertically after each new layer is deposited. It is the most popular technique in terms of availability for general consumer demand and quality.

1. Fused Deposition Modeling (FDM)

The patent for FDM (US Patent 5121329) was awarded on June 9, 1992, and it is trademarked by the company Stratasys. Similarly to other AM systems, the part in the FDM process is fabricated from the base up, using a layer-by-layer procedure.

FDM is a process in which a filament of a build material is extruded onto the part surface from a work-head. The work-head is controlled in the x-y plane during each layer and then moves up vertically by a distance equal to one layer in the z direction. The filament has a typical diameter ranging from 1 to 3 mm. It is unwound from a coil and fed to an extrusion nozzle into the work-head that heats the material to about 1°C above the filament's melting point so that it solidifies almost immediately after extrusion and cold welds to the previous cold layers within about 0.1 second (Figure 3.9).

While FDM is similar to all other 3D printing processes, as it builds layer by layer, it varies in the fact that material is added through a nozzle under constant pressure and in a continuous stream. This pressure must be kept steady and at a constant speed to enable accurate results. Material layers can be bonded by temperature control. It is often added to the machine in spool form, as shown in Figure 3.10.

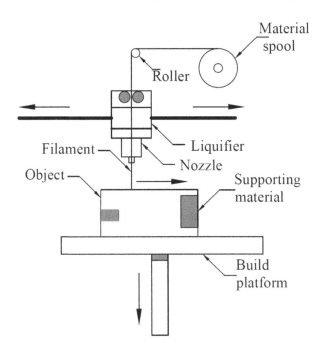

FIGURE 3.9 Scheme of FDM. (Courtesy of Dassault Systems, 2018.)

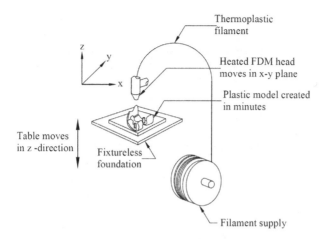

FIGURE 3.10 Schematic illustration of the fused deposition modeling process. (Courtesy of Stratasys, Inc.)

A wide variety of materials can be used in FDM. These have been expanded to include polymers, wax, metals, and ceramics. Wax is used to fabricate models for investment casting. Several types of polymers (thermoplastics), including polylactic acid (PLA), acrylonitrile butadiene styrene (ABS), medical-grade ABS, and E20 (a thermoplastic polyester-based elastomer), are available. ABS offers good strength, and more recently, polycarbonate and poly (phenyl) sulfone materials have been introduced, which extend the capabilities of this method further in terms of strength and temperature range. These materials are nontoxic, allowing the FDM machine to be set up in an office environment. High-impact polystyrene (HIPS), thermoplastic polyurethane (TPU), aliphatic polyamides (PA, also known as Nylon), and more recently, high-performance plastics such as polyether ether ketone (PEEK) or poly-etherimide (PEI) can also be used. FDM is favored when tough plastic components are needed. Part tolerances of ±50 μm can easily be achieved by this process.

The starting data of Stratasys Inc. was a CAD geometric model of software modules QuickSlice and SupportWork™. The first module is used to slice the model into layers, while the second is used to generate support structures if needed during the build process. In this case, a dual extrusion head and a different material are used to create the supports. The second material is cheaper, designed to be readily separated from the primary modeling material, and breaks away from the part without impairing its surface. A water-soluble support material that can be easily washed away is frequently used. Typical layer thickness varies from 0.18 to 0.36 mm. About 400 mm filament length can be deposited per second by the extrusion work-head. The width (road width) can be set between 0.25 and 2.5 mm.

The process requires many factors to be controlled in order to achieve a high-quality finish. The nozzle that deposits material will always have a radius, as it is not possible to make a perfectly square nozzle, and this will affect the final quality of the printed object. Accuracy and speed are low when compared with other processes, and the quality of the final model is limited by the material nozzle thickness. When using the process for components where a close tolerance must be achieved, gravity and surface tension must be accounted for (Gibson et al. 2015). As with most heat-related post-processing processes, shrinkage is likely to occur and must be taken into account if a close tolerance is required.

Complicated parts, such as those shown in Figure 3.11, necessitate the use of dual-head FDM systems. One head is used to extrude the modeling material, while the other is used to extrude the less dense supporting material.

With the expiration of this technology's patent, there is now a large open-source development community called RepRap (repeated rapid printing), which utilizes FDM technology. This has led to a measurable price decrease. However, the ME technique has dimensional accuracy limitations and is very anisotropic.

As it became possible to equip FDM with multiple extruders for speeding up the fabrication or opening multi-material capabilities, composite fabrication became possible. Composite filament fabrication (CFF) is one such technique. This term was coined by the company Markforged and uses two print nozzles. One nozzle operates following the typical ME process; it lays down a plastic filament that forms the outer shell and the internal matrix of the part. The second nozzle deposits a continuous strand of composite fiber (made with carbon, fiberglass, or Kevlar) on every layer. These continuous strands of composite fibers inside 3D printed parts add a strength to the built object that is comparable to parts made of metal.

Even composites can be 3D printed with the ME technique on machines equipped with only one extruder. The sole condition is that the base material (a thermoplastic)

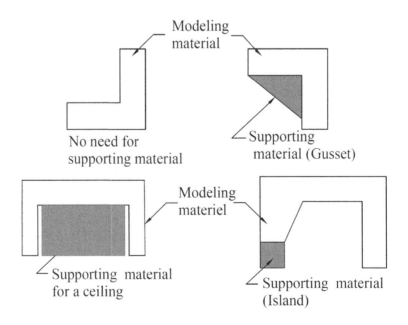

FIGURE 3.11   Supporting material for overhanging parts.

is present in sufficient quantities to guarantee fusion between layers. Therefore, a mix of two materials is possible within a single filament using wood 3D printing (wood particles embedded in PLA), metal 3D printing (metal particles embedded in thermoplastic), ceramic 3D printing (ceramic particles embedded in thermoplastic), and even carbon 3D printing (carbon fibers embedded in thermoplastic).

Advantages and limitations of FDM process:
Advantages: Advantages of the FDM process include:

- Use of readily available acrylonitrile butadiene styrene (ABS) plastic, which can produce models with good structural properties, close to a final production model
- In low-volume cases, FDM can be a more economical method than using injection molding
- The FDM process is characterized by compact equipment
- Low maintenance cost, quiet, nontoxic, cost-effective, and office-friendly process
- Fast for small shapes
- Produces strong parts
- Able to produce multi-colored parts using colored ABS

Limitations: FDM has some disadvantages, e.g.:

- Presence of seam line between layers
- Need for supports
- Long build time when making large-size models
- Relatively low accuracy and poor surface finish
- Not capable of making complicated parts
- Delamination caused by temperature fluctuation
- Accuracy and productivity are low when compared with other processes
- Accuracy of the final model is limited by material nozzle size
- Constant pressure of material is required in order to improve the surface quality

3.1.4 POWDER BED FUSION

Powder bed fusion (PBF) processes represent the first commercialized, and most widely used, AM processes for manufacturing intricate components using any engineering material such as metals, ceramics, polymers, and composites. PBF follows the basic principle of additively manufacturing the product layer by layer. All PBF processes share a basic set of characteristics. These include one or more thermal sources for inducing fusion between powder particles, a method for controlling powder fusion to a prescribed region of each layer, and mechanisms for adding and smoothing powder layers. When the selective fusion of one layer has been completed, the building platform is lowered by a predetermined distance (ranging from 20 to 200 μm), and the next layer of powder is deposited on the previous one. The process is then repeated with successive layers of powder until the part is completely built. This technique is widely used in many industrial sectors, such as aerospace, energy sector, transportation, medical industries, etc. The principle of the PBF system is presented schematically in Figure 3.12. The figure shows that the machine has two chambers: one is for spreading the powder layer using a roller or blade, and the second is the build chamber, where the product is built layer by layer.

The PBF process has been widely developed in the last few years because of its high-quality, low-cost products. In this process, no or minimum support is required, as the powder acts like a support structure. The powder used in the process can be recycled to produce more parts.

The major advantages of PBF over other AM methods are:

* PBF processes permit low cost per part and are cheap to run once purchased or rented
* Unused powder is easily recycled, which reduces overall material costs
* PBF is completely customizable and is only held back by geometric constraints, material choice, and imagination
* PBF can be used for rapid prototyping, visual prototyping, low-volume production, or full high-volume production runs
* PBF requires no support structures, reducing post-processing times
* The material choice across all PBF options is impressive; it is one of a few methods that can create reliable metal parts

The most notable limitations of PBF can be stated as follows:

* Most methods have relatively slow print times
* Printers of this type have high power usage, especially laser-driven printers

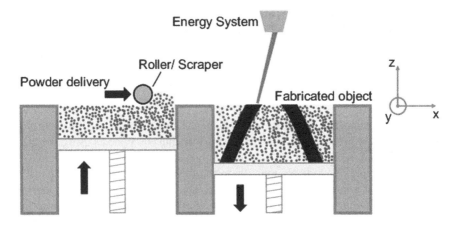

FIGURE 3.12 An overview of the principle of a PBF system. (From: Wiberg, A., *Towards Design Automation for Additive Manufacturing — A Multidisciplinary Optimization Approach*, Linköping University, Sweden, 2019.)

- The surface quality and overall integrity of a print rely heavily on the quality and grain size of its powder, as well as tight thermal management
- Most PBF parts are not specified for structural applications due to their layer-based manufacturing
- While recyclable, some of the unused powder will be wasted in the printing process
- Thermal distortion can shrink/warp parts (especially the case for polymer parts)

The most common and first used thermal sources for PBF were lasers, and the developed processes based on applying lasers are selective laser sintering (SLS), selective laser melting (SLM), and direct metal laser sintering (DMLS). Another highly efficient heat source for specific applications is electron beam, and the process is known as electron beam melting (EBM). This process involves significantly different machine architectures than laser sintering machines, by using a vacuum chamber. The vacuum chamber is filled with inert gas to prevent corrosion of the metal. A fifth PBF process that uses a lower-energy thermal printhead instead of a laser or electron beam for the fusion of powered thermoplastic is known as the selective heat sintering (SHS) process. Thus, PBF processes are sub-classified into these five different processes according to the ASTM classification (ASTM 2012). The main technologies involved in the development of these processes, their main features, and product ranges will be explained in the following sections.

1. **Electron Beam Melting (EBM)**

EBM uses a high-energy electron beam under high vacuum to induce fusion between metal powder particles. This process was developed at Chalmers University of Technology, Sweden, in 1997 and was commercialized by Arcam AB, Sweden, in 2001. It is a rapid manufacturing method and widely used in the aerospace, medical or orthopedic implant, and automotive industries. The process is three to five times faster than other additive technologies. The EBM parts are built in vacuum, which increases the energy efficiency and supports processing of reactive metal alloys such as titanium. Parts produced via EBM are near net shape, like those made via precision casting processes; thus, secondary infiltration processes are eliminated. The process is also capable of producing complex parts of multi-piece assemblies that exhibit exceptional strength-to-weight ratios, reduce manufacturing costs, and minimize assembly time. The EBM process can also produce hollow parts with an internal strengthening scaffold, which is impossible with any other method; thus, it can deliver the required mechanical strength with much less mass. This reduces the cost of raw materials and the weight of the component.

In EBM, the powder bed must be conductive. Thus, the process can only be used for conductive materials (e.g. metals) and cannot be applied for nonconductive materials such as polymers or ceramics. This is the major limitation of this process, despite its distinguished inherited features. As it is commonly applied for metallic materials, the details of this process and the electron gun machine will be presented in Chapter 5. Figure 3.13 presents examples of typical titanium products built by EBM, while Figure 3.14a shows a hip prosthesis with EBM-made stem and acetabular cup, and details of the porous structure of the stem are shown in Figure 3.14b.

2. Selective Laser Sintering (SLS)

laser sintering processes were originally developed to produce plastic prototypes using a point-wise laser scanning technique. This approach was subsequently extended to metal and ceramic powders. This technology was first developed at the University of Texas in Austin in 1992. Since polymer laser sintering (PLS) machines and metal laser sintering (MLS) machines are significantly different from each other, each will be addressed separately.

The PLS process fuses thin layers of powder (typically 0.075–0.1 mm thick) that have been spread across the build area using a counter-rotating powder leveling roller. The part building process takes place inside an enclosed chamber filled with nitrogen gas to minimize oxidation and degradation of the powdered material. The powder in the build platform is maintained at an elevated temperature just below the melting point and/or glass transition temperature of the powdered material. Infrared

Landing gear knuckle as-built (Ti-5Al-4V)   Warm air mixer as-built ((Ti-5Al-4V)

FIGURE 3.13   Typical titanium products built by EBM.

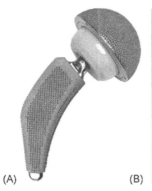   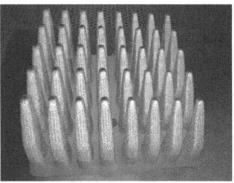

(A)   (B)

FIGURE 3.14   Typical EBM products. (Courtesy of Adler Ortho S.p.A., Italy.)

heaters are placed above the build platform to maintain and preheat the powder prior to spreading over the build area. This preheating of powder and maintenance of an elevated, uniform temperature within the build platform are necessary to minimize the laser power requirements of the process and to prevent warping of the part during the build due to non-uniform thermal expansion and contraction. After preheating of a powder layer, a focused CO_2 laser beam is directed onto the powder bed and is moved using galvanometers in such a way that it thermally fuses the material to form the slice cross section.

The surrounding powder remains loose and serves as a support for subsequent layers, thus eliminating the need for the secondary supports that are necessary for vat photopolymerization processes. This process repeats until the complete part is built. A cool-down period is typically required to allow the parts to uniformly come to a low enough temperature that they can be handled and exposed to ambient temperature and atmosphere; otherwise, parts may warp due to uneven thermal contraction. Finally, the parts are removed from the powder bed, loose powder is cleaned off the parts, and further finishing operations are performed, if necessary.

SLS uses a high-power laser to fuse small particles (polyamide, steel, titanium, alloys, ceramic powders, etc.). What sets sintering apart from melting is that the sintering processes do not fully melt the powder but heat it to the point where the powder can fuse together on a molecular level. The latest SLS machines offer laser powers from 30 to 200 W in a CO_2 chamber controlled machine.

Since the introduction of laser sintering (LS), each new technology developer has introduced competing terminology to describe the mechanism by which fusion occurs, with variants of "sintering" and "melting" being the most popular. These variants are discussed in Chapter 5.

3. Selective Laser Melting (SLM)

SLM was first developed by Fockele and Schwarze (F&S) with the help of Fraunhofer Institute of Laser Technology, Germany, in 1999. Compared with SLS, this technique is faster but requires inert gases for the laser. As with the SLS process, this process also fabricates products layer by layer with the help of high-energy laser beams on a powder bed. In this method, powder is melted rather than sintered. It is currently a very popular method for the fabrication of metal parts. In SLM, a laser is fitted on top with a set of lenses focused on the powdered material for melting the required areas of the layer. Once a layer is solidified, the build platform goes down, and the recoated arm makes a new layer on the top of the bed. This entire process will repeat until the whole part is produced. The minimum thickness of the powder material layer is 0.020 mm. Recently, high-powered lasers with the development of fiber optics have also been added in SLM to process different types of metallic materials, such as tungsten, copper, stainless steels, cobalt, chrome, titanium, and aluminum. This method has the advantages of producing high-precision and high-quality products. It is most useful for products with complex internal geometries, like spiral vents and nested cores, or for rare metals that would be prohibitively expensive to machine. A schematic of the SLS/SLM system is presented in Figure 3.15.

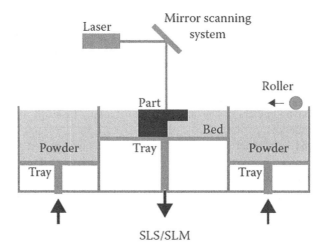

FIGURE 3.15   Schematic of a SLS/SLM system. (From Gardan, J., *International Journal of Production Research*, 54, 10, 3118–3132, 2016.)

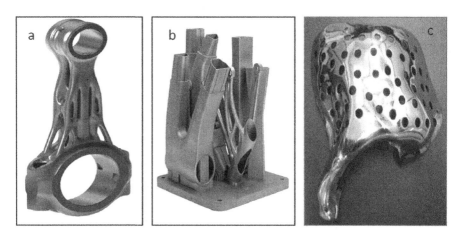

FIGURE 3.16   Typical SLM products. (a) Titanium Connecting Rod; (b) small batch manufacturing of a bicycle frame; (c) frontal customized implant as identical replica of the software planning.

Figure 3.16 shows some metallic products made by SLM. It should be noted that the geometrical shapes of these products cannot be directly manufactured using conventional processes. These improved design features exhibit higher mechanical and physical performance characteristics.

4. Direct Metal Laser Sintering (DMLS)

DMLS is similar to SLS with some differences. The technology is a PBF process that melts the metal powder locally using the focused laser beam. A product is manufactured layer by layer along the z axis, and the powder is deposited via a scraper moving in the x-y plane. The DMLS process was first developed by EOS, Germany, in 1995. The first generation of EOS machines included a 200-W laser source, and the second generation (EOSINT M 280) was launched with a 400-W fiber laser. The trend shows an increase in laser power and also an increase in work chamber size. DMLS often refers to the process that is applied to metal alloys for the manufacturing of direct parts in industry, including aerospace, dental, medical, and other industries that have small to medium-size, highly complex parts, and the tooling industry to make direct tooling inserts. Support structures are required for most geometries because the powder alone is not sufficient to hold in place the liquid phase created when the laser is scanning the powder. The rapid manufacturing of parts by the DMLS process requires the use of a powder that is composed of two types of particles. One type has a low melting point, and the other has a high melting point. The high–melting point particles generate a solid matrix, whereas the particles with the low melting point bind the matrix after being melted by the laser energy input (liquid-phase sintering and partial melting). Titanium and its alloys are widely used for various implants in the orthopedic and dental fields because of its good corrosion resistance and mechanical strength. By now, the concept of the DMLS technique for implantation biomaterial manufacturing is well accepted. Also, aluminum powder, in particular, broadens the range of possible applications of DMLS to lightweight structural components.

5. **Selective Heat Sintering (SHS)**

SHS is a plastics AM technique similar to SLS, the main difference being that SHS employs a less intense thermal printhead instead of a laser, thereby making it a cheaper solution and able to be scaled down to desktop sizes.

3.1.5 BINDER JETTING (BJ)

The original name for BJ was three-dimensional printing (3DP), and it was developed at Massachusetts Institute of Technology (MIT) in 1993. The process involves selectively adding binder where it is needed on top of powder layer by layer to form part cross sections. A typical layer comprises particles of about 100 µm. Hence, in BJ, only a small portion of the part material is delivered through the printhead. A wide range of polymers, composites, metals, and ceramic materials have been manufactured, but only a subset of these are commercially available. Some BJ machines contain nozzles that print color, not binder, enabling the fabrication of parts with many colors. Several companies licensed the 3DP technology from MIT and became successful machine developers, including ExOne and 3D Systems.

A novel continuous printing technology has been developed recently by Voxeljet that can, in principle, fabricate parts of unlimited length. Most of the part material is comprised of powder in the powder bed. Typically, binder droplets (80 µm in diameter) form spherical agglomerates of binder liquid and powder particles as well as providing bonding to the previously printed layer. Once a layer is printed, the powder bed is lowered, and a new layer of powder is spread onto it (typically via a counter-rotating rolling mechanism). A schematic of the architecture of almost all commercially available BJ machines is shown in Figure 3.17. An array of print heads is mounted on an x-y translation mechanism. If the process is capable of printing colored parts, some printheads are dedicated to printing binder material, while others are dedicated to printing color. Typically, the printheads used are standard, off-the-shelf printheads that are found in machines for 2D printing of posters, banners, and similar applications. Since the process can be economically scaled by simply increasing the number of printer nozzles, the process is considered a scalable, line-wise patterning process. Such embodiments typically have a high deposition speed at a relatively low cost (due to the lack of a high-powered energy source).

The BJ–DP process shares many of the advantages of powder bed processes. Parts are self-supporting in the powder bed, so that support structures are not needed. Similarly to other powder bed processes, parts can be arrayed in one layer and stacked in the powder bed to greatly increase the number of parts that can be built at one time. Finally, assemblies of parts and kinematic joints can be fabricated, since loose powder can be removed between the parts.

The printed part is usually left in the powder bed after its completion in order for the binder to fully set and for the green part to gain strength. Post-processing is necessary to reach a final functional product. This post-processing involves removing the part from the powder bed, removing unbound powder via pressurized air, and infiltrating the part with an infiltrant to increase strength and possibly, to impart other mechanical properties. Other post-processing procedures may be required,

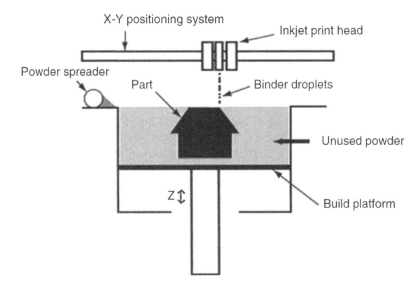

FIGURE 3.17 Schematic of binder jetting 3DP process (BJ–3DP). (With kind permission from Springer Science+Business Media: *Additive Manufacturing Technologies*, 2015, Gibson, I., Rosen, D., and Stucker, B.)

such as sintering (for metallic parts), machining, or polishing, which will protect the integrity of the print and ensures it reaches final part specifications for its application. Nevertheless, infiltration remains the most commonly applied process.

Examples of commonly used infiltrates include molten wax, varnish, lacquer, cyanoacrylate glue (super glue), polyurethane, and epoxy. Figure 3.18 presents the main parameters of both BJ and the process of infiltration.

Applications of BJ–3DP processes are highly dependent upon the material being processed. Low-cost BJ machines use a plaster-based powder and a water-based binder to fabricate parts. Polymer powders are also available. Some machines have color printheads and can print visually attractive parts. With this capability, the market has developed colorful figures from various computer games, as well as personal dummies or sculptures, with images taken from cameras. With either the starch or polymer powders, parts are typically considered visual prototypes or light-duty functional prototypes. In some cases, particularly when using elastomeric infiltrant, parts can be used for functional purposes. With polymer powders and wax-based infiltrant, parts can be used as patterns for investment casting, since the powder and wax can burn off easily. For metal powders, parts can be used as functional prototypes or for production purposes, provided that the parts have been designed specifically for the available metal alloys. Molds and cores for sand casting can be fabricated by some BJ machines that use silica or foundry sand as the powder. This is a sizable application in the automotive and heavy equipment industries.

Powder handling and spreading systems are similar to those used in PBF processes. Differences arise when comparing low-cost visual model printers (for plaster or polymer powders) with metal or sand printers. For the low-cost printers, powder containers (vats) can be hand-carried. In the latter cases, however, powder beds can weigh hundreds or thousands of kilograms, necessitating different material handling

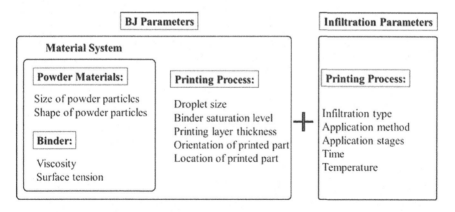

FIGURE 3.18 Main parameters of both BJ and the process of infiltration. (With kind permission from Springer Science+Business Media: Post-process Influence of Infiltration on Binder Jetting Technology. *Advanced Structured Materials*, 65, 2017, Garzón, E. O, Lino, J. Alves, and Neto, R. J.)

and powder bed manipulation methods. For sand printers, the vats utilize a rail system for conveying powder beds to and from de-powdering stations, and cranes are used for transporting parts or molds.

Common Applications and Materials for PB–3DP:

BJ is commonly applicable in such industries as aerospace and heavy industry, where thermally stable and wear-resistant parts are needed. It is also widely applicable for making casting patterns, cores, and molds for intricate products. Jewelry and decorative industries have also embraced BJ as a means of producing one-of-a-kind, colorful, and custom-made objects. An unusual yet promising application of BJ may also lie in the food industry using starch, cellulose, sugar, etc. Typical BJ–3DP products are presented in Figure 3.19.

Advantages of BJ Processes:

The BJ processes share many of the advantages of MJ. With respect to MJ, BJ has some distinct advantages.

- BJ can be faster, since only a small fraction of the total part volume must be dispensed through the printheads. However, the need to distribute powder adds an extra step, slowing down binder processes somewhat
- Due to its fast print speeds and relatively low costs, BJ is a cost-effective choice for 3D printing
- A range of materials (in powder form) can be used, giving the flexibility to create parts from materials like polymers, metals, ceramics, and even sandstone, and silica
- The combination of powder materials and additives in binders enables material compositions that are not possible, or not easily achieved, using direct methods
- The process requires no support structures because the part is surrounded by the unbound powder. This also means that several parts can be printed in the same powder bed at once, ideal for low to medium batch production
- A significant advantage of BJ over other 3D printing technologies is that no additional heat is applied to the part (unlike FDM, SLS, and DMLS processes), which eliminates distortions such as warping of the part
- Slurries with higher solids loadings are possible with BJ compared with MJ, enabling better-quality ceramic and metal parts to be produced
- BJ processes lend themselves readily to printing colors onto parts

Limitations of Binder Jetting:

Although there are valuable benefits to BJ technology, there are also some limitations that are worth considering:

- As a result of the process, parts produced with BJ technology may have limited mechanical properties, or poorer accuracies and surface finish, requiring additional post-processing to strengthen them

FIGURE 3.19 Typical products manufactured using BJ machines. (a) A full-color print (Sandstone), (b) an oil and gas stator printed from stainless steel and infiltrated with bronze, (c) resin pattern for casting, (d) turbine component, SS 420 infiltrated with bronze (design: Airbus Deutschland GmbH), (e) binder jetted reflector antenna from SS 316 and infiltrated with copper, (f) BJ shell structure sand mold.

- Additional post-processing can add much more time and cost to the overall production process
- For metal BJ, the parts are produced in their green state, and infiltration or even sintering post-processing is necessary

Plaster-Based 3DPrinting Process (PB–3DP)

This process is a subset of BJ technology for specific types of products that are based on certain types of powder such as plaster, sandstone, sand, or gypsum. Typically, a layer of the powder is laid on the machine build platform, and the binding ink is ejected by the nozzles of the printhead onto the powder so as to harden it. Once a single layer of powder is completely bound, the additional powder is added via a rake-like instrument over the previous film, and the process is repeated until the entire object fabrication is completed.

The most commercially available powder from 3D Systems is plaster based (calcium sulfate hemihydrate), and the binder is water based. Printed parts are fairly weak, so they are typically infiltrated with another material. In general, parts with any infiltrant are much stiffer than typical thermoplastics or resins, but are less strong and have very low elongation at break (0.04–0.23%).

The production of large sand casting patterns and molds is one of the most common applications for plaster-based BJ technology. The low cost and speed of the process make it an excellent solution for elaborate pattern designs that would be very difficult or impossible to produce using traditional techniques. The cores and molds are generally printed with sand or silica. After printing, the molds are generally immediately ready for casting. A typical example is shown in Figure 3.20, which recommends this technique to revolutionize the traditional casting technology.

FIGURE 3.20 Multi-part sand casting assembly used to cast an engine block. (Courtesy of ExOne)

3.1.6 Directed Energy Deposition (DED)

DED is an AM process in which a focused thermal energy source such as a laser or electron beam or plasma arc is used to fuse materials by melting while they are deposited. In the case of electron beam–based systems, the process must be performed in a vacuum to prevent the electrons interacting with or being deflected by air molecules. On the other hand, laser-based systems require a fully inert chamber when working with reactive metals. Alternatively, it is possible to use a shroud of shielding gas, which is sufficient to protect the metal being deposited from contamination. DED is, typically, used for metallic parts; however, it can be used with polymers and ceramics. Generally, any weldable metals can be fabricated using DED, including aluminum, Inconel, niobium, stainless steel, tantalum, titanium and titanium alloys, and tungsten. BED works by depositing material onto a base or component under repair through a nozzle mounted on a 4- or 5-axis arm. The metallic material is fed through the nozzle in either powder or wire form. As it is being deposited, the heat source melts the material simultaneously. This procedure is done repeatedly until the layers have solidified and created or repaired an object. DED can be used to fabricate parts but is generally used for repair or to add material to existing components. Generally-speaking, the applications for DED fall into three categories: near net shape parts, feature additions, and repair.

The electron beam additive manufacturing (EBAM) process is shown in Figure 3.21, with computer numerically controlled (CNC) motion and a wire feeder within a vacuum chamber allowing movement in x-y or tilt orientations to fuse a deposited bead of metal one bead at a time, layer by layer. Using this approach, with high vacuum, very large structures can be deposited. ₁₁High deposition rates are

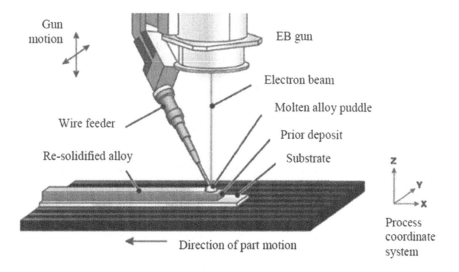

FIGURE 3.21 Electron beam additive manufacturing (EBAM™) process. (Courtesy of Sciaky, Inc., United States)

possible using a wide range of available wire alloys and sizes and choice of deposition parameters. Deposition rates up to 6.8–18 kg/h can be realized. The electron beam–based process can provide advantages over current laser beam systems in beam power, power efficiency, and deposition rate (Lachenburg and Stecker 2011).

The advantages of DED include:

- Control of the grain structure, which makes it suitable for the repair of parts
- Production of large parts with minimal tooling
- Production of components with hybrid structures using multiple materials with different compositions

The disadvantages include:

- Finish depends on the material used and may require post-finishing operations
- Material used is still relatively limited

The print quality of DED depends on the following factors (www.xometry.com/):

1. Porosity: Internal defects in the deposited material lead to weak points that should be minimized by drying the powder to keep moisture out and by the proper use of shielding gas
2. Scanning speed affects the size of the melt pool, the cooling rate, and consequently, the grain structure
3. The optimum speed depends on the material used and the desired grain structure
4. The power provided by the energy source has a direct effect on the melt and is related to the scan speed
5. Energy transferred to the component must be sufficient to properly melt the host material even as the DED printhead moves along the build path

DED technology is complex and challenging to implement due to the following (www.xometry.com/):

1. It requires a large initial investment to set up
2. Skilled and experienced operators are necessary to run the DED system efficiently and accurately
3. When designing components to be built using DED technology, the printhead motion must be considered, which adds effort and work hours each time
4. DED is relatively new; there is little standardization in digital information management, design, or manufacturing processes for these systems

DED covers two main terminologies. These are laser metal deposition (LMD) and laser engineered net shaping (LENS).

1. **Laser Metal Deposition (LMD)**

LMD is an AM process combining high-power laser and powder processing used for the manufacture of high-precision near net shape components from powders. As shown in Figure 3.22, a laser beam forms a melt pool on a metallic substrate, into which powder is fed. The powder melts to form a deposit that is fusion-bonded to the substrate. The required geometry is built up layer by layer. Both the laser and the nozzle that delivers the powder are manipulated using a CNC system or a robotic arm. LMD enables the building of whole geometries or the repair of an already existing but damaged part.

During this process, a surface layer is created by means of melting and simultaneous application of any given material. The laser usually heats the workpiece with a defocused laser beam and melts it locally. At the same time, an inert gas, mixed with a fine metal powder, is introduced. The metal powder melts at the heated area and is then fused to the surface of the workpiece. Due to the high cost of the powder used, laser equipment, and protective atmosphere, LMD is regarded as an expensive but versatile process. Therefore, it has been used in high–value added aerospace and medical applications that require the creation or repair of high-value or bespoke parts, especially those on a larger scale. Figure 3.23 shows typical LMD application for shape generation.

LMD can be used to generate or modify 3D geometries. The laser can also perform repairing or coating processes with this method. In the aerospace industry, it is used to repair turbine blades. In tool construction and mold making, broken or worn edges and shaping functional surfaces are repaired or even armored locally. Bearing points, rollers, or hydraulic components in energy technology or petrochemistry are coated to protect against wear and corrosion (Figure 3.24).

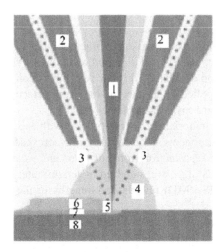

1-Laser beam
2-Powder nozzle
3-Powder
4-Shieldgas area
5-Weld pool
6-Weld bead
7-Bond zone
8-Substrate

FIGURE 3.22   Laser metal deposition (LMD). (Courtesy of CHIRON Group SE, Germany)

FIGURE 3.23   Typical LMD application. (www. trumpf.com)

FIGURE 3.24   High-speed laser metal deposition (HS-LMD). (www. trumpf.com)

Advantages of LMD:

- Creates rough and very fine structures
- Achieves high building rates in comparison to other methods in AM
- Several powders can be conveyed and supplied at the same time, which makes it possible to develop dedicated alloys or switch between materials
- Structures can be easily applied to 3D parts and uneven surfaces
- Changes to the geometry are easily possible
- Change between different materials in a work process with ease
- Used to repair tools or components, for example in the aerospace industry
- Used to apply thin layers for protection from corrosion and wear
- Achieves maximum speeds of several hundred meters/minute: high-speed laser metal deposition (HS-LMD) reveals its strengths in the coating of brake disks
- Offers a low-heat, high-strength method of laser AM, minimizing dilution, heat effect zones, and stress distortions
- Produces fully customized, functionally graded parts for demanding applications, or restores existing parts to their original strength

- Highly precise and fast, allowing the creation of rough to very fine structures
- The material range is extensive, offering single-component metal powders, tungsten carbides, alloy powder, and even custom powder blends to create sandwiched bimodal structures or new alloys. Material is also minimally wasted, since it is added as needed
- Effective in fabrication, repair, welding, and cladding, making it a multi-purpose manufacturing process
- Used on existing and uneven surfaces and is not held back by a build platform
- Bonds are truly metallurgical and not mechanical, making welds much stronger than traditional welds or plating techniques

Disadvantages Include:

- Capital cost is very high
- Material price is also high
- Because of the low deposition rate, it is suited mostly to small components
- The process is very complex, requiring trained operators
- The bimodal nature of the microstructure produced by LMD can become an issue, especially if the laser is not properly adjusted or materials do not play well
- Oxygen contamination is more common in LMD than in other laser-based AM processes, increasing shield gas wastage and the risk of brittleness/part failure
- Surface finishes can be rough and porous

Typical Applications of LMD Technology Include:

- Repairs of tooling surfaces, high-value parts, engine components, turbine blades, etc.
- The surfacing of oil and gas drills
- Repairs of sintered tools/metal components
- New alloy development
- Medical implants
- Rapid prototyping
- Deposition of vanadium carbide tool steels and titanium alloys onto high-stress components
- Fabrication of high-value or bespoke parts for aerospace, oil and gas, power, and tooling applications

2. Laser Engineered Net Shaping (LENS)

A high-power laser is used to melt metal powder supplied coaxially to the focused laser beam through a deposition head. The laser beam travels through the center of the head and is focused to a small spot by one or more lenses. As shown in Figure 3.25, the x-y table is moved in raster fashion to fabricate each layer of the

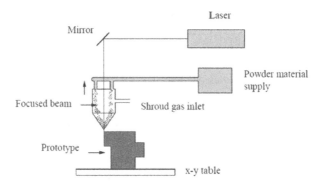

FIGURE 3.25 LENS process.

part. The head is moved up vertically after each layer is completed. Metal powders are delivered and distributed around the circumference of the head either by gravity or by using a pressurized carrier gas. In order to control print properties and promote layer-to-layer adhesion by providing better surface wetting, an inert shroud gas is often used to shield the melt pool from atmospheric oxygen.

A variety of materials, such as stainless steel, Inconel, copper, titanium, and aluminum, can be processed. Most systems use powder materials, but there has also been work done with material provided as fine wires. In such a case, the material is fed off-axis to the laser beam. The building area is usually contained within a chamber both to isolate the process from the ambient surroundings and to shield the operators from possible exposure to fine powders and the laser beam. The laser power used varies from a few hundred watts to 20 kW or more, depending on the particular material, feed rate, and other working parameters. The objects fabricated are near net shape, which generally requires finish machining. Critical build parameters include laser power, scan speed, powder feed rate, hatch distance, laser focal distance, and deposition pattern. Ultimately, these parameters and their interplay dictate the overall quality of the final build and require optimization to ensure build integrity.

Typical applications of LENS include the fabrication of parts for applications in key industries such as medical and aerospace. LENS creates functionally graded porous structures, which have been exploited for orthopedic implant applications. The process is suited to applications in environmental control, which eliminates issues relating to part oxidation during manufacturing, as the parts are processed in an inert environment. Figure 3.26 shows a repaired gear using the same process. Figure 3.27a shows a part after deposition and before machining, and Figure 3.27b shows the part after the spin test; the leading edge thickness was around 0.025″.

Advantages of Laser Engineered Net Shaping

* Typical build volume is comparatively large, possibly $900 \times 1500 \times 900$ mm, which allows the creation of much larger parts while simultaneously reducing relative powder demands

FIGURE 3.26 The repaired gear with LENS 3D printed. (https://3dprintingindustry.com/)

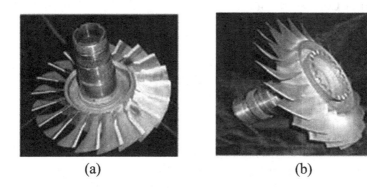

(a)   (b)

FIGURE 3.27 (a) Part after deposition and before machining. (b) After the spin test, the leading edge thickness was around 0.025″. (www.multistation.com/)

- LENS machines generate smaller melt pools, allowing higher solidification/cooling rates as compared with conventional manufacturing processes like casting
- LENS is able to fabricate multiple materials and create functionally graded materials due to the multiple powder feed lines
- LENS can be used to fabricate products using materials with high melting temperatures
- Manufactured parts are found to be more robust, less brittle, and less prone to cracking at low stress
- Produces a more defect-free interfacial bond between the layers for optimization within a given process window
- Creates complex porous structures. Porosity can be introduced not only through design architecture but also inherently within the bulk of the processed material itself
- Manufacturing can be performed without the need for a powder bed

Disadvantages of Laser Engineered Net Shaping

- The residual stress is significantly higher, which affects part precision and may lead to part collapse during deposition
- Due to the uneven heating and cooling rate, there is non-uniformity in microstructure and macrostructure and mechanical properties with respect to the height of build
- The build accuracy is entirely related to the optimization of build process parameters for a given metallic powder to produce a respective model at a desired layer height
- More time is generally needed for post-processing as compared with PBF technologies
- There is a staircase effect that affects surface roughness and negatively affects the mechanical properties of the finished part

3.1.7 Sheet Lamination (SL)

SL is used to produce colored parts in highly detailed resolution. Accordingly, a thin layered aluminum foil or paper-based filaments are cut into shaped layers, often by lasers or a very sharp blade. The layers are coated with adhesive and glued together layer by layer. The precision of the results and the final quality depend on the thickness of layered materials. Printed objects may be modified by machining. SL uses a wide variety of materials, such as paper, polymer, ceramic, and metal; each material uses different binding methods.

- Paper is the most common, with pre-applied adhesive, where heat and pressure are used to activate the adhesive layer
- Polymers use heat and pressure without the adhesive, since it relies on melting the sheets together
- Metal sheets are bound using ultrasonic welding
- Fiber-based material and ceramic use thermal energy in the form of oven baking to combine the layers together

SL can be categorized into the following seven types:

- Laminated object manufacturing (LOM)
- Selective lamination composite object manufacturing (SLCOM)
- Plastic sheet lamination (PSL)
- Computer-aided manufacturing of laminated engineering materials (CAM-LEM)
- Selective deposition lamination (SDL)
- Composite-based additive manufacturing (CBAM)
- Ultrasonic additive manufacturing (UAM)

Every type of SL works marginally differently from the others, although the main principle is the same. A schematic overview of SL is shown in Figure 3.28 of the original LOM, which was the first commercialized, in 1991.

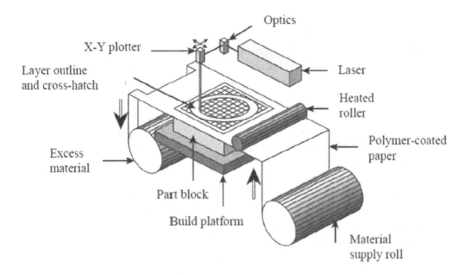

FIGURE 3.28 Sheet lamination overview. (Courtesy of Additive Manufacturing Technologies)

Four steps are followed for SL:

1. First, a thin sheet of the material is fed from the roller or placed onto the build platform
2. Depending on the type of SL, the next layer may or may not be bonded to the previous sheet
   - SDL and UAM bond the layers together and then cut the 3D shape at the end
   - CAM-LEM cuts the layers into shape and then bonds the layers together
3. This process is continued until it completes all the layers to achieve the full height of the part
4. The print block is removed, and all the unwanted outer edges are removed to reveal the printed 3D object

Advantages of SL Include:

- Faster printing time
- Ease of material handling
- Ceramic (CAM-LEM) and composite fiber (SLCOM) parts can be manufactured
- OEM components, such as sensors, wires, etc., can be embedded into the part
- Relatively low cost
- No support structures are needed
- Larger working area

- Full-color prints (LOM/SDL)
- The material state does not change during or after the process
- Multi-material layers are possible (UAM)
- Cut material can be easily recycled

Disadvantages of SL Are:

- Post-processing is required
- Layer height cannot be changed
- Finishes can vary depending on paper or plastic material
- Limited material options available
- Time-consuming and difficult to remove the excess material after the laminating phase
- Hollow parts are difficult to produce in some types of sheet lamination processes, such as "bond then form"
- Bonding strength depends on the laminating technique used
- Adhesive bonds are not good enough for long-term use of the product: strength and integrity
- Material waste is high if the part being made is smaller than the build area or the sheet size

Figure 3.29 shows carbon fiber sheet laminated parts. The main two types of SL are laminated object modeling and ultrasonic consolidation.

1. Laminated Object Manufacturing (LOM)

In LOM, the paper sheets are heated by a roller that also presses them against the build plate to improve the adhesion. The surrounding boundaries of the part are usually created with cross-hatching to provide easier extraction when the process is

Sheet Lamination

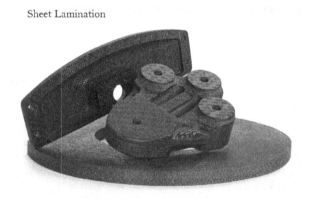

FIGURE 3.29   Carbon fiber sheet laminated parts. (Courtesy of envisiontec)

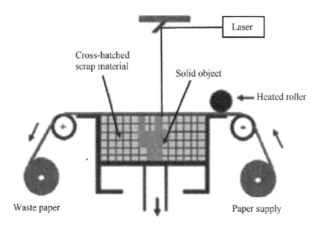

FIGURE 3.30 Laminated object modeling.

finished. A sub-process, selective deposition lamination (SDL), uses colored printed sheets of paper to create full-color 3D objects.

In LOM (Figure 3.30), profiles of object cross sections are cut from paper or other material using a laser beam. The paper is unwound from a feed roll onto the stack and first bonded to the previous layer using a heated roller, which melts a plastic coating on the bottom side of the paper. The profiles are then traced by an optical system that is mounted to an x-y stage.

After the laser has cut each profile, the roll of paper is advanced, a new layer is glued onto the stack, and the process is repeated. After fabrication, some trimming, hand finishing, and curing are needed. In LOM, the surface finish, accuracy, and stability of paper objects are not as good as for materials used with other AM methods. However, material cost is very low, and objects have the look and feel of wood and can be worked and finished in the same manner to form patterns for sand castings. While there are limitations on materials, work has been done with plastics, composites, ceramics, and metals. Some of these materials are available on a limited commercial basis.

LOM is mainly used for rapid prototyping. Applications include creating visual prototypes for demonstrations, color matching, and other purposes. LOM is a very fast and inexpensive way to 3D print objects in several kinds of materials. Sheets of material are bonded together and cut in the right geometry according to the 3D model. LOM is mainly used for rapid prototyping processes, not for production (Figure 3.31).

Advantages of the LOM technique

- Cheaper process
- No significant requirement of support materials
- No residual stress development

FIGURE 3.31 LOM paper based technology allows multicolor 3D printing. (Courtesy of MCor)

- No variation in the material properties
- Demonstrates the scope for manufacturing bigger components

Disadvantages of LOM

- LOM can only use materials that can be cut and bonded by heat or glue, which limits the range of colors, textures, and properties of the models
- LOM models also tend to have rough edges and surfaces, as the material is cut by the laser or the knife

2. Ultrasonic Consolidation

Metal foil is rolled out one by one under a sonotrode, which is a cylindrical tool that welds the foil against the previous layers of foil using ultrasonic vibration under high pressure. This sequence is repeated until the target surface is covered by a layer of metal foil.

Specific applications for ultrasonic additive manufacturing (UAM) include (www .insidemetaladditivemanufacturing.com/):

- Injection molding dies
- Parts with embedded channels
- Incorporation of second-phase material
- Components with complex geometries, which are difficult to produce through conventional manufacturing techniques

3.2 CONCLUSIONS CONCERNING THE AM TECHNOLOGIES

In this chapter, the seven AM categories with their derived processes have been comprehensively discussed. AM has evolved and blossomed into a plethora of processes (Figure 3.1), which have been presented and discussed based on process categories according to the ASTM F42.

Table 3.1 summarizes these categories along with their processed materials, industrial applications, advantages, and limitations (Jandyal et al. 2022). Though AM started in the early 1980s, its wider application in industry is still promising and competing well.

AM has revolutionized the production of various engineering materials, including polymers, composites, metals/alloys, and ceramics. Each AM technique comes with its own set of advantages, limitations, and material compatibility. The choice of material and technique depends on the desired properties, resolution, strength, and intended application of the final product. Table 3.2 outlines the compatibility of different engineering materials (polymers, composites, metals/alloys, and ceramics) with various AM techniques. This table provides a summarized overview of the compatibility OK between different engineering materials and various AM techniques. It's important to note that materials listed under each technique might not cover the entire spectrum of what's possible; ongoing research and advancements continually expand the range of materials compatible with each AM process.

AM of polymers, composites, metals/alloys, and ceramics will be treated in a detailed manner in Chapters 4, 5, and 6. In these chapters, the basics, applications, advantages, and limitations of the adopted techniques will be presented.

3.3 REVIEW QUESTIONS

3.3.1 SOLVED PROBLEMS

1. For an SLA system, estimate the time required to produce a hollow cylindrical polymeric model of the following dimensions.
   - Outside diam. = 220 mm
   - Inside diam. = 180 mm
   - Height = 50 mm
     Assume
   - Average scanning speed of the laser beam at the surface v = 1500 mm/s
   - Spot size diameter of the laser beam at the surface D = 0.5 mm
   - Repositioning time between layers Tr = 8 seconds
   - Layer thickness t = 0.1 mm

Solution:

Time to complete each layer Ti = Ai/vD
Ai = $(\pi/4)$ $(Do^2 - Di^2)$ = 12,571 mm²
Ti = 12,571/1500 × 0.5 + 8 = 24.76 seconds
Number of layers n = 50/0.1 = 500 layers

TABLE 3.1

AM Categories, Applications, Materials, Benefits, and Drawbacks

Ser. No	AM-Category	Materials	Applications	Advantage	Limitation
1	Vat Polymerization	A photo-active resin monomers, polymer–ceramics	Biomedical models	High surface finishHigh accuracyComplicated parts can be easily manufacturedHigh thermal durabilityAM prints made by this process can serve as patterns for casting	Less surface area is exposed to laser (about 0.15 mm), which makes it a slow processHigh initial investment cost. Overhanging parts are difficult to manufactureThe photosensitive resin is difficult to handle
2	Material Jetting	HDPE: high density polyethylene PS: polystyrenes PMMA: polymethyl methacrylate PC: polycarbonate ABS: acrylonitrile butadiene styrene HIPS: high-impact polystyrene EDP: environmentally degradable plastic Metals Ceramics	Electronics applications	Can achieve outstanding accuracy and surface finishesParts are good for use in patterns for casting	Limited number of materialsSlow build processParts are fragile and of poor mechanical properties due to wax-like materials
3	Material Extrusion	Continuous filaments of thermoplastic polymers, fiber-reinforced continuous polymerics	Rapid prototyping of advanced composite parts and toys	Less initial investment costComplex shapes can be easily madeHigh flexibility	A slow process, although the time taken depends on the part to be manufacturedQuality is not as good as vat polymerization or Powder Bed Fusion.

(continued)

TABLE 3.1 (CONTINUED)

AM Categories, Applications, Materials, Benefits, and Drawbacks

Ser. No	AM-Category	Materials	Applications	Advantage	Limitation
4	Powder Bed Fusion	Compressed fine powder components, limited polymerics, metals and alloys	Medicinal, electronic, aviation, and lightweight structures	Low costNo external support is requiredWide material choicePowder recyclingComplex parts can be manufactured. Suitable for mass productionGood accuracy and precision	Post-processing is requiredWeak structural material properties. Time-consuming processHigh cost of manufacturingRequires post-processingLarge surfaces, tiny holes are difficult to manufacture accurately
5	Binder Jetting	Metals, sand, and ceramics that are granular in shape	Fabrication of full-color prototype and wide sand casting cores and molds	High resolutionHigh surface finishNo need for post-processingPrinting can be done over a large areaMultiple printings at one time	Limited materials are availableLow part strengthSubstrate is required for printingLow densityShrinkage without infiltration
6	Directed Energy Deposition	Alloys and metals in the form of wire or powder, polymers and ceramics	Aerospace, retrofitting, repair, cladding, biomedical	Denser print creation is possible. Allows directional solidification, which enhances featuresUtilized effectively for repairing and refurbishing components	Time-consuming processLow accuracy, poor resolution and surface finishLimited material is availableLimitation for complex printing with fine details and shapes
7	Sheet Lamination	Polymer, metal-filled tapes, ceramics, metal rolls, and composites	Paper making, foundry sector, smart structures	External support is not required. InexpensiveQuick processSuitable for large parts	Post-processing is requiredPoor dimensional accuracyPoor surface finishComplex parts are difficult to manufacture

Source: Jandyal, A. et al., in *Sustainable Operations and Computers*, 3, 33–42, 2022.

TABLE 3.2

Compatibility of Different Engineering Materials (Polymers, Composites, Metals/Alloys, and Ceramics) with Various Additive Manufacturing Techniques

Engineering Materials	Vat Photopolymerization	AM Categories					
		MJ	BJ	PBF	ME	DED	SL
Thermosets: Epoxies and acrylates	x	x					
Thermoplastics: Polyamide, ABS, PPSF		x	x	x	x		x
Wood, Paper							x
Metals/Alloys: Steels, stainless steels, Ti-alloys, Co-alloys, Cr-alloys, non-ferrous			x	x		x	x
Ceramics							
Industrial: Zirconia, Alumina, Silicon nitride	x		x	x			x
Structural: Cement, Foundry sand		x	x	x			
Composites	Combinations of the above materials can also be printed by selected AM technologies as indicated in Chapter 4.						

Source: Adapted from ISO-DIS 17296-2 Additive manufacturing—General principles—Part 2 Overview of process categories and feedstock.

Total time of AM = 500 × 24.76 = 12,380 seconds = 206 minutes = 3.44 hours

2. A prototype of a tube with a square cross section is to be fabricated using stereolithography. The outside dimension of the square = 100 mm and the inside dimension = 90 mm. The height of the tube (z direction) = 80 mm. Layer thickness = 0.10 mm. The diameter of the laser beam (spot size) = 0.25 mm, and the beam is moved across the surface of the photopolymer at a velocity of 500 mm/s. Estimate the time required to build the part, if 10 s are lost each layer to lower the height of the platform that holds the part. Neglect the time for post-curing.

Solution:

Given:

• Outside dimension of the square = 100 mm
• Inside dimension of the square = 90 mm
• Height = 80 mm
• Average scanning speed of the laser beam at the surface v = 500 mm/s
• Spot size diameter of the laser beam at the surface D = 0.25 mm

- Repositioning time between layers Tr = 10 seconds
- Layer thickness t = 0.1 mm

Time to complete each layer Ti = Ai/vD
Ai = (Do² − Di²) = 1900 mm²
Ti = 1900 / 500 × 0.25 + 10 = 25.2 seconds
Number of layers n = 80/0.1 = 800 layers
Total time of AM = 800 × 25.2 = 20,160 seconds = 336 minutes = 5.6 hours

3. Solve Problem 1 except that the layer thickness = 0.40 mm.

Solution:
As problem 1 except the layer thickness t = 0.4 mm
Time to complete each layer Ti = Ai/vD
Ai = (π/4) (Do² − Di²) = 12,571 mm²
Ti = 12,571/1500 × 0.5 + 8 = 24.76 seconds
Number of layers n = 50/0.4 = 125 layers
Total time of AM = 125 × 24.76 = 3095 seconds = 51.58 minutes = 0.86 hours

4. The part in Problem 2 is to be fabricated using FDM instead of stereolithography. Layer thickness is to be 0.20 mm, and the width of the extrudate deposited on the surface of the part = 1.25 mm. The extruder work-head moves in the x-y plane at a speed of 150 mm/s. A delay of 10 s is experienced between layers to reposition the work-head. Compute an estimate for the time required to build the part.

Solution:
Given

- Outside dimension of the square = 100 mm
- Inside dimension of the square = 90 mm
- Height = 80 mm
- The extruder work-head moves in the x-y plane at a speed v = 150 mm/s
- The width of the extrudate deposited on the surface of the part = 1.25 mm
- Repositioning time between layers Tr = 10 seconds
- Layer thickness t = 0.2 mm

The part is to be fabricated using FDM instead of stereolithography.
Time to complete each layer Ti = Ai/vD
Ai = (Do² − Di²) = 1900 mm²
Ti = 1900/150 × 1.25 + 10 = 20.13 s
Number of layers n = 80/0.2 = 400 layers
Total time of AM = 400 × 20.13 = 8052 seconds = 134 minutes = 2.24 hours

5. Solve Problem 2, except using the following additional information. It is known that the diameter of the filament fed into the extruder work-head is 1.25 mm, and the filament is fed into the work-head from its spool at a rate

of 30.6 mm/second while the work-head is depositing material. Between layers, the feed rate from the spool is zero.

Solution:
Given:
Outside dimension of the square = 100 mm
Inside dimension of the square = 90 mm
Height = 80 mm
The filament is fed into the work-head from its spool at a rate of 30.6 mm/second, which equals the extruder work-head speed
The width of the extrudate deposited on the surface of the part = 1.25 mm
Repositioning time between layers T_r = 10 seconds
Layer thickness t = 0.2 mm
Time to complete each layer T_i = A_i/vD
$A_i = (D_o^2 - D_i^2) = 1900$ mm²
$T_i = 1900/30.6 \times 1.25 + 10 = 59.67$ seconds
Number of layers n = 80/0.2 = 400 layers
Total time of AM = 400 × 59.67 = 23,868 seconds = 398 minutes = 6.63 hours

6. A solid cone-shaped part is to be fabricated using stereolithography. The radius of the cone at its base = 35 mm and its height = 40 mm. The layer thickness = 0.20 mm. The diameter of the laser beam = 0.22 mm, and the beam is moved across the surface of the photopolymer at a velocity of 500 mm/s. Estimate the time required to build the part, if 10 seconds are lost each layer to lower the height of the platform that holds the part. Neglect post-curing time.

Solution:
Given:

- Radius of the cone at its base = 35 mm
- Its height = 40 mm
- Average scanning speed of the laser beam at the surface v = 500 mm/s
- Spot size diameter of the laser beam at the surface D = 0.22 mm
- Repositioning time between layers T_r = 10 seconds
- Layer thickness t = 0.2 mm
- Cone diameter D = 70 mm
- Cone height H = 40 mm
- No. of layers n = H/n = 40/0.2 = 200 layers

Time to complete each layer T_i = A_i/vD
$A_{mean} = (\pi/4) (D_o^2 - 0)/2 = (\pi/4) (70)^2/2 = 1925$ mm²
$T_i = 1925/500 \times 0.22 + 10 = 27.5$ seconds
Total time of AM = 200 × 27.5 = 5500 seconds = 91.7 minutes = 1.53 hours

3.3.2 UNSOLVED QUESTIONS

1. What are the three types of starting materials in AM?
2. What industry branch preferably uses stereolithography?
3. What materials can be processed with metal laser sintering or laser melting processes?
4. Besides the starting material, what other feature distinguishes the AM technologies?
5. Of all of the current AM technologies, which one is the most widely used?
6. Describe the AM technology called SGC.
7. Describe the AM technology called LOM.
8. What is the starting material in FDM?
9. Compare between DLP and SLA.
10. What are the fields of application of DLP?
11. What are the advantages and limitations of the following AM categories?
    ME, PBF, DED, BJ, MJ
    What are advantages and limitations of the following AM processes?
    SLA, DLP, CLDP, SGC
12. SGC is one of the vat photopolymerization processes. Explain why this process is no longer used to build parts.
13. What technologies are related to PBF?
14. Find out the chronological order of MJ evolution relative to other 3D printing processes.
15. Compare the effect of the two possible droplet formation technologies on the productivity of the MJ system.
16. Make a list of advantages and limitations of each of the AM processes you know.
17. Consider the case of a coffee mug, provided by a long vertical handle connected horizontally from its ends, being produced by the SLA process. How can the top of the handle be manufactured, since there is no supporting material beneath the arch?
18. What are advantages of directed energy deposition (DED)?
19. Show by line sketch the principles of laser metal deposition (LMD).
20. What are the benefits behind using LMD?
21. What are the main applications of laser engineered net shaping (LENS)?
22. What are the categories of sheet lamination (SL)?
23. Explain how sheet lamination can be used for a wide variety of materials.
24. State the main advantages and limitations of sheet lamination.
25. Show, using a line diagram, the principles of ultrasonic addition manufacturing (UAM).
26. Compare powder bed fusion vs. binder jetting.
27. What are the most popular vat photopolymerization AM technologies? Briefly describe them.
28. Describe the AM technology called laminated object manufacturing.

29. What is the starting material in FDM?
30. Construct a table illustrating the engineering materials that can be printed by different AM categories.
31. What is 4D additive manufacturing?

3.3.3 MULTIPLE CHOICE QUESTIONS

1. Which of the following AM processes starts with a photosensitive liquid polymer to fabricate a component (two correct answers): (a) fused deposition modeling, (b) selective laser sintering, (c) solid ground curing, and (d) stereolithography?
2. Which one of the following AM technologies uses solid sheet stock as the starting material: (a) fused deposition modeling, (b) laminated object manufacturing, (c) solid ground curing, or (d) stereolithography?
3. Which of the following AM technologies uses powders as the starting material (two correct answers): (a) fused deposition modeling, (b) selective laser sintering, (c) solid ground curing, and (d) three-dimensional printing?
4. AM technologies are never used to make production parts: (a) true or (b) false?
5. Which of the following are problems with the current material addition AM technologies (three best answers): (a) inability of the designer to design the part, (b) inability to convert a solid part into layers, (c) limited material variety, (d) part accuracy, (e) part shrinkage, and (f) poor machinability of the starting material?
6. Of all the current material addition AM technologies, which one is the most widely used: (a) three-dimensional printing, (b) fused deposition modeling, (c) selective laser sintering, (d) solid ground curing, and (e) stereolithography?

BIBLIOGRAPHY

ASTM. (2009) *International Committee F42 on Additive Manufacturing Technologies (ASTM F2792–10)*. West Conshohocken, PA: Standard Terminology for Additive Manufacturing Technologies.

American Society for Testing and Materials (ASTM) Standard. (2012) *Standard Test Method for Crimp Frequency of Manufactured Staple Fibers*. ASTM D, 3937-12, 256-

Gardan, J. (2016) Additive Manufacturing Technologies: State of the Art and Trends. *International Journal of Production Research*, 54(10), 3118–3132. https://doi.org/10.1080/00207543.2015.1115909

Garzón, E. O, Lino, J. Alves, and Neto, R. J. (2017) Post-process Influence of Infiltration on Binder Jetting Technology. *Advanced Structured Materials*, 65. https://doi.org/10.1007/978-3-319-50784-2_19

Gibson, Ian, Rosen, David, and Stucker, Brent. (2015) *Additive Manufacturing Technologies*. 2nd ed. New York: Springer.

Gibson, Ian, Rosen, David, Stucker, Brent, and Khorasani, Mahyar. (2021) *Additive Manufacturing Technologies*. Springer. https://doi.org/10.1007/978-3-030-56127-7

Groover, Mikell P. (2010) *Fundamentals of Modern Manufacturing-Materials, Processes, and Systems*. 4th ed. Hoboken, New Jersey, USA: John Wiley & Sons, Inc.

https://make.3dexperience.3ds.com/
https://3dprintingindustry.com/
https://www.insidemetaladditivemanufacturing.com/
https://www.multistation.com/
https://www.trumpf.com/
https://www.xometry.com/
ISO-DIS 17296–2. Additive manufacturing - General principles - Part 2 Overview of process categories and feedstock.
Jandyal, Anketa, Chaturvedi, Ikshita, Wazir, Ishika, Raina, Ankush, and Haq, Mir Irfan Ul. (2022) 3D Printing – A Review of Processes, Materials and Applications in Industry 4.0. *Sustainable Operations and Computers*, 3, 33–42. https://doi.org/10.1016/j.susoc.2021.09.004
Janusziewicz, R., Tumbleston, J. R., Quintanilla, A. L., Mecham, S. J., and DeSimone, J. M. (2016) Layerless Fabrication with Continuous Liquid Interface Production. *Proceedings of the National Academy of Sciences USA*, 113(42), 11703–11708. https://doi.org/10.1073/pnas.1605271113
Kalpakjian, S. and Schmid, S. R. (2003) *Manufacturing Processes for Engineering Materials.* Upper Saddle River: Pearson Education, Inc.
Lachenburg, K. and Stecker, S. (2011) Nontraditional Applications of Electron Beams. In *ASM Handbook Welding Fundamentals and Processes*, edited by T. Babu, S. S. Siewert, T. A. Acoff, and V. L. Linert, 540–544. Metals Park: ASM International.
PagacMarek, HajnysJiri, MaQuoc-Phu, JancarLukas, JansaJan, StefekPetr, and Mesicek, Jakub. (2021) A Review of Vat Photopolymerization Technology: Materials, Applications, Challenges, and Future Trends of 3D Printing. *Polymers*, 13, 598. https://doi.org/10.3390/polym13040598
Wiberg, Anton. (2019) *Towards Design Automation for Additive Manufacturing — A Multidisciplinary Optimization Approach.* Sweden: Linköping University.

4 Technologies of Additive Manufacturing of Polymers and Composites

4.1 POLYMERS

Additive manufacturing (AM) uses a wide range of materials, including clay, metals, polymers, and composite materials to build complex parts that may not be possible using traditional manufacturing techniques in a layer-by-layer fashion. This chapter concentrates on the polymer-based AM technologies. The versatility of polymer-based AM, combined with low machine, processing, and feedstock costs, has led to its use for a range of biomedical applications such as orthodontic braces and hearing aids. In this case, the polymers used must have suitable biocompatibility and be able to withstand sterilization.

The word "polymer" comes from the Greek roots *poly* (many) and *meros* (parts); these parts are known as monomers. Monomers are made up of organic materials in which atoms of carbon are joined in covalent bonds with other atoms of hydrogen, oxygen, nitrogen, silicon, chlorine, fluorine, or sulfur. Natural polymers such as silk, shellac, wood, rubber. and cellulose were available to ancient people. However, synthetic or semisynthetic polymers, manufactured as industrial products, are recently developed materials. The term "plastics" is a general common name given to most synthetically developed polymers that may contain other constituents to improve their performance and/or reduce product cost. Plastics can be cast, pressed, or extruded into a variety of shapes, such as films, fibers, plates, tubes, bottles, boxes, and many more (Youssef et al. 2023).

There is an ever-increasing demand for replacing metallic materials in machinery and equipment with plastics for improving the strength-to-weight ratio, safety, and appearance, and reducing noise and cost. Currently, most of the monomers are obtained by fractional distillation of petroleum or natural gas. These monomers are linked to form polymers through chemical reactions known as polymerization. The resulting polymers can be categorized into three main categories according to their properties and applications: these are known as: thermoplastics, thermosets, and elastomers, as shown in Figure 4.1.

At low temperatures, amorphous polymers are rigid, hard, and glassy. When heated, molecules are excited, exhibiting a slow rate of increase in the specific

DOI: 10.1201/9781003451440-4

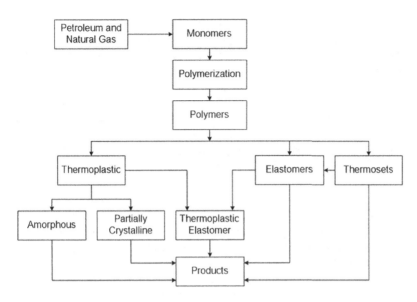

FIGURE 4.1 Classification of polymeric materials. (From Youssef, H. et al., *Manufacturing Technology: Materials, Processes and Equipment*, CRC Press, California, 2023. With permission of CRC)

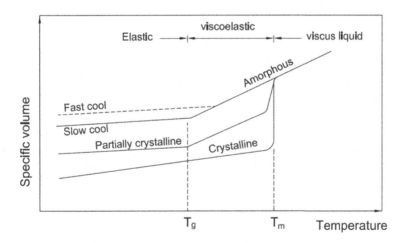

FIGURE 4.2 Change of characteristics of thermoplastic polymers. (From Youssef, H. et al., *Manufacturing Technology: Materials, Processes and Equipment*, CRC Press, California, 2023. With permission of CRC)

volume up to a certain temperature known as the glass transition temperature T_g (Figure 4.2), where the material changes to a rubbery state, exhibiting viscous as well as elastic behavior. Beyond T_g, the rate of increase of the specific volume is higher. Consequently, with further heating, the material changes gradually to a liquid without showing a clear melting point T_m or a change of the rate of increase of the specific volume in the liquid phase. Figure 4.2 also shows that faster rates of cooling from the melt lead to higher glass transition temperatures. On the other hand, crystalline and partially crystalline polymers exhibit a relatively sharp decrease of the specific volume at a distinctive temperature when cooled from the melt to the viscoelastic condition, indicating a melting point similar to metallic materials. The figure also shows that partially crystalline materials have a glass transition temperature. These characteristics made polymers good candidates for AM.

Thermoplastics are solid materials at room temperature, but when heated above the glass transition temperature T_g (or melting point T_m), they deform in a highly viscous manner, which allows them to be easily shaped in a mold or die into products. The process is reversible on subsequent heating and cooling, i.e. they can be reshaped without significant degradation. Therefore, it is customary to form these polymers into sheets, or films, and ship them to other AM manufacturers to form them into final products. Table 4.1 shows the structure and properties of the major families of thermoplastics.

Thermosetting polymers, known as thermosets, are characterized by a highly crosslinked structure extending in a three-dimensional arrangement. This crosslinking (curing) reaction is irreversible, i.e. when reheated, the part tends to degrade and burn without melting. Therefore, thermosets do not exhibit a sharply defined glass transition temperature or a melting point. Table 4.2 presents the structure and properties of the major families of thermosets.

4.2   TECHNOLOGIES OF ADDITIVE MANUFACTURING OF POLYMERS

Due to their tremendous synthetic versatility and adaptability, the availability of a variety of preformed processible options, and the ability to result in a wide range of properties, polymers have become the most sought and researched materials for AM technologies. Polymer AM offers numerous advantages over conventional polymer processing techniques. Thermoplastics, hydrogels, thermosets, functional polymers, elastomers, polymer blends, and biological systems are the range of polymers used in AM. ASTM International classifies AM technologies into seven categories (see Figure **3.1**). The majority of these technologies, except directed energy deposition (DED), are frequently used for polymer AM. DED uses focused thermal energy in the form of laser, electron beam, or plasma arc to melt and bind materials fed in the form of powder or wire as they are being deposited. DED has the capacity to produce large volume deposition rates and is predominantly used only for metals. It is used for repair and to generate compositional gradients.

4.2.1   VAT PHOTOPOLYMERIZATION

The most common types of this technology include:

1. Stereolithography (SLA)
2. Digital light processing (DLP)

TABLE 4.1

Structure and Properties of the Major Families of Thermoplastics

Thermoplastic	Abbreviation	Composition	Tensile strength, MPa	Modulus of elasticity, MPa	Elongation, %	Sp. gravity	T_g, °C	T_m, °C
Polyethylene	PE	$(C_2H_4)_n$						
Low Density	LDPE		9.6–7.2	140	100–500	0.91–0.94	−100	105–115
High Density	HDPE		30	700	20–100	≥0.941	−115	135
Ultrahigh molecular weight	UHMWPE		40		≥350	0.93–0.935	–	–
Polypropylene	PP		35	1400	10–500	0.9	−20	176
Polyvinylchloride	PVC	$(C_2H_3Cl)_n$	50–80	2900–3300	2 (rigid form)	1.39	82	212
Polystyrene	PS	$(C_8H_8)_n$	50	3000–3600	3–4	1.05	95	240
Acrylonitrile butadiene styrene	ABS	$[(C_3H_3N)_x \cdot (C_4H_6)_y \cdot (C_8H_8)_z]$	50	2100	10–30	1.06		200
Polymethyl-methacrylate	PMMA	$(C_5H_8O_2)_n$	55	2800	5	1.15–1.19	105	
Polyamide: Nylon 6,6	PA 6,6	$[(CH_2)_6(CONH)_2(CH_2)_4]$	70	700	300	1.14	50	260
Polyethylene terephthalate	PET	$(C_2H_4 \cdot C_8H_4O_4)_n$	55	2300	200	1.3	70	265
Polycarbonate	PC	$[C_3H_6(C_6H_4)_2CO_3]_n$	65	2500	110	1.2–1.22	150	267
Cellulose acetate	CA	$(C_6H_9O_5 \cdot COCH_3)_n$	30	2800	10–50	1.3	105	306
Polyoxymethylene	POM	$(CH_2O)_n$	70	2500	25–75	1.42	−80	175
Polytetrafluorethylene	PTFE	$(C_2F_4)_n$	20	425	100–300	2.2	127	327

Source: Youssef, H. et al., *Manufacturing Technology: Materials, Processes and Equipment*, CRC Press, California, 2023. With permission of CRC.

TABLE 4.2

Structure and Properties of the Major Families of Thermosets

Thermoset	Abbreviation	Composition	Tensile strength, MPa	Modulus of elasticity, MPa	Elongation, %	Sp. gravity
Phenolic resins:Phenol-formaldehyde		Phenol (C_6H_5OH)-formaldehyde (CH_2O)	70	7000	<1	1.4
• Amino resins:			50	9000	<1	1.5
• Urea-Formaldehyde	UF	$[CH_2\text{-}N\text{-}CO\text{-}N\text{-}CH_2]_n$				
• Melamine-Formaldehyde	MF	$(C_3H_6N_6)\text{-}(CH_2O)$				
Epoxy resins	–	–	70	7000	0	1.1
Polyester resins		$(C_4H_2O_3)+(C_2H_6O_2) + (C_8H_8)$	30	7000	0	1.1
Silicone resins		$(CH_3)_6\text{-}SiO)_n$	30	-	0	1.65

Source: Youssef, H. et al., *Manufacturing Technology: Materials, Processes and Equipment*, CRC Press, California, 2023. With permission of CRC.

3. Continuous digital light processing (CDLP)
4. Daylight polymer printing (DPP)

1. *Stereolithography:* SLA uses photosensitive monomer resins that react to the radiation of ultraviolet (UV) light. Upon irradiation, these materials undergo a chemical reaction (photopolymerization) to become solid. A solid 3D polymeric parts can be produced by curing one layer over a previous layer. This method of 3D printing has some advantages over other methods, including achieving a higher amount of detail in objects and creating less waste.

SLA offers a cost-effective way to produce parts with a high degree of accuracy and smooth surface finish using a range of materials (Figure 4.3). SLA parts are generally brittle and not suitable for functional prototypes. Their mechanical properties and visual appearance degrade over time when the parts are exposed to sunlight. Polymer stereolithography is a well-known AM process already used in bioMEMS and lab-on-chip applications.

2. *Digital Light Processing*: DLP is a 3D printing process similar to SLA, which offers the advantage of higher surface quality and accuracy. The quality of the 3D printed components is comparable to that of conventional injection molding (Figure 4.4); part made from biocompatible material by DLP for dental technology). Design guidelines for DLP include (www.jellypipe.com/):

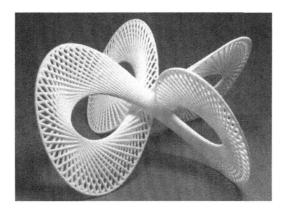

FIGURE 4.3 SLA produces parts with a high degree of accuracy and smooth surface finish. (Courtesy of amfg.ai, 2018)

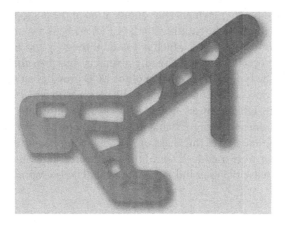

FIGURE 4.4 Part made from biocompatible material by DLP for dental technology. (www .jellypipe.com/)

- Minimum wall thickness: 1 mm
- Lettering and surface details: min. 0.5 mm
- Minimum diameter of holes: 1 mm
- Tolerances: ±0.2%, min. ±0.1 mm depending on geometry and material
- Components must not have any closed cavities, otherwise the resin cannot drain off
- Support structures are used for overhangs
- The surface may be of lower quality

3. ***Continuous Digital Light Processing*:** CDLP 3D printing technology is a vat polymerization AM process that enables high-volume and scalable part production; highly accurate process rendering of the external and

internal geometry of bone tissue engineering scaffolds; loading of internal pore spaces with cells, bioreactor-delivered nutrient and growth factor circulation, and scaffold resorption. It may be necessary to render resorbable polymer scaffolds with 50 µm or greater accuracy to achieve these goals. Figure 4.5 shows schematics of SLA and CDLP. SLA typically requires a deep vat of resin; as parts are built, they attach to an elevator, which moves downward through the polymer resin as each layer is rendered at the surface by a moving laser. In contrast, CDLP systems render parts by projecting an image through a clear basement plate into a tray containing the resin, curing at the bottom surface rather than the top surface. The parts attach to a build platform that moves upward, away from the basement plate, after each layer is projected (Dean et al. 2012).

4. **Daylight Polymer Printing:** DPP uses photocentric instead of a laser or a projector to cure the polymer. This technique, also called LCD 3D printing, uses unmodified liquid crystal display (LCD) screens and a specially formulated daylight polymer. The solidification of the photopolymer occurs using a low-power light-emitting diode (LED) array as a source for the blue light (400 nm wavelength), which is passed through a thin LCD panel that works as an optical shutter for the light source, thus providing whole-layer exposure. The typical 2K resolution of the LCD provides an xy pixel size of 42 µm (600 dpi) over a build volume of $12.1 \times 6.8 \times 16\ cm^3$ with 25 µm layer thickness. In DPP, a lower-energy, longer-wavelength light, which travels further and ensures more reliable polymerization than UV, is used. In a DPP 3D printer, a layer thickness of 100 µm and an exposure time of 80 s for each layer were used to 3D print the composites and unfilled photopolymer. To increase the crosslinking in the formulated composites and obtain

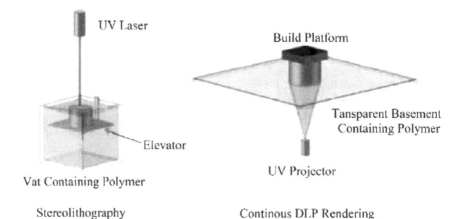

FIGURE 4.5   SLA and CDLP. (From: Dean, D. et al., *Virtual and Physical Prototyping*, 7, 1, 13–24, 2012, Taylor and Francis. With permission)

consistent 3D printing, the exposure time was extended to 100–110 s(Malas et al. 2019). The DPP process allows the use of cheaper materials than those used in other photopolymer 3D printing technologies. It allows simultaneous exposure of the entire cross section of the object, which is not possible in the case of point techniques such as SLA 3D printing. In the case of parts printed on the basis of a light source (LCD screen), the key parameter is the size of a single pixel, which determines the correct mapping of the surface curvatures (Kroma et al. 2021).

4.2.2 MATERIAL JETTING

As previously mentioned in Chapter 3, to facilitate jetting, materials that are solid at room temperature must be heated so that they liquefy. For high-viscosity fluids, the viscosity of the fluid must be lowered to enable jetting. The most common practices are to use heat or solvents or other low-viscosity components in the fluid. The limitation on viscosity quickly becomes the most problematic aspect for droplet formation in MJ. The most commonly used polymers are polypropylene, high density polyethylene (HDPE), polystyrenes (PS), polymethyl methacrylate (PMMA), polycarbonate (PC), acrylonitrile butadiene styrene (ABS), environmentally degradable plastic (EDP), and high-impact polystyrene (HIPS). These materials are formulated to have a viscosity of 18–25 cP and a surface tension of 24–29 dyn/cm at a printing temperature of 130°C.

MJ boasts one of the highest levels of accuracy, since with tiny droplets of material, layers can be printed as thin as 0.013 mm, allowing parts with a very smooth surface finish and enabling parts with small but highly accurate features to be produced. MJ can be used for full-color and more notably, multi-material 3D printing, as the printhead used in the printing process typically incorporates multiple nozzles. These nozzles can deposit different materials and/or different colors in a single printing process. Objects created through the MJ process can possess a range of material properties such as rigidity, flexibility, opaqueness, and translucency. MJ requires support structures that can be easily dissolved in an ultrasonic bath. When dissolved correctly, the supports don't leave marks on the surface after removal.

Objects produced by MJ are typically weaker, particularly when compared with other 3D printing techniques like SLS. This makes material jetted parts generally unsuitable for functional applications. MJ is typically used to produce parts for which look matters more than function. MJ is somewhat constrained by the speed of the printing process. Because small droplets of material are deposited over a small part of the build area at a time, the process takes more time to create a part. For MJ, typically, viscous materials can be successfully printed. Applications of MJ include:

- Production of realistic anatomical models both for educational purposes and for pre-surgical planning and training.
- Printing full-color anatomical models that look like real body parts.
- Low-volume production of molds and casting patterns.

MJ has two variations used for polymers: the Drop on Demand (DOD) stream and PolyJet or Multi-jet Modeling (PJM/MJM).

1. ***Drop on Demand***: DOD methods can deposit droplets of 25–120 μm at a rate of 0–2000 drops per second to form a 3D structure. The polymers that can be used with this method are limited. The process is characterized by:
   - The method is precise and affordable.
   - Common applications include full-color product prototypes; injection mold–like prototypes; low run injection molds; medical models.
   - Dimensional accuracy: ±0.1 mm is possible.
   - Best surface finish; full color and multi-material available.
   - Parts are brittle, not suitable for mechanical parts.
   - Higher cost than SLA/DLP for visual purposes.
2. ***PolyJet***: PJ/MJM is a powerful 3D printing technology that produces smooth, accurate parts, prototypes, and tooling. With microscopic layer resolution and accuracy down to 0.014 mm, it can produce thin walls and complex geometries using the widest range of materials available with any technology. Table 4.3 shows PolyJet technical data, while Figure 4.6 shows PolyJet printed parts. The process is able to:
   - Create smooth, detailed prototypes that convey final-product aesthetics.
   - Produce accurate molds, jigs, fixtures, and other manufacturing tools.
   - Achieve complex shapes, intricate details, and delicate features.
   - Incorporate the widest variety of colors and materials into a single model for unbeatable efficiency.

MJM works like an inkjet printer. The printing head creates successive layers of the part by dispensing droplets of either UV-curable photopolymers or casting waxes. Multi-jet can achieve resolutions and layer thicknesses similar to those of PolyJet

TABLE 4.3
PolyJet Technical Data

Accuracy	0.1–0.3 mm
Layer resolution	16 μm
x/y resolution	43 μm
Wall thickness (minimum)	1 mm
Feature size (minimum)	0.3 mm
Part dimensions (normal resolution)	48.5 × 39 × 20 mm
Part dimensions (maximum)	99 × 79 × 49 mm
Shore hardness range	27 to 95 A
Finishes	Matte, glossy, and custom finishes depend on the type of materials

Source: https://jiga.io/resource-center/materials/polyjet/

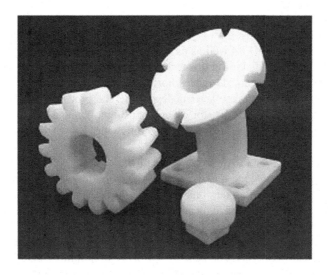

FIGURE 4.6　PolyJet printed parts. (https://jiga.io/resource-center/materials/polyjet/)

TABLE 4.4
PolyJet vs. Multi-Jet Comparison

Attribute	PolyJet	Multi-jet
Uses photopolymer	Yes	Yes
Prints composite material parts	Yes	No
Easy post-processing	No	Yes
Used for functional prototypes	Yes	Yes
Used for small-to-medium production volumes	No	Yes
Need for support materials	Yes	Yes
Can form complex and detailed geometries	Yes	Yes
Costly	Yes	Yes

Source:　www.xometry.com/

printing—as small as 16 µm. Unlike the PolyJet process, however, the MJM technology cannot print multiple materials simultaneously. Table 4.4 compares PolyJet and MJM.

4.2.3　MATERIAL EXTRUSION

Material extrusion (ME) is an AM technique in which a polymer filament is continuously fed into the printing area through extruding nozzles, which are heated (usually in the range of 150–250 C), and then deposited onto the building platform layer by layer. ME can accommodate the use of thermoplastics like ABS, aliphatic polyamides (PA, aka nylon), HIPS, polylactic acid (PLA), and thermoplastic polyure-thane (TPU). More recently, plastic materials such as polyether ether ketone (PEEK) and polyetherimide (PEI), as well as paste-like materials, including ceramics, have been successfully extruded using this technique. Typical layer thickness varies from 0.178 to 0.356 mm. The nozzle moves horizontally, and a platform moves up and down vertically after each new layer is deposited. The process is the most popular technique in terms of availability for general consumer demand and quality.

ME machines can be designed to print the following (www.xometry.com/):

1. Polymers such as PLA, ABS, acrylonitrile styrene acrylate (ASA) and nylon are common and receive a wide range of additives. Some others are technically waxes, but they function much the same as polymers.
2. Ceramics are 3D printed to form (generally artistic) pottery. The nozzles for this purpose are 10–50 times the size of plastic extruders.
3. Concrete and various construction materials have garnered recent attention as 3D printing options. The idea is to print entire buildings, and the process differs from fused deposition modeling (FDM) primarily just in scale.

Examples of ME technology include FDM and fused filament fabrication (FFF).

1. *Fused Deposition Modeling (FDM):* Figure 4.7 shows a desk top vise printed using FDM, which is characterized as follows:
   • The process leaves visible seam lines between layers.
   • Overhanging parts need supports to be printed.
   • Build times are long.
   • The resolution is low compared with other AM techniques.

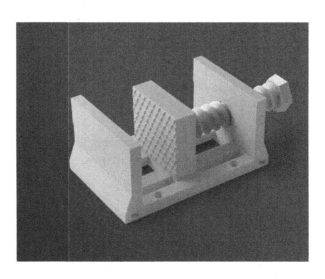

FIGURE 4.7   Desk top vise using FDM. (https://quickparts.com/)

- Parts do not require chemical post-processing, which is highly desirable for biomedical applications.
- The materials used are cost-effective.
- The equipment is simple to use and low-cost as compared with other 3D printing techniques.
- Low mechanical strength compared with injection molded parts.
- Poor surface finish.
- Poor layer adhesion.
- Layer delamination may occur.
- The process generates void spaces between deposition lines, which adversely affects the parts' stiffness and strength.

These problems can be resolved by post-processing methods such as microwave treatment, vapor smoothing, metal plating, cold welding, gap filling, polishing, and dipping, The use of resin to fill the voids, optimization of the process parameters, and structural optimization are addressed by Devine Declan (2019). Part tolerances of ±50 μm can easily be achieved by this process.

2. *Fused Filament Fabrication (FFF):* FFF employs the same process as FDM. It applies layers of filaments to a flat printing bed using a heated nozzle or extruder. The main difference between FFF and FDM is the lack of a heated print environment. FDM's heated chamber helps control the part's temperature and reduce residual stresses in the finished part (Devine Declan 2019).

Table 4.5 compares FDM and FFF. The critical advantages of FDM over FFF are: the resolution of the build (better-looking and more accurate models), and the reduced risk of distortion and porosity that results from a haphazard temperature environment.

4.2.4 BINDER JETTING

The process and technology for plastic BJ involves plastic powder and a liquid binding agent similar to the metal version, but there are several variations and patented

TABLE 4.5
Comparison of FDM vs. FFF

Attribute	FDM	FFF
High model strength	Yes	No
Higher resolution	Yes	No
Reduces residual stress and warping	Yes	No
Lower operational cost	No	Yes
High CAPEX cost for non-professional users	Yes	No

Source: www.xometry.com/

technologies that make plastic BJ a difficult category to define. Polymer BJ begins with a polymer powder (usually a type of nylon) spread across a build platform in a thin layer. Then, inkjet heads dispense a binder-like glue precisely where the polymer should be joined on each layer. In some methods, there's a heating unit attached to the inkjet head or on a separate carriage that fuses the parts of the layer that receive the fluid. The methods that include this heating step create stronger parts than the ones that don't, because the polymer powder is essentially melted together rather than only glued together (https://all3dp.com/).

The printhead moves horizontally along the x and y axes of the machine and deposits alternating layers of the powder material and the binding material. After each layer, the object being printed is lowered on its build platform. Due to the method of binding, the material characteristics are not always suitable for structural parts, and despite the relative speed of printing, additional post-processing can add significant time to the overall process. The process is characterized as follows:

- Parts are self-supporting in the powder bed, so that support structures are not needed.
- Parts can be arrayed in one layer and stacked in the powder bed to increase the number of parts that can be built at one time.
- Assemblies of parts and kinematic joints can be fabricated, since loose powder can be removed between the parts.

The most relevant parameters of the liquid binder material are the viscosity and surface tension. Increased liquid binder viscosity delays spreadability. Viscosity reduction is achieved by lowering the solids loading or adding dispersants. The printed part is usually left in the powder bed after its completion in order for the binder to fully set and for the green part to gain strength. Post-processing involves removing the part from the powder bed, removing unbound powder via pressurized air, and infiltrating the part with an infiltrant to increase strength and possibly to impart other mechanical properties.

After printing, a typical green body can be 30–75% vol. powder, 10% vol. binder, and the rest void space. Infiltration takes place when a selected liquid is drawn into the open pores of the printed part through capillary action and solidifies. Both low- and high-temperature infiltration are possible, depending on the part material and binding mechanism. The first occurs at or slightly above room temperature, and the second at temperatures between 6 and 10°C of the infiltrate's melting point. Examples of commonly used infiltrates include molten wax, varnish, lacquer, cyanoacrylate glue (super glue), polyurethane, and epoxy (Youssef et al. 2023).

BJ is commonly applicable in such industries as aerospace and heavy industry, where thermally stable and wear-resistant parts are needed. It is also widely applicable for making casting patterns, cores, and molds for intricate products. The jewelry and decorative industries have also embraced BJ as a means of producing one-of-a-kind, colorful, and custom-made objects. An unusual yet promising application of BJ may also lie in the food industry using starch, cellulose, sugar, etc. Typical BJ–3DP products are presented in Figure 4.8.

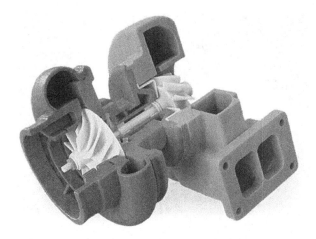

FIGURE 4.8 Parts made using colorJet printing. (Courtesy of 3D Systems)

4.2.5 POWDER BED FUSION (PBF)

PBF processes involve the spreading of the powder material over previous layers using a roller or a blade. A hopper or a reservoir below or beside the bed provides a fresh powder supply. Polymer powders such as polyamides (PA), polystyrenes (PS), thermoplastic elastomers (TPE), and polyaryletherketones (PAEK) are used.

Selective Laser Sintering (SLS): Selective laser sintering allows the processing of almost any material that consolidates upon heating, including polymers, metals, and even ceramics. It allows the processing of polymer powders such as PA, PS, TPE, and PAEK that consolidate upon heating. The high requirements of heat sources during the melting of metals and ceramic powders hinder the widespread use of this method. However, this challenge is decreased when using polymer-based systems, as these can be sintered at temperatures that are lower than those of metals or ceramics (Devine Declan 2019). Figure 4.9 shows a selection of laser-sintered parts made of polymers.

The quality of printed parts is affected by a multitude of factors, including powder composition and morphology, laser energy input, scan spacing and speed, and processing (or powder bed) temperature. There are only a few grades of polymers that are commercially available for use in the SLS process. These include amorphous, semi-crystalline, reinforced, or filled polymers.

Selective Heat Sintering (SHS): SHS technology was invented and patented by a Danish company (Tjellesen and Hartmann) in 2008 and was launched at Euromold in 2011 as a 3D printer available for small businesses. Their latest machine, Blueprinter M3, includes smoother lines and an increased build volume. It has also been highly optimized for daily use and to be much more user-friendly, less noisy, and to use less power. An increased build volume (200 mm × 157 mm × 150 mm) allows the production of multiple parts in a single print run as well. SHS is a less well-known type of PBF process. It works by using a thermal printhead to apply heat to layers of powdered thermoplastic. When a layer is finished, the powder bed

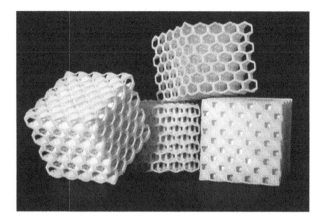

FIGURE 4.9   Selective laser-sintering of polymers. (www.brazilianplastics.com/)

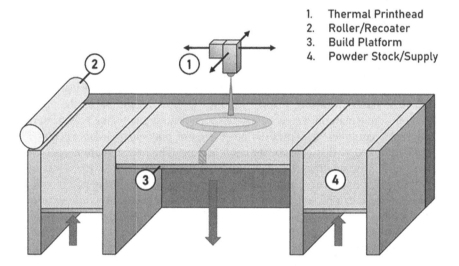

FIGURE 4.10   Functional diagram of SHS 3D printers. (Courtesy of Christian Cavallo Consulting, LLC)

moves down, and an automated roller adds a new layer of material, which is sintered to form the next cross section of the model. The printer's thermal printhead moves back and forth across the powder bed for melting cross sections of the powder into a layer of solid plastic. Therefore, the system does not need sophisticated scanning devices. The object is built without the need for any support structure whatsoever, and thus, this technology gives freedom to build almost any shape. SHS is best for manufacturing inexpensive functional prototypes for concept evaluation, fit/form, and functional testing. Figure 4.10 shows the functional diagram of SHS. SHS technology provides finished, high-quality parts as shown in Figure 4.11.

FIGURE 4.11 3D printed plastic parts by SHS. (https://additive-x.com/)

4.2.6 SHEET LAMINATION

Sheet lamination or laminated object manufacturing (LOM) is a process in which feedstock in the form of sheets of synthetic polymers (or paper) is bonded together or cut, respectively, to form an object. The sheets constitute a layer of the AM process. Various thermoplastics, including PC and PMMA, and polymer-based composites have been utilized to build objects using the LOM technique. LOM has also been used to fabricate complex multilayer ceramic components with the help of polymer additives. Figure 4.12 shows the plastic sheet lamination process. Polymers use heat and pressure without the adhesive, as the process relies on melting the sheets together. Layers are combined using an adhesive or through thermal welding, which is more desirable than adhesive bonding for biomedical applications, as the adhesives may be biologically toxic. The process is often referred to as automated tape placement (ATP). Most ATP systems utilize a continuous fiber thermoplastic composite tape to generate high-strength composite parts. The tape is laid down in layers with pressure applied with a heated roller during bonding to ensure good contact between layers (Figure 4.13).

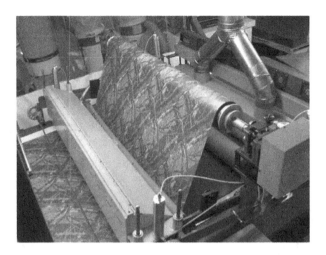

FIGURE 4.12   Plastic sheet lamination. (https://positroncorp.com/)

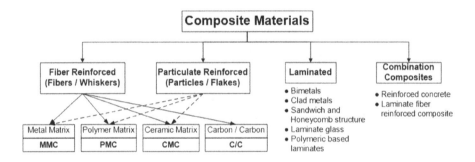

FIGURE 4.13   Classification of composite materials. (From Youssef, H. et al., *Manufacturing Technology: Materials, Processes and Equipment*, CRC Press, California, 2023. With permission of CRC)

4.3   COMPARISON AND CONCLUSIONS REGARDING AM OF POLYMERS

Regarding the AM of polymers, the following points can be concluded:

1. Liquid resin–based systems like SLA and PolyJet are capable of the best printing resolution. However, the resins that are approved for use in biomedical implants are limited. These systems are generally based on thermosetting polymer resins that, following curing, do not melt on application of heat.
2. Thermoplastic resin–based systems can be used in conjunction with FFF. The challenge of delamination between weld lines is a concern; however, these effects can be decreased with post-production processing.

TABLE 4.6

AM Technologies for Polymers

Category and Technologies	Abbreviation	Power Source	Materials
VAT photopolymerization			
• Stereolithography	SLA	Ultraviolet light	Photosensitive resin
• Digital light processing	DLP		
• Continuous liquid interface production	CLIP		
• Day light polymer printing	DLPP		
Material jetting	MJ		Photopolymer resins
• Drop on demand	DOD	Thermal energy	
• PolyJet	PJ		
Material extrusion			
• Fused Deposition Modeling	FDM	Thermal energy	Thermoplastics (ABS, PLA, PS,
• Fused Filament Fabrication (FFF)	FFF		nylon)
Binder jetting			
• Binder jetting	BJ	Binder/thermal energy	Polymer
Powder bed fusion			
• Selective laser sintering	SLS	Laser, electron beam	Polymer powder such as polyamides (PA), polystyrenes (PS), thermoplastic elastomers (TPE), and polyaryletherketones (PAEK)
Sheet lamination			
• Laminated object manufacturing (LOM)	LOM	Laser	Plastic foil

3. SLS and BJ are both based on powdered polymers, but they differ in how the powders are bound. With SLS, thermal energy is used to sinter the polymer particles together, leaving a potentially porous construct, whereas BJ utilizes adhesives, which may be toxic. However, if this can be overcome, both have potential for use in areas where surface roughness is desirable, such as in orthopedic implants.

4. LOM uses sheets of polymers, which enables the production of parts from a wide variety of polymer resins.

Table 4.6 summarizes and compares the viable AM processes (Devine Declan 2019).

4.4 ADDITIVE MANUFACTURING OF COMPOSITES

4.4.1 COMPOSITES

The term "composite materials" signifies that two or more materials are combined on a macroscopic scale to form a new useful material. The combinations are among any two or more of the known material families, i.e. metals, ceramics, and polymers. The key is the macroscopic examination (by the naked eye or at low magnification) of the new material to identify its constituents. This differentiates composites from other combinations such as alloys, where different constituents come together on the microscopic scale, and the material is macroscopically homogeneous, i.e. components cannot be distinguished by the naked eye. The main objective of a composite material is that it utilizes the most distinguished qualities of its constituents to produce a material with the characteristics needed to perform design requirements. The most significant characteristics of composites are the strength-to-weight ratio and the stiffness-to-weight ratio, which can be several times greater than steel or aluminum. Other improved properties include: fatigue life, toughness, wear resistance, corrosion resistance, appearance (aesthetic aspect), thermal conductivity or insulation, and acoustic insulation. The unique advantage of composites is the possibility of achieving combinations of properties that are not achievable with metals, polymers, or ceramics alone. Nevertheless, it should be noticed that the inherent structure of composites implies that they are heterogeneous and anisotropic.

Composite materials have a long history, starting with human existence on earth, where people used and are still using wood for everyday life. Recorded history showed that Ancient Egyptians used straw to strengthen mud bricks. They also used plywood when they realized that wood layers could be rearranged to achieve superior strength, as well as resistance to thermal expansion, and to swelling due to moisture absorption. Medieval swords and armor were made of layers from different metals (Japanese swords or sabers were made of steel and soft iron for high flexural and impact resistance). Reinforced concrete (cement and gravel, reinforced with steel rods) is a well-known construction material. More recently, fiber-reinforced polymer matrix composites have become main constituents in aircraft and aerospace vehicles due to their superior strength-to-weight and stiffness- to-weight-ratios.

According to the structure of the composite materials, they can be classified to four commonly accepted types:

1. Fiber-reinforced composites that consist of fibers or whiskers of one material embedded in a matrix material
2. Particulate reinforced composites that consist of particles of one material in a matrix material
3. Laminated composites that consist of layers of two or more materials
4. Combinations of some or all of the first three types, such as reinforced concrete

This classification is presented in Figure 4.13.

Table 4.7 presents the specific weight, strength, and modulus of elasticity (stiffness) for commonly known fibers in comparison with common structural metallic wire materials. The table also shows the strength–specific weight and stiffness–specific weight ratios, which are used as indicators of the effectiveness of fibers, especially in weight-sensitive applications such as aircraft and space vehicles. Notice

TABLE 4.7

Effective Properties of Commonly Known Fibers as Compared with Structural Metallic Materials

Fiber or metal wire	Diameter, μm	Specific weight, kN/ m^3	Tensile strength, GPa	Modulus of elasticity E, GPa	Strength/ Specific weight, km	E/Specific weight, mm
E-Glass	16	25.0	3.4	72	136	2.9
S-Glass	10	24.4	4.8	86	196.7	3.5
Carbon		13.8	1.7	190	123.2	13.8
Graphite	7	13.8	1.7	250	123.2	18.1
Kevlar	12	14.2	2.9	130	204	9.2
Beryllium	–	18.2	1.7	300	93.4	16.5
Boron	100	25.2	3.4	400	134.9	15.9
Silicon	14	21.6	–	95	–	4.4
Aluminum	–	26.3	0.62	73	23.6	2.8
Titanium	–	46.1	1.9	115	41.2	2.5
Steel	–	76.6	4.1	207	53.5	2.7

Source: Youssef, H. et al., *Manufacturing Technology: Materials, Processes and Equipment*, CRC Press, California, 2023. With permission of CRC.

in the table that Kevlar fibers offer the highest strength-specific weight ratio, and graphite provides the highest stiffness-specific weight ratio.

4.4.2 3D PRINTING OF COMPOSITES

There are several techniques for AM of composites: FFF, liquid deposition modeling (LDM), SLA, LOM, composite-based additive manufacturing (CBAM), and SLS. The most widespread and relatively simple technique is FFF/FDM. The material that is widely used is CF PA-12 filament (polyamide 12 matrix composite, reinforced with carbon fibers), which allows the printing of highly durable components characterized by high rigidity and tear strength values (www.compositesportal.com/).

Composite 3D printing materials with reinforcing fibers can be divided into two distinct categories: continuous fiber composites and chopped fiber composites. While the two different composites might contain exactly the same constituents, they might perform in a very different way depending on whether they have continuous fibers or chopped fibers. Continuous fibers are longer, unidirectional strands of reinforcing material that when integrated into a thermoplastic matrix, provide vastly superior strength compared with chopped fibers. This is because a strand can absorb and distribute loads across its entire length, so a longer continuous length has greater load-bearing capacity than a tiny chopped strand. Continuous fiber composite printing is more expensive than chopped fiber composite printing, as dedicated hardware is required (www.3erp.com/).

The base thermoplastics for FDM 3D printing composites range from polymers like PLA and ABS at the cheaper end of the scale, to high-performance polymers like PEEK at the premium end. Polyamide/nylon (PA)is the main material used for composite SLS powders (as it is the main material used in laser sintering generally), but high-performance materials like PAEK can also be used. Other materials include polycarbonate (PC), polyetherimide (PE), and polyphenylene sulfide (PPS). On the other hand, reinforcements can be offered using carbon fiber, fiberglass, glass beads, Kevlar, and graphene (www.3erp.com/).

PLA and polyurethane-type filaments as the thermoplastic matrices and an epoxy as the thermoset matrix are widely used in industry. The PLA filament and CF are smoothly extruded from a nozzle for thermoplastic composites, as presented in Figure 4.14a, while the thermoset composite system is shown in Figure 4.14b. (https://amchronicle.com/).

1. *3D Printing of thermoplastic composites with continuous fiber:*

Among a variety of polymer 3D printing techniques, extrusion-based printing methods show great promise in the fabrication of continuous fiber reinforced composites. An FDM printer head is used to fabricate continuous fiber composites based on the in-nozzle impregnation (Figure 4.15a). The PLA filamen° and the continuous carbon fiber are supplied separately to the printer head. When the nozzle is heated, the PLA is melted and fused to the fiber bundle. After the extrusion of the filament, the PLA matrix is quickly solidified and adheres to the previous layer.

In 2016, Markforged Inc. (Watertown, MA) released the first 3D composite FDM desktop printer to print continuous fiber thermoplastic composites. The printer features two separate extrusion nozzles for plastic filaments and continuous fiber supply, as shown in Figure 4.15b). The continuous fiber is pre-impregnated with plastic

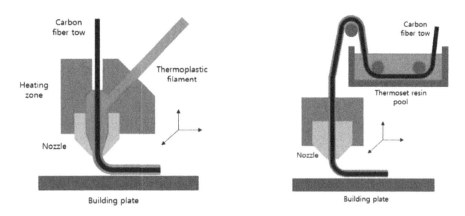

FIGURE 4.14   3D printing system of continuous carbon fiber-reinforced polymer composites for (a) thermoplastics and (b) thermosets. (https://amchronicle.com/)

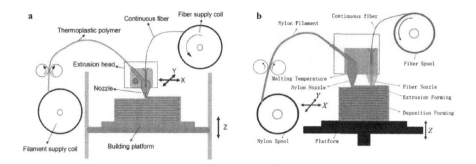

FIGURE 4.15 Schematic representation of the FMD 3D printing process for continuous fiber composites with (a) a single nozzle; (b) dual nozzles. (Panel (a): Reprinted from *Composites Part A: Applied Science and Manufacturing*, 88, Tian, X., Liu, T., Yang, C., Wang, Q. and Li, D., Interface and Performance of 3D Printed Continuous Carbon Fiber Reinforced PLA Composites, 198–205, Copyright (2016), with permission from Elsevier. Panel (b): Reprinted from *Additive Manufacturing*, 27, Mei, H., Ali, Z., Yan, Y., Ali, I. and Cheng, L, Influence of Mixed Isotropic Fiber Angles and Hot Press on the Mechanical Properties of 3D Printed Composites, 150–158, Copyright (2019), with permission from Elsevier.)

so that it can stick to the previous layer during the extrusion. Unlike the single-nozzle system, the dual-nozzle design enables the individual printing layer to be selectively reinforced at different locations. The printer is able to print composites with different thermoplastic matrix (nylon or nylon with chopped carbon fiber) and reinforcement fibers (carbon fiber, Kevlar, fiberglass) (Yu 2023).

3D printed nanocomposites can be created using filaments that incorporate nanoparticles. Shrinking the scale size to the nano-range can change the properties of the materials. A variety of nanomaterials, such as carbon nanotubes, graphene, and metal nanoparticles, are used in 3D printing for various applications (Ivanova et al. 2013). Nanocomposites signify numerous advantages due to their properties and improve the homogeneity of the product. The high viscosity of the thermoplastics can be further increased by the addition of nanofillers, which affects the printability of the nanocomposite filament.

2. *3D Printing of thermosetting composites with continuous fiber:*

Printing with these thermosets with continuous fiber is challenging because the viscous resins cannot solidify after the filament extrusion to provide a persistence force. As shown in Figure 4.16,, a fiber bundle is first impregnated with resin in a tank and then extruded through a squeezing nozzle at 130°C. Due to the high molecular weight of the epoxy resin, it remains in a nearly solid state at room temperature and thus, can be printed in a process similar to that of FDM. Finally, the printed composite is subjected to vacuum heating to complete the curing of the epoxy matrix. Figure 4.17 shows a typical 3D printed composite.

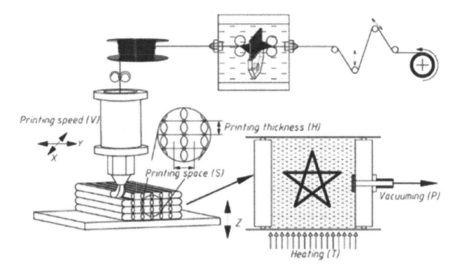

FIGURE 4.16   Schematic illustration of the 3D printing process for continuous fiber composites with epoxy matrix. (From Yu, K., Design and 3D Printing of Continuous Fiber Composites: Status, Challenges, and Opportunities. *iMechanicalJournal Club*)

FIGURE 4.17   3D printed composite. (www.3erp.com)

3. ***3D Printing of short carbon fiber–reinforced thermoset polymer composites:***

Chopped fibers are tiny strands of reinforcing material like carbon or Kevlar. Typically measuring less than a millimeter in length, these strands can be easily mixed into a thermoplastic matrix (PLA, ABS, etc.), giving the ordinary plastic increased strength and stiffness. Chopped fibers are incredibly useful because they are highly versatile: they can be mixed with a wide variety of thermoplastics, and the resulting composites can be printed on ordinary 3D printers. However, when

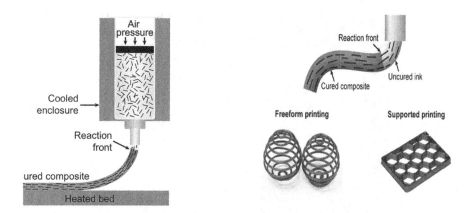

FIGURE 4.18 3D printing of short carbon fiber–reinforced thermoset polymer composites. (Adapted from Ziaee, M. et al., *American Chemical Society (ACS) Appl. Mater. Interfaces*, 14, 14, 16694–16702, 2022)

chopped fibers are mixed into a base material, each individual fiber takes on a random orientation, which makes the material less strong than a material with continuous fibers (www.3erp.com/).

As shown in Figure 4.18, upon extrusion and deposition of the composite ink from a printing nozzle, the ink is cured via frontal polymerization, leading to rapid printing of high-quality composites. Tailoring the processing conditions allows freeform or rapid, supported printing of 3D composite objects with zero void content and highly oriented carbon fiber reinforcements,

4.4.3 LAMINATED OBJECT MODELING OF COMPOSITE SHEETS

LOM is an automated process used for the fabrication of large composite structures in the aeronautical industry. Most commonly, the ATP system utilizes a continuous fiber thermoplastic composite tape to generate high-strength composite parts.

4.4.4 SELECTIVE LASER CENTERING (SLS)

SLS is a plastic 3D printing process suitable for the production of composite parts. Due to the cost and complexity of SLS systems, this technology is mostly employed by industrial users. Materials are a mix of thermoplastic powders (often nylon) and reinforcing elements like chopped fibers or glass beads.

4.4.5 COMPOSITE FILAMENT FABRICATION (CFF)

With composite filament fabrication, two printhead nozzles are used at once, with one following the usual material extrusion process to create the outer shell and internal matrix of the part, while the other deposits a continuous strand of composite fiber inside the printed parts to add strength. This creates parts with a strength comparable

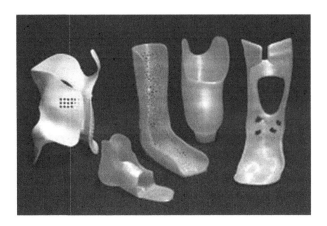

FIGURE 4.19   Polypropylene used in custom orthoses and orthopedic aids. (Courtesy of: PPprint)

TABLE 4.8
AM Technologies for Composites

Category and Technologies	Abbreviation	Power Source	Materials
Fused Deposition Modeling	FDM 3DP	Thermal energy	Polylactic acid (PLA)
3D Printing with Continuous Fibers	3DP	Binder/thermal energy	filament and the continuous carbon fiber Short Carbon
3D Printing of Short Carbon Fibers			Fiber Reinforced Thermoset Polymer
Sheet lamination	LOM	Laser	Continuous fiber
Laminated object manufacturing			thermoplastic tape
Selective laser sintering	SLS	Laser, electron	Thermoplastic powders
Selective laser sintering		beam	(often nylon) and chopped fibers or glass beads

to metal. Figure 4.19 presents polypropylene used in custom orthoses and orthopedic aids. CFF is a term that was coined by the company Markforged and uses two print nozzles. Table 4.8 presents the AM technologies for composites.

4.4.6   Ionic Polymer–Metal Composites (IPMCs)

IPMCs are composed of an ionic polymer like Nafion or Flemion, whose surfaces are chemically plated or physically coated with conductors such as platinum or gold. The unique actuation and sensing properties of ionic polymer–metal composites (IPMCs) are exploited in 3D printing to create electroactive polymer structures for application in soft robotics and bio-inspired systems. The process begins with extruding a precursor material (non-acid Nafion precursor resin) into a thermoplastic

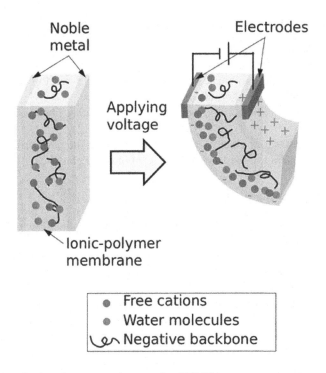

FIGURE 4.20 Ionic polymer–metal composites (IPMCs).

filament for 3D printing. The filament is then used by a custom-designed 3D printer to manufacture the desired soft polymer structures layer by layer (Carrico et al. 2015). Under an applied voltage (1–5 V), ion migration and redistribution due to the imposed voltage across a strip of IPMCs result in a bending deformation, as shown in Figure 4.20 Also, IPMCs can be in the form of an ionic hydrogel that is immersed in an electrolyte solution and connected to the electric field indirectly. If the plated electrodes are arranged in a non-symmetric configuration, the imposed voltage can induce a variety of deformations, such as twisting, rolling, torsion, turning, twirling, whirling, and non-symmetric bending deformation. Alternatively, if such deformations are physically applied to an IPMC strip, they generate an output voltage signal (a few millivolts for typical small samples) as sensors and energy harvesters. IPMCs are electroactive polymers that work well in a liquid environment as well as in air. As shown in Figure 4.20, when a voltage (electric field) is applied to the electrodes, positively charged conjugated and hydrated cations in the membrane molecular network are repulsed by the anode and migrate toward the negative electrode (the cathode), carrying the hydrated water molecules with them. This migration creates an osmotic pressure gradient across the membrane, causing the IPMC strip to bend or deform in a spectacular manner. On the other hand, mechanically bending or deforming the IPMC strips causes the conjugated cations to move around, and this creates an electric potential, output voltage, and transient current.

COMPARISON AND CONCLUSIONS REGARDING AM OF COMPOSITES

1. FDM builds 3D objects from PLA filament and continuous carbon fiber filament using thermal energy.
2. 3D printed models are made using short carbon fiber–reinforced thermoset polymer with binder/thermal energy.
3. LOM uses a laser to build objects from continuous fiber thermoplastic tapes.
4. SLS uses a laser or electron beam to build composite parts from thermoplastic powders (often nylon) and chopped fibers or glass beads.

4.5 REVIEW QUESTIONS

1. Of all of the current AM technologies, which one is the most widely used for polymers?
2. What is the starting material in fused deposition modeling (FDM)?
3. What industry branch preferably uses stereolithography?
4. Compare DLP versus SLA.
5. Differentiate between inkjet printing (IJP) and 3DP techniques.
6. Compare powder bed fusion vs. binder jetting.
7. List five types of polymeric materials that can be material jetted, and arrange them according to the ease of jetting from the viscosity–temperature viewpoint.
8. Explain how sheet lamination can be used for polymeric materials.
9. What are the processes used for additive manufacturing of polymers?
10. What are the processes used for additive manufacturing of composites?
11. Explain how short fibers are used in 3D printing of composites.
12. Show the process steps used for composite-based additive manufacturing (CBAM) technology.

BIBLIOGRAPHY

ASTM. (2009) *International Committee F42 on Additive Manufacturing Technologies (ASTM F2792–10)*. West Conshohocken, PA: Standard Terminology for Additive Manufacturing Technologies.

ASTM. (2012) *Standard Terminology for Additive Manufacturing Technologies (Withdrawn 2015, ASTM F2792-12a)*. West Conshohocken, PA: ASTM International.

Carrico, James D., Traeden, Nicklaus W., Aiureli, Matteo, and Leang, Kam K. (2015) Fused Filament 3D Printing of Ionic Polymer-metal Composites (IPMCs). *Smart Materials and Structures*, 24(12), 125021. https://doi.org/10.1088/0964-1726/24/12/125021

Dean, D., Wallace, J., Siblani, A., Wang, M. O., Kim, K., Mikos, A. G., and Fisher, J. P. (2012) Continuous Digital Light Processing (cDLP): Highly Accurate Additive Manufacturing of Tissue Engineered Bone Scaffolds. *Virtual and Physical Prototyping*, 7(1), 13–24. https://doi.org/10.1080/17452759.2012.673152

Devine Declan, M. (2019) *Polymer-Based Additive Manufacturing: Biomedical Applications*. Cham, Switzerland: Springer Nature Switzerland AG.

https://www.3erp.com/

https://additive-x.com/
https://all3dp.com/
https://amchronicle.com/
http://www.brazilianplastics.com/
https://www.compositesportal.com/
https://www.jellypipe.com/
https://jiga.io/resource-center/materials/polyjet/
https://positroncorp.com/
https://quickparts.com/
https://www.xometry.com/

Ivanova, O., Williams, C., and Campbell, T. (2013) Additive Manufacturing (AM) and Nanotechnology: Promises and Challenges. *Rapid Prototyping Journal*, 19(5), 353–364. https://doi.org/10.1108/RPJ-12-2011-0127

Kroma, A., Mendak, M., Jakubowicz, M., Gapiński, B., and Popielarski, P. (2021) Non-Contact Multiscale Analysis of a DPP 3D-Printed Injection Die for Investment Casting. *Materials*, 14(22), 6758. https://doi.org/10.3390/ma14226758

Malas, Asish, Isakov, Dmitry, Couling, Kevin, and Gibbons Gregory, J. (2019) Fabrication of High Permittivity Resin Composite for Vat Photopolymerization 3D Printing: Morphology, Thermal, Dynamic Mechanical and Dielectric Properties. *Materials*, 12(23), 3818. https://doi.org/10.3390/ma12233818

Mei, Hui, Ali, Zeeshan, Yan, Yuekai, Ali, Ihtisham, and Cheng, Laifei. (2019) Influence of Mixed Isotropic Fiber Angles and Hot Press on the Mechanical Properties of 3D Printed Composites. *Additive Manufacturing*, 27, 150–158. https://doi.org/10.1016/j.addma.2019.03.008

Pou, Juan, Riveiro, Antonio, and Davim, J. Paulo. (2022) *Additive Manufacturing*. Netherlands: Elsevier Inc.

Tian, Xiaoyong, Liu, Tengfei, Yang, Chuncheng, Wang, Qingrui, and Li, Dichen. (2016) Interface and Performance of 3D Printed Continuous Carbon Fiber Reinforced PLA Composites. *Composites Part A: Applied Science and Manufacturing*, 88, 198–205. https://doi.org/10.1016/j.compositesa.2016.05.032

Youssef, H., El-Hofy, H., and Hamed, M. (2023) *Manufacturing Technology: Materials, Processes and Equipment*. California: CRC Press.

Yu, Kai. (2023) Design and 3D Printing of Continuous Fiber Composites: Status, Challenges, and Opportunities. *iMechanicalJournal Club*. https://imechanica.org/node/26442

Ziaee, Morteza, Johnson, James W., and Yourdkhani, Mostafa. (2022) Supporting Information for 3D Printing of Short Carbon Fiber Reinforced Thermoset Polymer Composites via Frontal Polymerization. *American Chemical Society (ACS) Appl. Mater. Interfaces*, 14(14), 16694–16702. https://doi.org/10.1021/acsami.2c02076

5 Technologies of Additive Manufacturing of Metallic Materials

5.1 METAL ADDITIVE MANUFACTURING PROCESSES

The new field of metal additive manufacturing (AM) has the ability to produce hard-to-manufacture components in complex structural shapes that are difficult or impossible to fabricate by conventional means, as a direct replacement for conventionally manufactured components. Metal AM is now finding acceptance for critical applications such as medical implants, aerospace, and in many other fields, with a clearly demonstrated ability to produce complex shapes.

The metal AM processes consolidate feedstock metallic materials such as powder, wire, or sheets into a dense metallic part by fusion, melting, and solidification with the aid of an energy source such as a laser, electron beam, or electric arc, or by the use of ultrasonic vibration in a layer-by-layer manner. Powder is used in the form of powder bed or powder feed, wire is applied in a wire-feed process, while rolled sheets or foils (typically in the thickness range of 0.1–1.0 mm) are stacked for the ultrasonic additive manufacturing (UAM) process.

Among the seven ASTM F42 standard categories of AM presented in Chapter 3, only four are applied for directly manufacturing metallic materials. These are powder bed fusion (PBF), direct energy deposition (DED), binder jetting (BJ), and sheet lamination (SL). Two of the other three categories specified in the standard (i.e. vat photo-polymerization and material extrusion) are devoted to polymeric materials, and the third (material jetting) is capable of being applied to metal technologies but is not currently economically viable. Two techniques of metal AM comprise several sub-processes, as shown in Figure 5.1. The present specific features of these techniques as related to AM of metals and alloys will be demonstrated in this section.

The basic principles of applying these processes to build the product layer by layer using a specific source of power, along with their general applications, advantages, and limitations, are discussed in Chapter 3. In this section, the explicit features of these processes relevant to manufacturing metallic materials are presented.

5.1.1 Powder Bed Fusion Processes (PBF)

PBF embraces processes whereby a focused thermal energy source, whether a laser beam or an electron beam, is used to selectively melt or sinter a layer of the powder

DOI: 10.1201/9781003451440-5

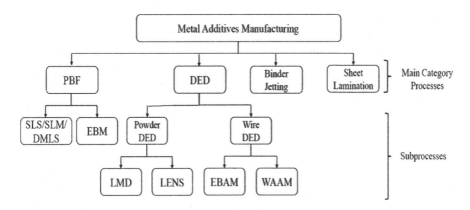

FIGURE 5.1 Generalized map of metal additive manufacturing processes and subprocesses. (Modified from Abdel-Aal, H. (2022) *Additive Manufacturing of Metals Fundamentals and Testing of 3D and 4D Printing*, McGraw Hill Co., New York, 2022)

bed. For metallic materials with relatively high melting points, high-power laser beam sources are applied for melting of the powder. This includes selective laser sintering (SLS), selective laser melting (SLM), and direct metal laser sintering (DMLS). On the other hand, the electron beam melting (EBM) process is used when the electron beam fully melts the selectively fused parts and the liquefied metal conforms to the surrounding metal powder in the build chamber. Although SLM and EBM use the same powder bed principle for layer-wise selective melting, there are significant differences in the hardware setup.

SLM is used for AM of metal powders (pure or alloy). This is different from SLS, which is an AM process whereby a laser sinters the particles of a polymer-based powder. Despite these similarities, the three main differences between SLM and SLS are the material used to build the parts, the process used to build them, and the cost. Since a metal powder is melted in SLM, the build process between the two machines is different. An inert gas (argon or nitrogen) must be pumped into the build chamber to facilitate melting. Consequently, the high temperatures present within SLM systems necessitate sufficient cooling for parts before they are removed. SLS printers have long lead times, and completed parts must cool for up to 12 hours, while SLM can reduce manufacturing lead times and cooling costs.

However, SLM requires increased material usage due to the need for support of overhangs on parts during printing.

5.1.1.1 Selective Laser Sintering (SLS) Process

SLS is an AM process based on the use of powder-coated metal additives. A continuous or pulsating laser beam is used as the heating source for the scanning and fusion of particles in sand shapes of predetermined size in the layers. The geometry of the scanned layers corresponds to the STL models established by computer-aided design (CAD). The process is repeated for subsequent layers from the bottom to the top until the product is complete.

In addition to its suitability for polymeric, ceramic, and composite materials, metallic materials used in SLS include various types of metals (Al, Cr, Ti, Fe, Cu), metal alloys (Fe–Cu, Fe–Sn, Cu–Sn), as well as bronze, nickel, and Inconel 625. Almost any material that can be combined with another material with a low melting point and acts as an adhesive can be selectively laser centered. The process takes place in a controlled atmosphere (Despa and Gheorghe 2011).

5.1.1.2 Selective Laser Melting (SLM) Process

SLS is the most widely applied and perhaps the most evolved AM metal technology. A range of metal alloys are available but generally limited to those engineering metal alloys optimized for powder bed fusion (PBF) AM. The high cost of powder is currently a limiting factor to its adoption; however, continuous development of the parameters of the laser beam and the powder characteristics encourage growing fields of application.

The laser system typically comprises a set of lenses and a scanning mirror or galvanometer to maneuver the position of the beam.

Among the various available industrial lasers, CO_2 and Nd:yttrium-aluminum-garnet (Nd:YAG) lasers are the two most investigated lasers for SLM/SLS. It is well known that the type of laser beam affects the final properties (i.e. mechanical properties, physical density, and surface texture) of the final part. Hence, the choice of laser is not independent of the material that has to be melted/sintered, as different lasers would have a varied effect on the same material. Some of the major differences among the various laser systems include wavelength, coherence, mode of operation, and beam diameter. Optimally, the laser wavelength should be adapted to the powder material, since laser absorption greatly changes with the material and repetition rate or wavelength of the laser light. CO_2 lasers have a wavelength of 10.6 μm and are well suited to sintering various polymers and ceramic oxide powders. Nd:YAG lasers have an active medium of neodymium in a YAG. These lasers have a short wavelength of 1.06 μm, which may outperform CO_2 lasers for metallic materials that absorb much better at short wavelength. They are commonly used for the fabrication of aluminum alloy/silicon carbide, titanium/silicon carbide, titanium/graphite, and single layers and multilayers of metal/ceramic composites.

Recently, a new generation of machines that employ lasers of enhanced beam quality are being developed using Yb (ytterbium)-fiber lasers that involve low laser powers and high scanning rates to achieve SLM processing of Al 6061, AlSi12, and Al–Cu alloy powders. In a fiber laser, the active gain medium is an optical fiber doped with rare-earth elements. The main advantage of fiber lasers over other types is that the laser light is both generated and delivered by an inherently flexible medium, which allows easier delivery to the focusing location and target. Table 5.1 shows the main characteristics and ranges of application for these types of laser.

Figure 3.18 (Chapter 3) shows some metallic products made by SLM. It should be noted that the geometrical shapes of these products cannot be directly manufactured using conventional processes. These improved design features exhibit higher mechanical and physical performance characteristics.

TABLE 5.1

Types of Laser Used in SLS/SLM Processes

Laser	CO_2 laser	Nd:YAG laser	Yb-Fiber laser
Wavelength (μm)	9.4 and 10.6	1.06	1.07
Efficiency (%)	5–20	10–20 (diode pump)	10–30
Output power (kW)	Up to 20	Up to 16	Up to 10
Pump source	Electrical discharge	Laser diode	Laser diode
Operation mode	CW[a] and pulse	CW and pulse	CW and pulse
Fiber delivery	Not possible	Possible	Possible
Maintenance periods (h)	2000	10,000 (diode life)	Maintenance free (25,000)

[a]CW: continuous wave laser.

5.1.1.3 Direct Metal Laser Sintering (DMLS) Process

Direct metal laser sintering (DMLS) is a 3D printing technology that has been recognized as one of the most effective forms of additive manufacturing. It requires the use of a powder, which is composed of two types of particles. One type has a low melting point, and the other has a high melting point. The high-melting-point particles generate a solid matrix, whereas the particles with the low melting point bind the matrix after being melted by the laser energy input (liquid-phase sintering and partial melting). The process is applied to metal alloys for the direct manufacturing of parts in industry, including aerospace, dental, medical, and the tooling industry. Support structures are required for most builds because the powder alone is not sufficient to hold in place the liquid phase created when the laser is scanning the powder. Titanium and its alloys are widely used for various implants in the orthopedic and dental field because of its good corrosion resistance and mechanical strength. By now, the concept of the DMLS technique for implantation biomaterial manufacturing is well accepted. Also, aluminum powder, in particular, broadens the range of possible applications of DMLS to lightweight structural components. A major difference between DMLS and SLS is the type of materials that can be used to "print" the required parts. SLS can be used with a variety of metals and non-metallic materials, whereas DMLS is designed to work solely with metals.

One big advantage of DMLS is that custom manufactured parts made using this process tend to be free of internal defects and residual stress. This ability to create defect-free parts is critical when parts are to be used in a high-stress environment such as the automotive or aerospace industries. One of the biggest downsides to DMLS is the cost of the 3D metal printer.

5.1.1.4 Electron Beam Melting Process

The EBM system is essentially a giant scanning electron microscope, which requires a filament, magnetic coils to collimate and deflect the position of the beam, and an electron beam column. The electron beam is emitted from a tungsten filament and is controlled by two magnetic fields that organize and direct the fast-moving electrons.

The first acts as a magnetic lens, which focuses the beam to the correct diameter. The second magnetic field deflects the focused beam to the target point on the powder bed. The focused electron beam scans across a thin layer of pre-laid powder, causing localized melting and solidification. The gun delivers high output (up to 4 kW) at very high scanning speeds, up to 1000 m/s. and fast build rates, up to 80 cm³/hour. The process is three to five times faster than other additive technologies.

The EBM parts are built in a vacuum to prevent the loss of energy that would be caused by the fast-moving electrons colliding with air or gas molecules. The vacuum has two advantages:

- The process is 95% energy-efficient, which is five to ten times greater than laser technology.
- The vacuum supports processing of reactive metal alloys such as titanium.

A schematic of the standard EBM system is shown in Figure 5.2.

The EBM machine produces components from special titanium alloy parts, common in the aerospace industry, which exhibit properties that match those of wrought materials and exceed those of investment castings. Directly from the EBM machine, the parts are 100% dense, eliminating the need for secondary infiltration processes. Parts produced via EBM are near net shape, like those made via casting processes. Since the electron beam fully melts the titanium, the liquefied metal conforms to the surrounding metal powder, which yields a surface finish similar to a precision sand casting. The process is also capable of producing complex parts of multi-piece assemblies that exhibit exceptional strength-to-weight ratios, reduce manufacturing costs, and minimize assembly time. The EBM process can also produce hollow parts with an internal strengthening scaffold, impossible with any other method; thus, it can deliver the required mechanical strength with much lower mass. This reduces the cost of raw materials and the weight of the component.

5.1.1.5   Sintering/Melting Mechanisms for PBF

What sets sintering apart from melting is that the sintering processes do not fully melt the powder but heat it to the point where the powder can fuse together on a molecular level.

Since the introduction of SLS/SLM and EBM, each new technology developer has introduced competing terminology to describe the mechanism by which fusion occurs, with variants of "sintering" and "melting" being the most popular. There are four different fusion mechanisms present in PBF processes. These include solid state sintering, chemically induced binding, liquid-phase sintering (LPS), and full melting. Most commercial processes primarily utilize LPS and melting. A brief description of each of these mechanisms and their relevance to AM follows.

a. Solid State Sintering
Sintering, in its classical sense, indicates the fusion of powder particles in their "solid state" at elevated temperature without melting. This occurs at temperatures in the range of 0.7°C to 0.9°C of the absolute melting temperature. The driving force for

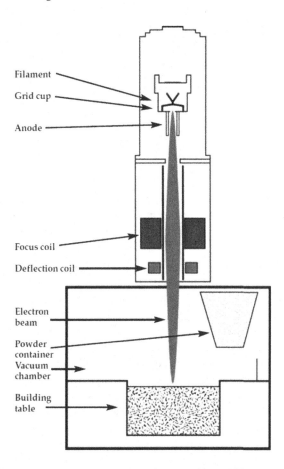

FIGURE 5.2   Schematic of a standard electron beam melting (EBM) system. (From https://ars.els-cdn.com/)

solid state sintering is the minimization of the total free energy of the powder particles, primarily by diffusion between powder particles, as presented in Figure 5.3.

To achieve very low porosity levels, long sintering times or high sintering temperatures are required. The use of external pressure, as is done with hot isostatic pressing (HIP), increases the rate of sintering. Smaller particles sinter more rapidly and initiate sintering at a lower temperature than larger particles. In general, diffusion-induced solid state sintering is the slowest mechanism for selectively fusing regions of powder within an LPS process. Since the time taken for fusion by sintering is typically much longer than for fusion by melting, few AM processes use sintering as a primary fusion mechanism.

b. Chemically Induced Sintering

Chemically induced sintering involves the use of thermally activated chemical reactions between two types of powders or between powders and atmospheric gases

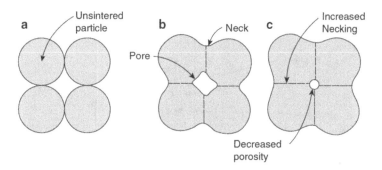

FIGURE 5.3   Solid state sintering. (a) Closely packed particles prior to sintering. (b) Particles agglomerate at high but below the absolute melting temperature. (c) As sintering progresses, neck size increases and pore size decreases (From Moshiri, M., PhD Thesis, 2020)

to form a by-product that binds the powders together. This fusion mechanism is primarily utilized for ceramic materials. Examples of reactions between powders and atmospheric gases include: laser processing of SiC in the presence of oxygen, whereby SiO_2 forms and binds together a composite of SiC and SiO_2, and laser processing of Al in the presence of N_2, whereby AlN forms and binds together the Al and AlN particles. For chemically induced sintering between powders, it is known that mixtures of high-temperature structural ceramic and/or intermetallic precursor materials can be made to react using a laser. Hence, raw materials that exothermically react to form the desired by-product are pre-mixed and heated using a laser. By adding chemical reaction energy to the laser energy, high-melting-temperature structures can be created at relatively low laser energies. One common characteristic of chemically induced sintering is part porosity. As a result, post-process infiltration or high-temperature furnace sintering to higher densities is often needed to achieve properties that are useful for most applications. The cost and time associated with post processing have limited the adoption of chemically induced sintering in commercial machines.

c. Liquid-Phase Sintering (LPS) or Partial Melting

LPS is possibly the most versatile mechanism for PBF. LPS is a term used extensively in the powder processing industry to refer to the fusion of powder particles when a portion of constituents within a collection of powder particles become molten, while other portions remain solid. In LPS, the molten constituents act as the glue that binds the solid particles together. As a result, high-temperature particles can be bound together without needing to melt or sinter those particles directly. LPS is used in traditional powder metallurgy to form, for instance, cemented carbide tools, where Co is used as the lower-melting-point constituent to glue together particles of WC. There are many ways in which LPS can be utilized as a fusion mechanism in AM processes. In many LPS situations, there is a clear distinction between the binding material and the structural material. The binding and structural materials can be combined in three different ways: as separate particles, as composite particles, or as coated particles, as shown in Figure 5.4. Darker regions in the figure represent the lower-melting-temperature binder material. Lighter regions represent the high-melting-temperature structural material. For indistinct mixtures, microstructural alloying eliminates distinct binder and structural regions.

d. Full Melting

Full melting is the mechanism most commonly associated with PBF processing of engineering metal alloys and semi-crystalline polymers. In these materials, the entire region of material subjected to impinging heat energy is melted to a depth exceeding the layer thickness. The thermal energy of subsequent scans of a laser (next to or above the just-scanned area) is typically sufficient to re-melt a portion of the previously solidified solid structure; thus, this type of full melting is very effective at creating well-bonded, high-density structures from metals and polymers.

5.1.1.6   Comparison between SLM and EBM

As compared with the SLM system, the EBM has higher build rates (up to 80 cm^3/hour because of the high energy density and high scanning speeds) but with inferior dimensional and surface finish qualities. In both the SLM and EBM processes, because of rapid heating and cooling of the powder layer, residual stresses are developed. In EBM, high build chamber temperature (typically 700–900°C) is maintained by preheating the powder bed layer. This preheating reduces the thermal gradient in the powder bed and the scanned layer, which reduces residual stresses in the part and eliminates the requirement for post-heat treatment. Preheating also holds powder particles together, which can act as a support for overhanging structural members. So, supports required in the EBM are only for heat conduction and not for structural support. This reduces the number of supports required and allows the manufacture of more complex geometries. A powder preheating feature is available in a very few laser-based systems, where it is achieved by platform heating. In addition, the entire EBM process takes place under vacuum, since this is necessary for the quality of the electron beam. The vacuum environment reduces thermal convection, thermal gradients, and contamination and oxidation of parts like titanium alloys. In SLM,

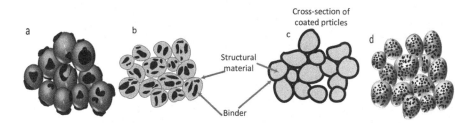

FIGURE 5.4   Liquid phase sintering variations used in powder bed fusion processing: (a) separate particles, (b) composite particles, (c) coated particles, and (d) indistinct mixtures. (With kind permission from Springer Science+Business Media: *Additive Manufacturing Technologies*, 3rd ed., 2021, Gibson, I., Rosen, D., Stucker, B. and Khorasani, M.)

part manufacturing takes place in an argon gas environment for reactive materials to avoid contamination and oxidation, whereas nonreactive materials can be processed in a nitrogen environment. So, EBM-manufactured parts would be expected to have a lower oxygen content than SLM-manufactured parts. In spite of having these advantages, EBM is not as popular as SLM because of its higher machine cost, low accuracy, and non-availability of large build-up volumes. The characteristic features of SLM and EBM are summarized in Table 5.2.

5.1.2 Direct Energy Deposition (DED) Processes

DED comprises all processes where focused energy generates a melt pool into which feedstock is deposited. The heat source for these processes can be a laser beam, an electron beam, or an electric arc. The feedstock used can be either powder or wire. The DED category of AM systems is presented in Chapter 3, including the principles, advantages, and limitations of the subprocesses: laser melting deposition (LMD), laser engineering net shaping (LENS), and the electron beam wire DED (EBAM).

5.1.2.1 Laser Metal Deposition (LMD)

LMD is an additive manufacturing process in which a laser beam forms a melt pool on a metallic substrate, into which powder is fed from hoppers through nozzles. The powder melts to form a deposit that is fusion-bonded to the substrate. The substrate can be positioned in a stationary position (three-axis systems) or on rotating axes (five-axis systems) to increase the ability of the machine to process more complex geometries. After solidification, the filler material forms single weld beads. Multiple weld beads placed next to each other form layers or volumes.

TABLE 5.2
Differences between SLM and EBM for Metal AM

	SLM	EBM
Thermal power source	Laser	Electron beam up to 3 kW
Build chamber atmosphere	Argon or nitrogen (inert gas)	Vacuum
Scanning	Galvanometers	Deflection coils
Energy absorption	Absorptivity-limited	Conductivity-limited
Powder preheating	Infrared or resistive heater 100–200°C	Electron beam (preheat scanning) up to 900°C
Scan speed	Limited by galvanometer inertia	Very fast (magnetically driven)
Maximum build rate (cm³/h)	20–35	80
Layer thickness (μm)	20–100	50–200
Powder particle size	Fine	Medium
Energy cost	High	Moderate
Surface finish	Excellent to moderate4–11 Ra	Moderate to poor50–200 Ra
Feature resolution	Excellent	Moderate
Materials	All (metals, polymers, ceramics)	Metals (conductive)

LMD can be used to generate or modify 3D geometries. The laser can also perform repairing or coating processes with this manufacturing method. It is commonly applied to repairing turbine blades, tool construction and mold making, broken or worn edges, and coating or armoring functional surfaces of bearing points, rollers, or hydraulic components in energy technology or petro-chemistry to protect against wear and corrosion.

A comparison of the two additive processes SLM and LMD is shown in Table 5.3. Because of their respective features, a combined additive process chain has the potential to benefit from high structural complexity with SLM while increasing build-up rates and material flexibility with LMD.

5.1.2.2 Laser Engineering Net Shaping (LENS)

LENS is the most common DED process, where powder delivery is controlled using up to four delivery nozzles. Salient features of LENS technology are the unconstrained powder feed, which in some LENS models can realize five-axis control. Additionally, the typical build volume is comparatively large, up to $900 \times 500 \times 900$ mm. Compared with most commonly used PBF technologies, this enables the creation of much larger parts while simultaneously reducing relative powder demands. Compared with many traditional metal AM build processes, LENS has many unique characteristics, which allow a range of interesting opportunities for part production. The first such attribute stems from the ability of LENS to create functionally graded porous structures, which has been exploited for orthopedic implant applications. LENS is also highly suited to repairing parts with complex geometry due to the high level of environmental control, which eliminates issues relating to part oxidation during manufacturing, as parts are processed in an inert environment (Mojtaba et al. 2020).

5.1.2.3 Electron Beam Additive Manufacturing (EBAM)

Electron beam (EB) processing has the distinct advantages of high energy density (high beam quality, e.g. small spot size) and high beam powers (multi-kilowatt) and is performed in a high-purity vacuum, parts per billion (ppb) oxygen, versus parts per million (ppm) levels as present in commercial welding grade argon. DED-EB

TABLE 5.3
Comparison of SLM and LMD

	Part dimensions	Structural complexity	Substrate	Material flexibility
SLM	Limited by the process chamber	High, e.g. lattice structures	Flat surfaces	Some powder for the whole process
LMD	Limited by the machine working area	Limited, e.g. walls	Arbitrary surfaces	In-process change of powder

Source: Graf, B., Schuch, M., Kersting, R., Gumenyuk, A. and Rethmeiera, M. (2015) in *Lasers in Manufacturing Conference, LiM*, Wissenschaftliche Gesellschaft Lasertechnik.

machines are produced by Sciaky and are referred to as electron beam additive manufacturing (EBAM). A DED-EB process developed by NASA is referred to as electron beam free form fabrication or EBF3. The higher-purity vacuum environment of these machines offers a primary advantage by enabling the deposition of highly reactive materials and those susceptible to contamination by oxygen or other contaminants picked up during solidification and cooling. These systems integrate a mobile EB gun, computer numerically controlled (CNC) motion, and a wire feeder within a large high-vacuum chamber allowing movement in the x–y or tilt orientations to trace out and fuse a deposited bead of metal, one bead at a time. Using this approach, very large vacuum chamber build environments can be created, allowing the deposition of very large structures. High deposition rates are possible using a wide range of available wire alloys and sizes and choice of deposition parameters. Near net shaped components display a distinctly stepped weld bead overlay shape that requires machining to create the final shape. One disadvantage is the slow cooling rate of the deposit within the vacuum environment and its potential effect on large grain growth and other metallurgical effects of the deposit. High degrees of distortion or residual stress may result when depositing large structures, requiring post-process heat treatment (Milewski 2017).

5.1.2.4 Wire and Arc Additive Manufacture (WAAM)

With the electric arc as energy-carrying beam, WAAM is an emerging advanced digital manufacturing technology using a layer-by-layer cladding principle to gradually form metal solid components from the line-surface-volume through the addition of the wire material according to the 3D digital model under the control of the program, which adopts electric arc welding machines, such as metal inert-gas welding (MIG) and tungsten inert-gas welding (TIG), or plasma arc welding (PAW) as the heat source (Figure 5.5), and the formed parts are composed of all welded joints with uniform chemical composition and high density.

The energy-carrying beam of WAAM has the characteristics of low heat flux density, large heating radius, and high heat source intensity, coupled with the fact that the instantaneous point heat source reciprocating during the forming process strongly interacts with the forming environment. The more stable the arc is, the more favorable it is to the forming process control, that is, the more favorable it is to the dimensional accuracy control of the formed shape. The stacking speed is high. The arc printing has a fast wire feeding speed and high stacking efficiency, which has obvious advantages in the efficient manufacture of near net shaped large-sized parts (1000–3000 mm) with modest complexity and a high rate of production (50–130 g/min). Many materials are available in the welding community at relatively low cost; thus, AM costs are much lower than for equivalent AM technologies, This technology is expected to become a new manufacturing technology for mass production in the civilian market.

Limitations of the process are low surface quality of the solid components and the lower ability to shape complex geometry than the laser. This is because the arc forming position is determined by the positions of the welding torch, the welding wire, and the robot. Further, residual stress and heat input distortion—deformations

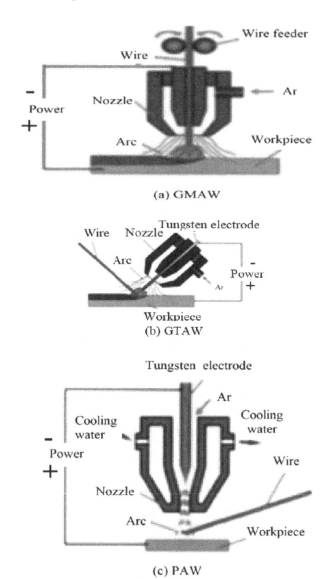

(a) GMAW

(b) GTAW

(c) PAW

FIGURE 5.5 Schematic diagram of three arc welding technologies applied in WAAM processes. (From *Results in Engineering*, 13, Li, Y., Su, C. and Zhu, J., Comprehensive Review of Wire Arc Additive Manufacturing: Hardware System, Physical Process, Monitoring, Property Characterization, Application and Future Prospects, 100330, Copyright (2022), with permission from Elsevier)

caused by residual stress—are a major cause of tolerance loss, poor dimensional accuracy, and surface finish in WAAM (Figure 5.6).

Wire arc additive manufacturing has unparalleled efficiency and cost advantages over other metal AM techniques in the forming of large-size structural parts.

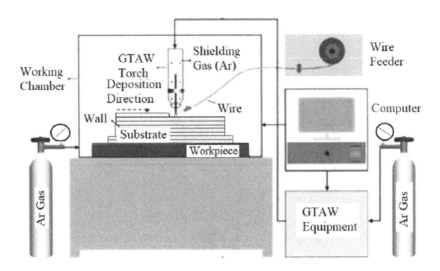

FIGURE 5.6 Schematic of the WAAM system constituents. (From https://miro.medium .com/)

It is called a low-energy, sustainable, and green manufacturing technology by the European Space Agency. Recently, plenty of research institutions have developed their WAAM system and successfully applied it to fabricate industrial products. Typical applications of these products are shown in Figure 5.7.

A comparison of the main differences in the processing parameters and product quality characteristics for the different techniques of the two main categories of AM of metallic materials, DED and PBF, is presented in Table 5.4.

5.1.3 Binder Jetting (BJ) Processes

For metals, the BJ process creates "green" parts made of bonded metal powder that are then cured, or dried, in an oven. The printed part is then removed from the powder bed (depowdering) and cleaned before final sintering in a high-temperature furnace, where the particles fuse together into a final metal object that is dense, accurate, and can be machined. Now, metal parts made with BJ deliver a final part density that is better than press-and-sinter, and as good as or better than parts produced with metal injection molding (MIM), depending on the specific metallic material. Additionally, parts made with this process routinely deliver improved density as compared with gravity or low-pressure metal castings. Any unbound powder left over from the printing process can be reused several times through a mixing and reconditioning process, ensuring an overall efficiency of material consumption of up to 96%. The process enables mass manufacturing at speeds and costs that compete with traditional manufacturing technologies, besides its applications in prototypes and batch production. It is one of the few solutions in that respect.

Metal binder jet systems are capable of processing more than 20 metals, including aluminum alloys, copper and copper alloys, carbides, nickel alloys, nitrides, oxides,

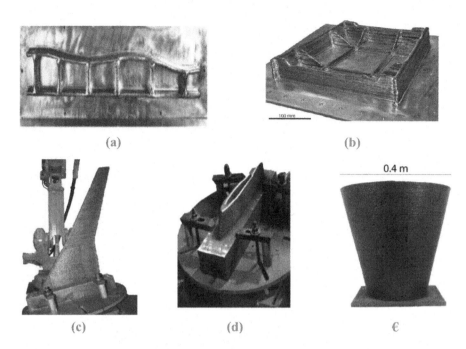

(a)   (b)

(c)   (d)   €

FIGURE 5.7   (a) 1.2 m, Ti–6Al–4V wing spar built for BAE Systems; (b) 24-kg Ti–6Al–4V external landing gear assembly, (c) high-strength steel wing model for wind tunnel testing, (d) particular of hollow structure, (e) mild steel truncated cone. (Panel (a) courtesy of BAE Systems; panels (b)–(e) modified from Williams, S.W., Martina, F., Addison, A.C., Ding, J., Pardal, G. and Colegrove, P., *Materials Science and Technology*, 32(7), 160208081556009, 2015)

stainless steels, titanium alloys, tool steels, refractory metals, and precision metal alloys. These materials require a variety of BJ systems, based on several factors, such as the approach to powder management and printhead types that can process certain binder chemistries that may not work with all materials. Some materials, such as aluminum and titanium alloys, may also require an inert or controlled atmosphere system.

Powders: The main property of powders is their depositability, which depends on the size and shape of the particles. For the dry state, preferred particles are less than 20 microns, while particles smaller than 5 microns can be deposited both dry and wet. Fine powders (~1 µm) tend to agglomerate due to the Van der Waals forces and moisture effect. To exploit the advantages of powders with small and large particles, multimodal powders are used. Larger particles allow the powder mixture to be easily spread in the dry state, while the smaller particles fill the interstices among the large particles to increase the density of the printed part. In both dry and wet deposition methods, the particle shape is less important than the size, but spherical powders are preferred for dry deposition because they tend to have a better flow and have low internal friction.

TABLE 5.4

Comparison of Different Techniques of DED and PBF Processes

	PED			PBF	
	Powder	Wire		Powder	
ProcessFeedstock	laser	E-beam	Electric arc	Laser	E-beam
Heat source	laser	E-beam	Electric arc	Laser	E-beam
Nomenclature	DED-L	DED-EB	DED-PA/DED-GMA	PBF-L	PBF-EB
Power (W)	100–3000	500–2000	1000–3000	50–1000	
Speed (mm/s)	5–20	1–10	5–15	10–1000	
Max. feed rate (g/s)	0.1–1.0	0.1–2.0	0.2–2.8	–	
Max. build size (mm × mm × mm)	200 × 1500 × 750	200 × 1500 × 750	5000 × 3000 × 1000	500 × 280 × 320	
Production rate	High	Medium	Low	High	
Dimensional accuracy (mm)	0.5–1.0	1.0–1.5	Intricate features are not possible	0.04–0.2	
Surface roughness (μm)	4–10	8–15	Needs machining	7–20	
Post-processing	HIP and surface grinding are seldom required	Surface grinding and machining are required to achieve better finish	Machining is essential to produce final parts	HIP is rarely required to reduce porosity	

Source: DebRoy, T., Wei, H.L., Zuback, J.S., Mukherjee, T., Elmer, J.W., Milewski, J.O., Beese, A.M., Wilson-Heid, A., De, A. and Zhan, W., Additive Manufacturing of Metallic Components – Process, Structure and Properties. *Progress in Materials Science*, 92, 112–224, Copyright (2018), with permission from Elsevier.

Types of binders: There are different ways of binding the powder. Common binding methods are organic liquids, in-bed adhesives, hydration systems, acid/base systems, inorganics, metal salts, solvents, phase changing materials, and sintering aids/inhibitors.

The most relevant parameters of the liquid binder material are the viscosity and surface tension. The spread characteristics are important because a liquid that wicks out considerably from the impact area results in a rougher surface texture. Increased liquid binder viscosity delays the spreadability, while viscosity reduction is achieved by lowering the solids loading or adding dispersants. The parameters inherent in the printing process are droplet size, binder saturation level, printing layer thickness, and orientation and location of the printed part, which influence the strength and surface quality of the 3D printing process.

The saturation level is the ratio of ink to bed pore volume and depends on the droplet size, droplet spacing, layer thickness, and bed packing density. The saturation level needs to be high enough that the ink penetrates to the previous layer to bind the part in the vertical direction, but not so high that excess ink wicks away from the impact zones and roughens the surface finish by binding extra powder.

Drying/Curing: After printing, the build box containing the printed parts—which are called "green" at this stage—is placed into an oven, where the binder is dried and the part essentially cures, building strength in the part. This curing step gives the green parts strength to be removed from the print bed. Some systems perform this curing step inside the printer, which involves an expensive setup. Lower-cost ovens can perform drying or curing separately out of the machine.

Depowdering: The process of removing green parts from the loose powder contained in the build box and preparing those parts for sintering is called "depowdering". During this process, parts are removed, and the unused loose powder that surrounds the parts is recycled for use in future builds. The degree of recyclability may vary between different BJ systems. Often, parts are cleaned with a simple air gun.

Sintering is performed in a furnace where the green state parts are heated to temperatures below the melting point of the powder to be fused and densified. Depending on the powder alloy, the method for densification can be either solid state sintering or liquid-phase sintering. Solid state sintering works by diffusion of the alloy at temperatures approaching the solidus point of the material to minimize porosity in the final parts. Liquid-phase sintering can be implemented by picking a temperature range between the solidus and liquidus points of the alloy. During both sintering mechanisms, the particles begin to "neck" or attach to one another, and the void spaces collapse, creating a dense part. Parts typically undergo shrinkage during the sintering process and potentially, distortion if not designed and supported properly. In some BJ systems, the software can predict and correct for shrinkage and distortion in part designs, delivering sinter-ready, printable geometries.

Infiltration is a way to achieve high-density parts without the large shrinkage associated with sintering to full density. Parts printed by BJ, especially in metallic and ceramic powders, are not strong enough to be used as functional parts such as those made by other AM technologies, so infiltration is necessary to improve the strength or final characteristics of the part. Infiltration takes place when a selected liquid is drawn into the open pores of the printed part through capillary action and solidifies. Both low- and high-temperature infiltration are possible, depending on the part material and binding mechanism. The first occurs at or slightly above room temperature, and the second at temperatures between −6 and −10°C below the infiltrate's melting point.

Inkjet Printheads: The industrial printhead is the engine of a BJ system. The two predominant printhead types used in BJ are thermal and piezoelectric. Both types of printheads are capable of delivering high-quality BJ results, though each type has its own advantages and disadvantages.

Thermal printheads work by heating a resistor, causing a small volume of water-based ink to vaporize, creating a bubble. As the bubble expands, it pushes a droplet of ink out of the printhead and onto the print medium. Thermal printheads are known for their high resolution and are commonly used in industrial printing applications. They virtually require a water-based binder to create a bubble; therefore, most thermal printheads are limited to water-based binder chemistries. This may limit the number of metals they can print.

Piezoelectric printheads, on the other hand, create droplets mechanically via displacement from the expansion and contraction of a piezoelectric crystal. They are known for their high speed and are commonly used in industrial printing applications such as wide-format printing and high-speed inkjet printing. They are typically more expensive and more complex to produce, but they may have higher resolution, better durability, and more performance options. For example, they can process a wider range of viscosities, including binders with particles in them, and deliver droplets or jet streams, enabling more sophisticated binder deposition strategies. Table 5.5 presents a comparison between both types.

Machine Design and Controls: The different designs of BJ systems have developed two distinct approaches: bed-to-bed simple design and the triple advanced compaction technology (ACT).

Bed-to-Bed Jetting: Bed-to-bed powder deposition is one of the earliest and least expensive forms of powder spreading. This approach, offered in the Shop System with a thermal printhead, is highly affordable and very reliable; binder jet systems first produced in 2003 using this approach are still printing commercial parts today (Figure 5.8a).

Triple Advanced Compaction Technology: Also known as Triple ACT, this approach is used with a piezoelectric printhead system. This approach uses different methods of dispensing, spreading, and compacting ultra-fine powders for the highest-density green parts. The hallmark of this approach is tight individual parameter control of particle dispensing, spreading, and compacting. This allows processing for a wide range of metals and ceramics (Figure 5.8b).

5.1.4 Sheet Lamination Processes

Sheet lamination (SL) uses stacking of precision-cut 2D metal sheets into slices to form a 3D object. After stacking, these sheets are either adhesively joined or

TABLE 5.5

Comparison between Thermal and Piezoelectric Printheads

	Thermal Inkjet Printheads	Piezoelectric Printheads
Operating mechanism	Ink chamber is heated to create a vapor bubble that expands and forces a precise drop of ink out of the nozzle	Electric charge is applied to piezoelectric elements (i.e. crystals) to deform the surface creating pressure that forces a precise drop of ink out of the nozzle
Printhead life	Shorter. Typically used as a consumable, but category offers a wide range of printhead life	Longer. Typically not a consumable but may need replacement every several years. Also may be repaired
Cost	Lower	Higher
Binder compatibility	Only works with water-based (aqueous) binders	Works with a wide variety of binders including water-based (aqueous) and solvent based, as well as higher viscosity binders including those that may contain particulate or non-liquid additives
Droplet size	Single; binary on or off	Variable: enables grayscaling and varying wave or droplet formations
Maintenance	Consumable; designed to be easily replaced in printer	More involved. Designed to be maintained over time
OEMs	HP, Canon, Memjet	Epson, Fujifilm, Xaar

OEMs: Original Equipment Manufacturers.

Source: From Desktop Metal Inc., (2023). *The Ultimate Guide to: Laser-Free Metal 3D Printing with Binder Jetting Technology.*

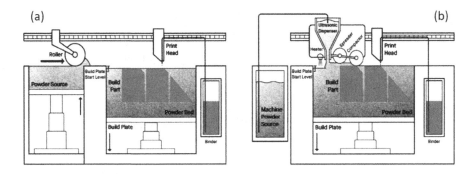

FIGURE 5.8 Two binder jetting design systems: (a) bed-to-bed design, and (b) triple advanced compaction technology (ACT). (From: Desktop Metal Inc. 2023). (From Desktop Metal Inc., *The Ultimate Guide to: Laser-Free Metal 3D Printing with Binder Jetting Technology*, 2023)

metallurgically bonded using brazing, diffusion bonding, laser welding, resistance welding, or ultrasonic consolidation (UC). A key feature of SL hardware is the order in which sheets are applied and cut/machined. Sheets may be either cut to the specified geometry prior to adhesion or machined post adhesion. The metal SL process ensures low geometric distortion (the original metal sheets retain their properties), ease of making large-scale parts, relatively good surface finish, and low costs. However, adhesively joined parts may not work well in shear and tensile loading conditions, geometric accuracy in the z direction is difficult to obtain due to swelling effects, and SL builds suffer anisotropy due to the type of joining processes.

In a brazing SL process, for example, sheets are coated with flux (or low-melting alloy), which acts as a brazing alloy for joining these sheets. In another process, special fixtures have to be developed for resistance welding SL to enable joining of layers. Due to these restraints, other solid state joining techniques between sheets have been considered. One innovative SL process joins sheets using an ultrasonic seam welding technique known as UAM or UC. It is a solid state welding process where high-frequency ultrasonic acoustic vibrations are applied through a rotating sonotrode pressed against the sheets to form a solid bond. UAM is currently one of the most often used technologies for metal SL.

Figure 5.9 exhibits the UAM system components and the directions of motion. During the bonding step, a cylindrical sonotrode is pressed at a constant normal

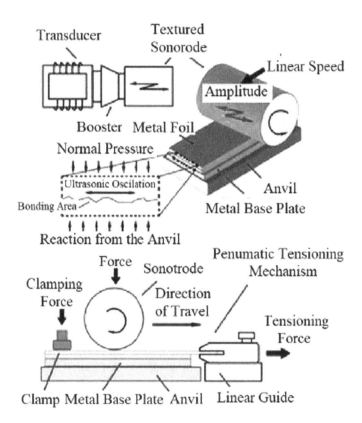

FIGURE 5.9 Ultrasonic additive manufacturing (UAM) process. (From Total Materia, 2021)

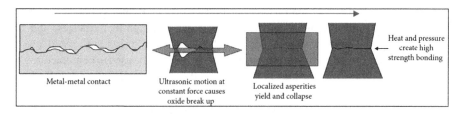

Metal-metal contact

Ultrasonic motion at constant force causes oxide break up

Localized asperities yield and collapse

Heat and pressure create high strength bonding

FIGURE 5.10 Steps to reach high-strength bonds via ultrasonic consolidation.

force against the metal foil, which is kept in place via a clamping or tensioning mechanism. The sonotrode rolls over the foil while oscillating at a constant frequency of approx. 20 kHz perpendicular to the direction of travel.

The process parameters have a direct effect on the quality of the final part. The important parameters are the ultrasonic amplitude (approximately 5–50 µm), the applied normal force (500–2000 N), the travel speed of the sonotrode horn (up to 50\mm/s), the texture of the sonotrode horn (Ra between 4 and 15 µm), and the substrate, which may be cold welded or preheated to a temperature 30 to 50% of the melting point. These parameters vary for different materials and need to be optimized for every material to yield a sound, high-strength bond via disruption of the surface oxide layer between welded parts (Figure 5.10). The placement of foils and their orientation also play a major role in deciding the final properties of the part. Sheet or foil thicknesses as low as 0.1 mm can be used to build a component.

In theory, any sheet metal that can be ultrasonically welded is a candidate material for the UAM process. Materials that have been successfully bonded using UAM include various aluminum alloys, nickel alloys, brass, steels, etc. The process has been applied also to distinct applications, including combinations of dissimilar metals, functionally graded materials, structurally embedded electronic systems, and fiber reinforced metal matrix composites.

Pairing of machining with the consolidation process is common and produces parts with a machined surface finish directly from the hybrid process. However, this hybrid process cannot manufacture complex overhangs, as no support material is deposited to provide mechanical support. Features may be additionally limited by the tool paths available for machining operations.

APPLICATIONS OF UAM

a. *Functionally graded materials (FGM):* The process can manufacture the FGM used for applications where high-level mechanical as well as physical properties are required, such as aerospace vehicle shields and rocket nozzles where the temperature reaches up to 2000°C. Other applications are heat exchanger plates, lightweight armor plates for defense applications, and heat engine components.

b. *Structurally embedded electronic systems:* In current industrial applications, the electric systems embedded in a solid enclosure or part are becoming essential so as to facilitate portability, robustness, and flexibility in functioning. Thus, UAM provides a single-stroke method to produce better electrical applications, such as embedded sensors, high-tech devices, and Internet of Things products.

c. ***Fiber reinforced metal matrix composites:*** UAM can be used to manufacture the MMC in which the matrix is a metal reinforced by silicon carbide (SiC) fibers. This composite manufactured by UAM is used in applications such as airplane wing frames, where low weight and higher strength are expected. The F-16 Falcon fighting plane uses silicon carbide fibers in a titanium matrix for structural components of the jet's landing gear (Choudhary et al. 2017).

5.2   FORMS OF METALS USED IN AM OF METALS (FEEDSTOCK MATERIALS)

The materials used for building products in metal AM processes take the form of powder (for all PDF and some of the DED processes), whereas EBAM and WAAM processes use thin wire, and the SL AM processes use thin sheets, as shown in Figure 5.11. The major characteristics of these forms are described in this section.

5.2.1   FEEDSTOCK POWDERS

These are commonly used as feedstock materials in the laser and EBAM techniques due to ease of feeding and controlled fusion/melting. Feeding a mixture of multiple alloy powders in a pre-set ratio further allows building a part with a composition/property gradient that is difficult to achieve by traditional processing. However, the manufacturing of high-quality powders remains a critical challenge due to their high surface area and susceptibility to oxidation. The main characteristics of powders that control product quality include the shape, size distribution, surface morphology, composition, and flowability of the powders. Scanning electron microscopy (SEM), X-ray, and computed tomography (CT) are used to examine the shape and the surface morphology of the powder particles. Laser diffraction and sieving methods are

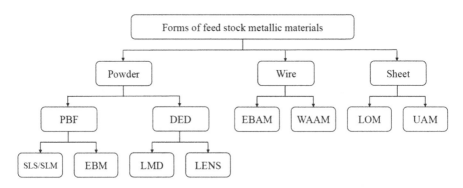

FIGURE 5.11   Forms of feedstock metallic materials and the processes using them. (Modified from Lewis, G., *World Journal of Engineering and Technology*, 10, 363–409, 2022)

used to ensure the size distribution of the powders. Flowability of the powders is measured by a Hall flow meter. Fine powder particles with uniform size distribution and smooth surface are able to provide uninterrupted flow through the feeder nozzles and promote small pool size under the concentrated beam. As a result, both DED and PBF processes prefer the use of alloy powders with good surface finish and size distribution. Details of the powder preparation techniques and characterization will be presented in Chapter 9. Powder preparation techniques enable diversity and flexibility of material properties to meet product performance requirements and to form components from refractory metals such as molybdenum, tantalum, and tungsten with high melting temperatures that exclude conventional casting processes. Alloys with immiscible phases, which will not form solutions under normal melting conditions, can be made to form alloy systems by the intimate mixture of particles prior to further processing, known as mechanical alloying. The cost of powdered metals is relatively high; however, this is offset by all the advantages that AM achieves.

5.2.2 Feedstock Wire

The use of filler wires can result in a greater rate of deposition compared with the powder particles. The wires of different alloys and sizes are manufactured by wire drawing and are relatively inexpensive compared with powders of the same alloy. However, the cost savings per kilogram are offset by the material lost to machining process waste, as wire-based AM processes often require extensive machining allowance to be removed to achieve the desired net shape. In addition, remaining material from a partially used spool may not be suitable for reuse, unlike the ability to reuse and recycle un-fused PBF process powder. Filler wires of diameters smaller than 0.8 mm are scarce, and thus, wire-fed AM processes require a larger melt pool size, resulting in a relatively rougher surface finish of the final part. However, the use of filler wires can result in a greater rate of deposition compared with the powder particles. As a benefit, wire feedstock has significantly less surface area per kilogram than powder product and is less likely to oxidize and absorb moisture or contaminants. Wire forms are easier to store and handle, and pose fewer hazards associated with environment, safety, and health when compared with metal powders.

Table 5.6 shows the difference in cost of wire compared with powder as common materials for metal AM. The cost of wire could be as low as 0.3 times the cost of wire for SS 316L.

TABLE 5.6
Approximate Cost of Different Metal Materials in Wire and in Powder

Feedstock	Ti-6Al-4V	Inconel 718	Inconel 625	Stainless steel 316L
Wire (£/kg)	120	58	49	12
Powder (£/kg)	280	80	80	40

Source: Xiong, J.T. et al., (2015). *Journal of Advanced Manufacturing Technology*, 58, 80–85.

5.2.3 Sheets and Foils

Aluminum alloys are amenable to the UAM process because solidification is avoided, copper alloys can be joined reliably, and multiple metal types can be joined together (such as Inconel and Al) to improve device performance. This multiple metal joining attribute also allows cladding, transition joints, and composite structures. Multiple metals can be joined reliably in UAM because intermetallic formation is largely suppressed due to the low processing temperature. This low processing temperature allows the inclusion of heat-sensitive electronics into the metal structure directly. Examples are including sensors into structures for improved control or health monitoring applications. This inclusion into the structure also provides protection to the sensor to avoid external wear or damage. The formed joint or deformation zone is near 10 μm in size and localized to the interface region, which leads to minimal heat generation in UAM. Formation temperatures have been measured to be near 150°C for aluminum and copper alloys. Also, the process has challenges when working with hard steels and nickels because the foil stock has an affinity to stick to the steel UAM tooling. Building layers of tantalum, molybdenum, and titanium using UAM for radiation shielding in structural panels has also been achieved.

High-quality metal powder and wire feedstock are very important for successful PBF in AM. A number of different metals are available in powdered form to suit exact processes and requirements. Titanium, steel, stainless steel, aluminum, and copper-, cobalt chrome-, titanium-, and nickel-based alloys are available in powdered form, as well as precious metals like gold, platinum, palladium and silver. Wire feedstock options are also wide-ranging; steel and stainless steel alloys as well as pure metals like titanium, tungsten, niobium, molybdenum, and aluminum are all available as wire feedstock. Table 5.7 represents the utilization of these metallic materials in different applications, followed by a brief account of the specific AM applications of these materials.

5.3 COMMON METALLIC MATERIALS USED IN AM

Table 5.8 summarizes the feedstock type, AM process, and mechanical properties of some of the most commonly used materials.

5.4 DEFECTS OF METALLIC PRODUCTS MADE BY AM PROCESSES

The common limitations of AM processes affect the geometrical, physical, and mechanical characteristics of their products. The regular defects that may affect metallic products include:

i. **Loss of alloying elements:** When the molten pool temperatures are very high, pronounced vaporization of alloying elements takes place. Since some elements are more volatile than others, selective vaporization can occur, causing changes in the overall composition of the alloy. These changes can affect solidification microstructure, corrosion resistance, and

TABLE 5.7

Different Metallic Materials Used in AM Applications in Different Industries

Alloy / Application	Aluminum	Maraging steel	Stainless steel	Titanium	Cobalt chrome	Nickel superalloy	Precious metals
Aerospace	•		•	•	•	•	
Medical			•	•	•		•
Energy, oil, and gas			•				
Automotive	•		•	•			
Marine			•	•		•	
Machinability and weldability	•		•	•		•	
Corrosion resistance			•	•	•	•	
High temperature			•	•		•	
Tools and molds		•	•				
Consumer products	•		•				•.

Source: Adapted from DebRoy, T., Wei, H.L., Zuback, J.S., Mukherjee, T., Elmer, J.W., Milewski, J.O., Beese, A.M., Wilson-Heid, A., De, A. and Zhan, W., Additive Manufacturing of Metallic Components – Process, Structure and Properties. *Progress in Materials Science*, 92, 112–224, Copyright (2018), with permission from Elsevier.

mechanical properties and can be a serious problem in producing high-quality components.

ii. **Porosity and lack of fusion voids** are common defects in AM that need to be minimized or eliminated due to their adverse effects on mechanical properties. Inert shielding gas in laser processes, and vacuum in EB processes, as well as control of the process parameters, can eliminate such defects.

iii. **Surface roughness** is one of the most important features of intricate components fabricated using AM. It is measured using a profilometer or by analyzing the surface morphology using SEM. Additively manufactured components for high-end applications require an average surface roughness less than 1 μm. However, as-deposited parts often exhibit rough surfaces and require post-processing such as surface machining, grinding, chemical polishing, post HIP, and shot peening to attain the required finish.

iv. **Cracking and delamination:** Solidification cracking can be observed along the grain boundaries of the build due to both solidification shrinkage and thermal contraction. Since the temperatures of the substrate or the previously deposited layers are lower than those of the depositing layer, the contraction of the depositing layer is greater than that of the lower layer, and the contraction of the solidifying layer is thus hindered by the substrate or the previously deposited layer. This results in the generation of a tensile stress at the solidifying layer. If the magnitude of this tensile stress

TABLE 5.8

Metallic Material Printing Methods, Processes, and Properties

Method	Feedstock type	Process	Material	Young's modulus (GPa)	Yield strength (GPa)	Ultimate strength (GPa)	Density (g/cm³)	Poisson's ratio
DED	Powder Fed	LENS	Ti6Al4V	113	945	979	4.41	0.34
DED	Wire Fed	EBF3	Inconel 718	202	1195	1372	8.2	0.288
LOM	Metal Sheet Fed	UAM	Al6061	68	217	225	2.7	0.33
LOM	Metal Sheet Fed	UAM	Al3003	68.9	262	266	2.73	0.33
PBF	Powder Bed Fed	EBM	Iron (Fe)	204	50	540	7.86	0.29
PBF	Powder Bed Fed	EBM	Copper (Cu)	110	33.3	210	8.96	0.343
PBF	Powder Bed Fed	SLM	Gold (AU)	79	80	900	19.3	0.415

Source: Das, S. et al., (2016). *MRS Bulletin*, 41, 10, 729–741.

exceeds the strength of the solidifying metal, cracking may be observed along those grain boundaries. Delamination is principally the separation of two consecutive layers, which is caused by the residual stresses at the layer interfaces exceeding the yield strength of the alloy. In AM components, cracking can be very long, spreading over several layers, or small, with a maximum length equal to the layer thickness.

v. **Residual stresses and distortions:** An inherent consequence of the deposition of liquid alloy powder on a relatively cooler substrate or prior deposited layers is the steep temperature gradient, thermal strain, and residual stresses. The residual stresses can lead to part distortion, loss of geometric tolerance, and delamination of layers during depositing, as well as deterioration of the fatigue performance and fracture resistance of the fabricated part. To mitigate residual stresses and/or stress-induced distortion, there must be tight control of the substrate preheat temperature, using shorter deposition length or scanning in smaller islands, an effective deposition strategy such as spiraling-in as against spiraling-out, increase in scanning speed, and decrease in layer height. It is noted that the substrate preheat temperature can reduce not only the final residual stresses post fabrication but also the stresses during building. The use of high preheat can also be important to mitigate solidification cracking of additively manufactured nickel-base superalloys.

REVIEW QUESTIONS

1. What are the distinguishing attributes of metallic materials compared with polymeric, ceramic, and composite materials in additive manufacturing processes?
2. Compare SLS and SLM as one ASTM category but two different AM processes.
3. Compare conventional lasers and the recently emerged fiber laser as power sources for PBF and DED processes.
4. Suggest applications for each of the four mechanisms of sintering/melting mechanisms for PBF processes.
5. What are the main differences between SLM and EBM for metal AM?
6. What distinguishes the electron beam head used in EBAM systems from that used in scanning electron microscopes?
7. Differentiate LMD and LENS as DED processes, showing typical applications of each.
8. What is the difference between GMAW, GTAW, and PAW as welding processes and as energy sources in the WAAM system in metal AM?
9. Discuss the differences between techniques of powder feeding and those of binder feeding in BJ systems.
10. What are the differences between sintering and infiltration of binder jet products?

11. What are the types of printheads in BJ systems and the role of each in the quality of products?
12. Compare the two technologies adopted in machine control of BJ systems: namely, the bed-to bed simple design and the triple advanced compaction technology (ACT).
13. Which specific sheet lamination process is used for manufacturing metal parts? What are the differences when compared with the LOM process?
14. What are the process parameters that lead to a high-strength bond in UAM?
15. What are the critical applications of UAM that cannot be achieved using conventional processes?. Give typical products for these applications.
16. Compare the effects of the different forms of feedstock metallic material on the selection of the AM processes.
17. Differentiate between the application of lasers and electron beams as heat sources on the selection of the feedstock form of metallic materials.
18. Give typical examples of AM industrial products based on the three forms of metallic materials.
19. What are the metal powder preparation methods and their effect on the quality of the powder required for AM?
20. Discuss the effects of the source of power and traversing speed on the production time and product quality in metal AM processes.
21. Why are aluminum alloy powders extensively used in AM processes?
22. What are the types of metallic materials suitable for orthopedic devices and implants?
23. What are the common defects of metallic products built by AM processes, and how can they be eliminated?

BIBLIOGRAPHY

Abdel-Aal, Hisham. (2022) *Additive Manufacturing of Metals Fundamentals and Testing of 3D and 4D Printing.* New York: McGraw Hill Co.

Choudhary, Suresh and Jondhale, Shivajirao S. (2017) Study of Ultrasonic Additive Manufacturing. *International Journal of Innovative Works in Engineering and Technology (IJIWET),* 3(3), ISSN: 2455-5797

Das, S., Bourell, D. L., and Babu, S. S. (2016) Metallic Materials for 3D Printing. *MRS Bulletin,* 41(10), 729–741. https://doi.org/10.1557/mrs.2016.217

DebRoy, T., Wei, H. L., Zuback, J. S., Mukherjee, T., Elmer, J. W., Milewski, J. O., Beese, A. M., Wilson-Heid, A., De, A., and Zhan, W. (2018) Additive Manufacturing of Metallic Components – Process, Structure and Properties. *Progress in Materials Science,* 92, 112–224. https://doi.org/10.1016/j.pmatsci.2017.10.001

Desktop Metal Inc. (2023) *The Ultimate Guide to: Laser-Free Metal 3D Printing with Binder Jetting Technology.*

Despa, V. and Gheorghe, I. Gh. (2011) Study of Selective Laser Sintering- A Qualitative and Objective Approach. *The Scientific Bulletin of VALAHIA University (Romania) – Materials and Mechanics,* 6, 150–155.

Gibson, Ian, Rosen, David, Stucker, Brent, and Khorasani, Mahyar. (2021) *Additive Manufacturing Technologies,* 3rd ed. Springer. https://doi.org/10.1007/978-3-030 -56127-7

Graf, Benjamin, Schuch, Michael, Kersting, Robert, Gumenyuk, Andrey, Rethmeiera, Michael. (2015) Additive Process Chain using Selective Laser Melting and Laser Metal Deposition. In *Lasers in Manufacturing Conference, LiM*, Wissenschaftliche Gesellschaft Lasertechnik.

https://ars.els-cdn.com/

https://miro.medium.com/

ISO / ASTM. (2015) Additive manufacturing - General principles-Terminology. 52900: 2015(E).

Lewis, Gladius. (2022) Aspects of the Powder in Metal Additive Manufacturing. *World Journal of Engineering and Technology*, 10, 363–409. https://doi.org/10.4236/wjet .2022.102022

Li, Yan, Su, Chen, and Zhu, Jianjun. (2022) Comprehensive Review of Wire Arc Additive Manufacturing: Hardware System, Physical Process, Monitoring, Property Characterization, Application and Future Prospects. *Results in Engineering*, 13, 100330. https://doi.org/10.1016/j.rineng.2021.100330

Milewski, J. O. (2017) *Additive Manufacturing of Metals, Springer Series in Materials Science, Volume 258*. Springer. https://doi.org/10.1007/978-3-319-58205-4

Mojtaba, Izadi, Aidin, Farzaneh, Mazher, Mohammed, Gibson, Ian, and Bernard Rolfe. (2020). A Review of Laser Engineered Net Shaping (LENS) Build and Process Parameters of Metallic Parts. *Rapid Prototyping Journal*, 26(6), 1059–1078. https://doi .org/10.1108/rpj-04-2018-0088

Moshiri, Mandanà. (2020) Integrated Process Chain for First-time-right Mould Components Production Using Laser Powder Bed Fusion Metal Additive Manufacturing. PhD Thesis. https://doi.org/10.13140/RG.2.2.17942.98880

Total Materia. (2021) Ultrasonic Additive Manufacturing (UAM): Part One.

Williams, S. W., Martina, F., Addison, A. C., Ding, J., Pardal, G., and Colegrove, P. (2016) Wire + Arc Additive Manufacturing. *Materials Science and Technology*, 32(7), 160208081556009. https://doi.org/10.1179/1743284715y.0000000073

Xiong, J. T., Geng, H. B., and Lin, X. (2015) Research Status of Wire and Arc Additive Manufacture and its Application in Aeronautical Manufacturing. *Journal of Advanced Manufacturing Technology*, 58, 80–85. https://doi.org/10.16080/j.issn1671-833x.2015 .23/ 24.080

6 Technologies for Additive Manufacturing of Ceramics (AMC)

6.1 INTRODUCTION

Along with extensive research on additive manufacturing (AM) of polymers and metals, additive manufacturing of ceramics (AMCs) is now the latest trend to come under the spotlight. Due to their various excellent properties, ceramics are used in a wide range of applications, including the chemical industry, machinery, electronics, aerospace, and biomedical engineering. The properties that make them versatile materials include high mechanical strength and hardness, good thermal and chemical stability, and viable thermal, optical, electrical, and magnetic performance.

Ceramic materials are usually the best choice for conditions of use in terms of temperature, pressure, corrosive environment, or a combination of all these. They are often described as materials of the future. Compared with metals, ceramics offer several advantages, including:

- chemical resistance to alkalis and acids
- lower density than metals
- insulating effect
- low thermal expansion
- high corrosion and temperature resistance
- extreme hardness and consequently, highly wear resistant
- high compressive strength
- good tribological properties
- excellent biocompatibility and thus, applicable in medical and food applications
- high E-modulus of elasticity at high temperatures
- have shiny and smooth surfaces
- can be produced with complex geometry

The following are the main applications of technical ceramics as linked to one or more of the characteristics listed above:

- Products used in luxury industry, such as watches and jewelry
- Wear parts, such as cutting and machining tools, tribology, wire-guides, and knives

DOI: 10.1201/9781003451440-6

- Parts for thermal protection, such as furnace bricks, space shuttle tiles, or ceramic coating on metal parts
- Chemically resistant products, found in chemical and agri-food industries
- Electrical insulators and superconductors
- Biocompatible implants and bone substitutes

And further applications for which ceramic materials appear to be most suitable. However, the high hardness of ceramics is responsible for their main drawbacks, such as low ductility and brittleness, which limit their use for structural applications.

This chapter focuses on different AMC techniques. Some of the recent developments over the past 20 years are addressed to understand this technology, keeping in mind that as traditionally known, ceramic parts are also manufactured from the powder form as starting material. The starting concept of AMC was first introduced by Chuck Hull in 1986 through stereolithography (SLA). Since then, several processes such, as selective laser sintering (SLS), fused deposition modeling (FDM), inkjet 3D printing (IJ3DP), laser engineered net shaping (LENS), and others, have been introduced to fabricate metals, ceramics, polymers, and composites (Bandyopadhyay and Bose 2016). Among these approaches, direct ink writing (DIW) or robocasting (RC) is also used to fabricate ceramic parts.

According to the form of the pre-processed feedstock prior to printing, these technologies can generally be categorized into slurry-based, powder-based, and bulk solid–based methods, as summarized in Table 6.1. This is also illustrated in Figure 6.1. Overviews of some successfully used AMC technologies will be given. The historical origins and evolution of each type of technology will be specifically emphasized. Moreover, a thorough comparison of important aspects of AMC technologies will be presented in this chapter.

TABLE 6.1
Ceramic 3D Printing Technologies

Feedstock form	Ceramic 3D printing technology type	Abbreviation
Slurry-based	Stereolithography	SLA
	Digital light processing	DLP
	Two-photon polymerization	TPP (2PP)
	Inkjet printing	IJP
	Direct ink writing (robocasting)	DIW (RC)
Powder-based	Three-dimensional printing	3DP
	Selective laser sintering	SLS
	Selective laser melting	SLM
Bulk solid-based	Laminated object manufacturing	LOM
	Fused deposition modeling	FDM

Source: Chen, Z. et al., (2018) *Journal of the European Ceramic Society*, 39, 4, 661–687.

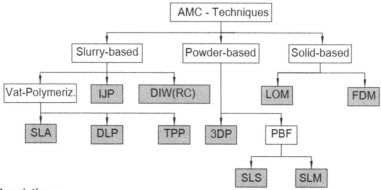

Abbreviations

AMC	Additive Manuact.of Ceramics	3DP	3D Printing
SLA	Stereolithography	PBF	Powder Bed Fusion
DLP	Digital Light Processing	SLS	Selective Laser Sintering
TPP	Two Photon Photopolimerization	SLM	Selective Laser Melting
IJP	Ink Jet Printing	LOM	Laminated Object Manufacturing
DIW	Direct Ink Writing	FDM	Fused Deposition Modeling
RC	Robocasting		

FIGURE 6.1 Classification of AMC techniques as based on slurry, powder, and bulk solid feedstock material.

The slurry-based technologies use ceramic/polymer mixtures with viscosities ranging from low-viscosity (~mPa·s) inks with a low ceramic loading (up to 30 vol%) to high-viscosity (~Pa·s) pastes with a much greater ceramic loading (up to 60 vol%). Powder-based ceramic 3D printing technologies mainly utilize powder beds, normally containing loose ceramic particles as feedstock. The ceramic particles are bonded either by spreading liquid binders or by powder fusion using thermal energy provided by a laser beam. Therefore, in this section, three types of powder-based 3D printing techniques are discussed, namely, three-dimensional printing (3DP), SLS, and selective laser melting (SLM). The first technique applies printheads to selectively jet liquid binder droplets onto the powders, whereas the latter two use the energy of laser beams to selectively sinter/melt ceramic powders. In the bulk solid–based section, there are two types, namely, laminated object manufacturing (LOM) and FDM.

Although all the seven AM categories according to ASTM International Committee F42 on Additive Manufacturing Technologies, 2009 (sheet lamination, powder bed fusion, directed energy deposition, material jetting, binder jetting, vat photopolymerization, and material extrusion), can theoretically handle ceramics, only some of them have been developed and are being successfully used to build ceramic parts today (Yang and Miyanaji 2017a, b).

Powder bed fusion (PBF) processes were among the first AM processes that were employed for ceramic fabrication. The PBF processes that utilize an electron beam

PRINT DEBINDING SINTERING FINAL PART

FIGURE 6.2 Steps of AM of ceramics. (Courtesy of Zetamix)

as a heat source have received little attention from the researchers for the fabrication of ceramic materials, likely due to the thermally induced challenges, such as cracking and structural distortion (Bourell et al. 2017). On the other hand, the applicability of the PBF processes with a laser beam heating supply has been studied by different researchers (Tang et al. 2003 and Castlino et al. 2002). The fabrication of yttria–zirconia powders via laser sintering/melting for dental applications was also experimentally investigated, although the resulting density of the structure was rather low (56%) (Bertrand et al. 2007).

LENS is a directed energy deposition (DED) process focused by the laser beam and produces molten material on the surface area. The powder material is fed through the feeding gas, which also acts as a coolant that solidifies the molten material as soon as it is deposited on the surface. LENS is much easier to control while achieving higher cooling rates. By using different powders carried in different nozzles and allowing them to react with different non-reactive gases, functionally graded materials can be produced. Due to the fact that LENS is used only to repair and not to produce parts, it will not be considered in this chapter.

While the development of AM technologies in the ceramic industry has been slower than in the polymer and metal industries, there is now considerable interest in developing AM processes capable of producing defect-free, fully dense ceramic components for different applications. A variety of AM technologies can be used to shape ceramics, but variable results have been obtained so far. Selecting the correct AM process for a given application depends not only on the requirements in terms of density, surface finish, size, and geometrical complexity of the product, but also on the nature of the particular ceramic material to be processed (Lakhdar et al. 2021).

It is important to note that all ceramic AM processes share a common characteristic: they use ceramic in its powder form. Moreover, they are indirect 3D printing processes. The printed parts need to go through debinding and sintering processes (Figure 6.2). The debinding process removes the matrix material that is used, while the sintering process is a high-temperature thermal treatment used to densify the part after the matrix material is removed.

6.2 SLURRY-BASED TECHNOLOGIES

6.2.1 Stereolithography of Ceramics (SLC)

SLC is a slurry-based vat polymerization process. It was first introduced in the early 1990s by Griffith and Halloran (1994) by mixing ceramic powders of silica (SiO_2)

and alumina (Al_2O_3) at a volumetric ratio of 45/55 and adding a UV-curable aqueous acrylamide solution (Figure 6.3), followed by curing the composite solution using a high-intensity UV laser.

SLC is perhaps the AM process that is the most preferred when it comes to ceramics. The process is based on traditional SLA of polymeric materials, in which a laser cures a liquid photopolymer in a vat layer by layer. In this case, the resin is a mix of photosensitive materials that is highly loaded with ceramic powder, giving it a pasty, viscous nature. Then, the parts are debinded and sintered. Because it works with a semi-liquid material, supports are required for all overhangs built in the same material. Advantages of SLA are that it can deal with small powder grains (that are easy to sinter) and produce finely detailed parts with very high accuracy and surface finish in a fairly wide range of available ceramic materials. Its main drawbacks are the photopolymer stability over time (when exposed to light), removal of the supports from complex structures, and the difficulty of developing new materials, which is linked to several factors inherent in this technology.

Existing machines are made by manufacturers such as Lithoz, 3DCeram, Prodways, Admatec, or DDM Systems. The size of the building platform ranges from about 50 to 500 mm. 3DCeram now has a wide range of AM machines, such as the C900, a hybrid press that can print more than one material at the same time,

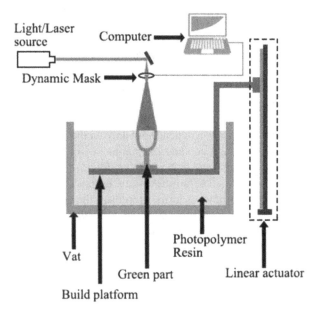

FIGURE 6.3 Schematic of ceramic stereolithography. (With kind permission from Springer Science+Business Media: Wang, Jia-Chang, Dommati, Hitesh, and Hsieh, Sheng-Jen (2019) Review of Additive Manufacturing Methods for High-performance Ceramic Materials. *The International Journal of Advanced Manufacturing Technology*, 103(1–4), 2627–2647)

dedicated to the electronics and energy industry, and two additional machines launched in 2019. These are C100 "KISS" (Keep It Smart and Simple) and C3600: the latter is specialized in technical ceramics (Figure 6.4).

In addition to the developments of the French company 3DCeram, there is an American company, Formlabs, which specializes in desktop SLA 3D printers. Figure 6.5 illustrates ceramic parts produced by the Formlabs SLA desktop. Dimensional accuracy can be lower than 10 μm for the complete controlled process (printing + debinding-sintering). Tethon 3D and Cerhum are specific ceramic materials providers for ceramic SLC processes. Figure 6.6a) illustrates a sintered alumina turbine blade fabricated using the SLC technique (Schwentenwein and Homa 2014). Figure 6.6b) illustrates a dense zirconia specimen with complex geometry produced using the SLC technique after post-processing (Chartier et al. 2002).

FIGURE 6.4   3DCeram's stereolithography machine C3600. (Courtesy of 3DCeram)

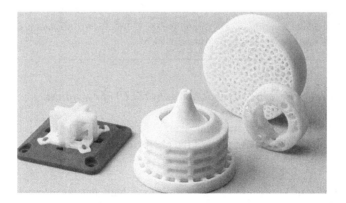

FIGURE 6.5   Typical ceramic parts produced by Desktop SLC 3D printers of Formlabs. (Courtesy of Formlabs)

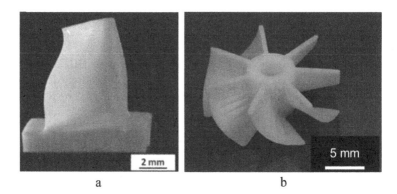

a   b

FIGURE 6.6 (a) Sintered alumina turbine blade fabricated using the SLC technique; (b) dense zirconia specimens with complex geometry were produced with no visible layer interfaces after post-processing. (Panel (a) from Schwentenwein, M. and Homa, J., (2014) *International Journal of Applied Ceramic Technology*, 12, 1, 1–7; panel (b) from Chartier, T. et al., (2002) *Journal of Material Science*, 37, 15, 3141–3147)

TABLE 6.2
Photocurable Polymers Used for Different Types of Ceramic Powder, Used in SLA

Ceramic powder	Photocurable polymer
Alumina (Al$_2$O$_3$)	Diacryl 101, Hexanedioldiacrylate (HDDA), Diacryl 101 and HDDA, Acrylamide, Acrylate, Acrylic and silicon acrylate
Silica (SiO$_2$)	Acrylamide, Acrylate, Acrylic and silicon acrylate
Lead zirconate titanate PZT	Diacryl 101 and HDDA
Barium titanate (BaTiO$_3$)	HDDA
Titanium oxide (TiO$_2$)	Epoxy resin
Hydroxyapatite (HA)	SL5180 resin (Huntsman)

Source: Adapted from Bandyopadhyay, A. and Bose, S., (2016) *Additive Manufacturing*, CRC Press, Boca Raton, London, New York. With permission

Parameters that affect the stereolithography (SLC) of ceramics:

- Ceramic powder ratio and grain size
- Chemistry and concentration of the photocurable polymer
- Power intensity of the UV lamp

Different types of photocurable polymers along with the corresponding ceramic powder types that have been used in SLA of ceramics are listed in Table 6.2 and Figure 6.7.

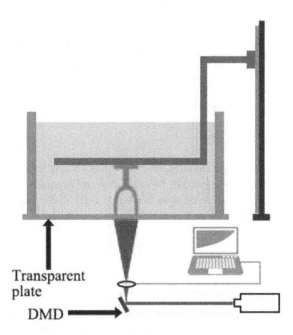

Transparent
plate
DMD

FIGURE 6.7 DLP of ceramics. (With kind permission from Springer Science+Business Media: Wang, Jia-Chang, Dommati, Hitesh, and Hsieh, Sheng-Jen (2019) Review of Additive Manufacturing Methods for High-performance Ceramic Materials. *The International Journal of Advanced Manufacturing Technology*, 103(1–4), 2627–2647

Steps to fabricate alumina ceramic using SLA:

1. The alumina powder is added to a selected photocurable polymer, such as acrylate, to form a generous ceramic polymer suspension. This polymer suspension should be stable and homogeneous, and its viscosity should be similar to conventional SLA resins in order to achieve proper flow during layer-by-layer processing (Griffith and Halloran 1994). Moreover, the suspension should be photocurable with high cure depth and low cure width to achieve high resolution during manufacturing.
2. Ceramic particles are then cured with photocurable polymer using a UV laser beam.
3. After curing is completed, the organic phase is removed thermally. The cured green part must be of high density to inhibit crack formation, deformation, and significant shrinkage after removing the polymer.

Applications, advantages, and limitations of ceramic SLC process:
Applications:
The first advantage of SLC is that it's compatible with a wide range of ceramics, including alumina, zirconia, zirconia-toughened alumina/alumina-toughened

zirconia (ZTA/ATZ), Cordierite, silicon-based ceramics, and hydroxyapatite/tricalcium phosphate (HAP/TCP). Second, stereolithography enables the printing of not only very dense parts but also very small details, thanks to the high resolution its thin layers provide. Ceramic SLA also made it possible to manufacture electric devices in small dimensions and complex structures. Moreover, SLC 3D printers are an excellent choice for jewelry, dental, and other high-precision casting applications.

Advantages:

- The process is efficient, flexible, and of high readiness level
- The process is characterized by repeatability and accuracy
- It enables the fabrication of near net shape fully dense parts with a high geometrical complexity and an excellent surface finish

Limitations:

- The limited types of ceramic materials that can be produced using the photopolymerization of ceramic slurries
- The processes may be limited in terms of maximum wall thickness that can be manufactured without the formation of porosity or cracks
- Large amounts of organic compounds such as dispersants and photoinitiator are added; this first results in very significant shrinkage that can lead to part deformation, and second, tends to complicate the debinding stage due to the release of gaseous species
- Supporting material is needed when producing parts of complex geometry
- Parts produced are of low density and of low mechanical strength
- The process is not capable of manufacturing large ceramic parts

Future developments of ceramic SLC process:
Attempts should be made to increase resolution and repeatability and to decrease the processing time. Recently, biocompatible ceramics such as hydroxyapatite (HA) and bioglass have been processed via this technology for biomedical implants for tissue engineering scaffolds. With the development of this technology, it will become a more promising AM method for manufacturing products for different applications.

6.2.2 Digital Light Processing (DLP) of Ceramics

The use of DLP as an AM technique for ceramics fabrication has been widely explored. DLP is another slurry-based vat polymerization process. High-density (97–99%) zirconia and alumina structural parts have been produced using this process. Since 2012, considerable efforts have been made by a research team at Vienna University of Technology using DLP for the production of complex ceramic structures with very fine features, with resulting relative densities above 90% and mechanical strength comparable with conventionally processed samples (Gmeiner et al. 2015). Other ceramic materials, such as zirconia and β-tricalcium phosphate (β-TCP), were also successfully printed with solid loadings of up to 50 vol% (Schwentenwein and Homa 2014).

It is worth noting that based on the above work, the team commercialized the DLP ceramic printing technique, which they called lithography-based ceramic manufacturing (LCM), by founding Lithoz GmbH for further development of the 3D printing of advanced fine ceramics (Schwentenwein et al. 2014). The higher efficiency of DLP over the conventional sheet lamination (SL) process makes it a promising 3D printing technique for ceramics fabrication.

6.2.3 Two-Photon Photopolymerization

Two-photon photopolymerization (2PP or TPP) is yet another lithography-based AM process characterized by its extreme resolution and accuracy, making it ideal for the manufacture of microscopic structures. 2PP has also been used to fabricate micro devices and sensors. Powder-based micro-stereolithography uses the SLA process to produce complex three-dimensional components as small as possible. Pham et al. (2006) reported the fabrication of complex SiCN ceramic microstructures with a 210-nm resolution via nano-stereolithography of a pre-ceramic polymer. The process is based on the two-photon absorbed crosslinking of a 2-isocyanatoethyl methacrylate–functionalized polyvinylsilazane precursor and its subsequent pyrolysis at 600°C under a nitrogen atmosphere. However, highly anisotropic shrinkage was obtained, which is detrimental to the accuracy of the process (Lakhdar et al. 2021).

6.2.4 Inkjet Printing (IJP) of Ceramic (MJ-Technology)

IJP is a slurry-based material jetting technology. Material jetting uses inkjet printers with nozzle diameters in the range of 20–75 μm to print low-viscosity ceramic particle suspensions termed "inks" (Figure 6.8). There are two main methods of material jetting: continuous stream (CS) and Drop on Demand (DOD). The DOD mode is preferred because of its higher positioning accuracy and smaller droplet size, and can be realized by squeezing ink through either thermal excitation or the piezoelectric effect. The range of ink materials has been extended to include polymers, metals, or ceramics for many applications. Material jetting of ceramic materials is difficult because ink formulation must follow conflicting requirements, such as the ink needing to be of low viscosity to avoid nozzle clogging while also containing enough ceramic content that the green body can be sintered to full density. Material jetting enables the use of ceramic powders of particle size less than 100 nm that are well dispersed as inks. Using these ultrafine powders promotes sintering to enable full-density parts after post-processing.

In recent years, there has been growing interest in the IJP of ceramic inks, where ceramic particles are well dispersed within a liquid solvent for direct and selective deposition onto a substrate through a printhead. The application of IJP to printing ceramic components was first described in the literature by Blazdell and co-workers, beginning in 1995, using ZrO_2 and TiO_2 ceramic inks, both of volume fractions as small as 5 vol.%. Such low volume fractions would lead to long drying times.

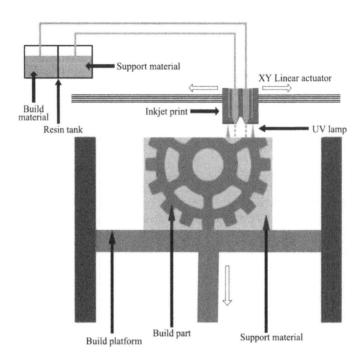

FIGURE 6.8 Diagram illustrating the material jetting technique. (With kind permission from Springer Science+Business Media: Wang, Jia-Chang, Dommati, Hitesh, and Hsieh, Sheng-Jen (2019) Review of Additive Manufacturing Methods for High-performance Ceramic Materials. *The International Journal of Advanced Manufacturing Technology*, 103(1–4), 2627–2647

However, their earlier attempts at ceramic printing generated only simple multilayer structures, which also exhibited poor surface quality. Another group then successfully printed small-sized pillar arrays based on submicron ZrO_2 particles loaded at 14 vol.% in the ink (Zhao et al. 2003). Seerden et al. (2001) reported the use of ceramic ink prepared with an Al_2O_3 loading of up to 40 vol%, resulting in the fabrication of ceramic parts with a feature size of less than 100 μm.

Applications, advantages, and limitations of IJP of ceramics:
Inkjet printing of ceramics has been frequently used in tissue engineering applications. Hydroxyapatite (HA) scaffolds have been printed using water-based solutions as binders (Leukers et al. 2005). Tetracalcium phosphate (TTCP), dicalcium phosphate, and tricalcium phosphate (TCP) have also been fabricated using citric acid as a binder (Vorndran 2008).

The IJP printed ceramics need to be sintered to enhance densification and mechanical properties. Post-processing is important in IJP of ceramics to achieve handleability and higher mechanical strength, especially in bone tissue engineering (Bose et al. 2012). Due to the low mechanical properties of porous 3D printed ceramic scaffolds, the effects of sintering (Tarafder et al. 2013), dopant addition

(Fielding and Bose 2013), and polymer infiltration Tarafder and Bose 2014 on the final properties of the part have been well studied.

IJP is without a doubt the most precise ceramic AM technology, but also the most expensive one. The process itself is quite slow, and it is unable to print large parts. It's particularly adapted for 3D printed electronics or for 3D printing very thin details, such as surgical tools with tiny, internal intricacies.

Nanoparticle jetting of ceramics:
This technique is based on material jetting technology and was developed by the company XJET. After years of work and more than 80 patents, in 2016, nanoparticle jetting was launched as a new technology and developed for AM of metals and ceramics. Two machine sizes are available, which work at a really high speed, with a very high accuracy and surface quality. The first XJET machines were the Carmel 1400 (Figure 6.9) and Carmel 700 AM systems. These are of a material jetting rather than a binder jetting technology. They print liquid ink containing a suspension of nanoparticles of metal or ceramic (only zirconia at this time) on a plate. The printed layers are heated to evaporate the solvent and pre-sinter nanoparticles. Green parts are then removed from the platform and sintered in a separate furnace.

The liquid suspensions are delivered and installed in sealed cartridges without problems. The precision of the inkjet printheads and the use of ultra-thin layers will create a super-sharp z resolution, enabling parts to be extremely neat, achieving excellent shape and high accuracy and surface quality.

6.2.5 Direct Ink Writing (DIW) or Robocasting (RC)

DIW, also known as robocasting (RC) or robotic material extrusion, was first filed as a patent by Cesarano and co-workers at Sandia National Laboratories in 1997. It is another slurry-based, extrusion ceramic AM process. The technique was originally developed for processing concentrated materials such as ceramic slurries with little organic content. DIW is an additive manufacturing technique in which a filament of

FIGURE 6.9 Carmel 1400 machine, based on nanoparticle jetting (MJ) of ceramics. (Courtesy of XJet)

a paste (known as an "ink" by analogy with conventional printing) is extruded from a small nozzle while the nozzle is moved across a platform. The object is thus built by "writing" the required shape layer by layer.

The material exits the nozzle in a liquid-like state but retains its shape immediately, exploiting the rheological property of shear thinning (Figure 6.10). It is distinct from FDM, as it does not rely on solidification or drying to retain its shape after extrusion.

Figure 6.11 shows the structure of the green parts after the robocasting process. The robocast part is also sintered up to 1650°C, resulting in parts with greater than

FIGURE 6.10   Schematic diagram of DIW (RC) for ceramics AM. (Courtesy of Ezardo)

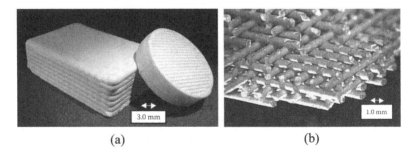

            (a)   (b)

FIGURE 6.11   (a) and (b) Examples of robocasting process by mullite suspension. (Data from Stuecker, J.N. et al., (2003) *Journal of Materials Processing Technology*, 142, 2, 318–325.)

96% density (Stuecker et al. 2003(. More recently, a kind of concentrated, aqueous colloidal ceramic slurry consisting of SiC, Al_2O_3, and Y_2O_3 has been used to build parts with complex geometry via robocasting. After robocasting, drying, and calcining processes, green parts were fired at 1700°C in argon. The sintered structures display average grain size around 1–2 μm and above 97% of theoretical density, which is outstanding compared with other AM techniques (Cai et al. 2012). With the development of this new technology, robocasting has also been applied to construct medical devices and scaffolds for tissue engineering. Calcium phosphate is one of the most common ceramics used for robocasting because of its excellent biocompatibility.

The deposition manifests itself as extrusion of the materials through nozzles, whose openings are required to be much larger than those of IJP nozzles because of the higher viscosity of the materials used in DIW. Objects are built up by moving nozzles to directly "write" the designed shape layer by layer until the part is complete. The part is typically very fragile and soft after printing. Drying, debinding, and sintering usually follow to give the part the desired mechanical properties and to ensure that it is free of organics.

DIW enables a cheaper and faster manufacturing process compared with photocuring AM techniques. The use of semi-liquid pastes allows shape retention thanks to the high solid loading and viscoelastic properties. Therefore, DIW is able to fabricate free-standing structures with high–aspect ratio walls or spanning parts without the need for supports, which is not possible with other AM techniques. It is well suited to the fabrication of tailored porous ceramic structures possessing periodic features, with little surface quality/resolution needed. In this sense, dense engineering ceramics are difficult to process using DIW, thus limiting its applications.

DIW can produce non-dense ceramic bodies that are very fragile and must be sintered before they can be used for most applications. A wide variety of different geometries can be formed using the technique, from solid monolithic parts to intricate microscale "scaffolds" that can be formed quite easily and allow bone and other tissues in the human body to grow and eventually replace the transplant.

6.3 POWDER-BASED PROCESSES

6.3.1 THREE-DIMENSIONAL PRINTING (3DP) OF CERAMIC POWDER-BASED BINDER JETTING

The abbreviation "3DP" is retained for the individual 3D printing technique as defined hereafter, whereas "3D printing" is now as understood, a general and popular term representing the assembly of AM technologies.

3DP is a powder-based binder jetting process for ceramic printing. As previously mentioned in Chapter 3, this process was first devised by Sachs et al. (1993) at the Massachusetts Institute of Technology (MIT), 1993, who filed a patent for it in 1989. Although 3DP can be considered an indirect ceramic inkjet printing process, its distinctive feature is the use of powder beds, which is why 3DP is classified as a powder-based technique in this connection. In the process of 3DP, organic binder solution in droplet form is sprayed through printheads onto selected regions of a powder bed surface. A new layer of powder is then supplied and spread on the previous layer to repeat the building process until the part is formed. After this, loose powder is removed to reveal the part. A schematic diagram of 3DP is depicted in Figure 6.12.

Binder jetting is the only ceramic AM technology that allows the production of large parts at a high processing rate. However, this method requires a large initial investment, even if the raw material itself is quite affordable. The process cannot print small details or make parts that are hollow. Furthermore, ceramic binder-jetted parts have a relatively rough surface and porous structure. Another major drawback with binder jetting is that it usually produces high levels of porosity (typically around 20% to 30%) after sintering. This is linked to the low powder density of the print bed as well as the large size of the powder particles being used. Thus, binder jetting is a good choice when producing objects that need to be rough and porous, such as ceramic cores, filters, crucibles, or other refractory products. Figure 6.13 shows full color binder-jetted ceramic models manufactured by Johnson Matthey.

Applications of 3DP of ceramics:
One useful application is the fabrication of ceramic mold shells and cores for metal casting. The method has also been used to fabricate structural components with even larger sizes of up to several meters. More recently, the 3DP method has been extended to other commercial applications due to its considerable flexibility with respect to materials and geometries. In particular, promising explorations have been made in the biomedical field, including the fabrication of components for tissue engineering, which generally require less precision in resolution and surface finish, as well as porous features of the printed parts for cultivation purposes. Biocompatible ceramics such as HA, CP, and TCP are often used in 3DP to print scaffolds for bone replacement.

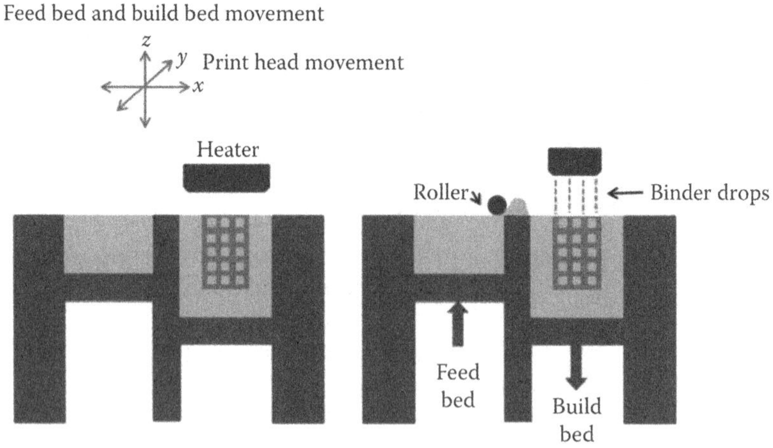

FIGURE 6.12   Schematic of PB-3DP of ceramics. (Reprinted from Bose, S., Vahabzadeh, S. and Bandyopadhyay, A., Bone Tissue Engineering Using 3D Printing. *Materials Today*, 16, 12, 496–504, Copyright (2013), with permission from Elsevier)

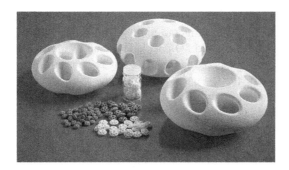

FIGURE 6.13 Full color binder-jetted ceramic models printed by 3DP-binder jetting. (Courtesy of Johnson Matthey)

Advantages and limitations of 3DP of ceramics:
Advantages:

- High flexibility of geometrical design without the addition of supports, as the powder bed supports the structure while printing.
- The process is of simple technology.
- Best suited to the fabrication of porous ceramic parts.
- Possible for a wide range of materials. It works with almost any material available as a powder.
- The process does not demand liquids or photocurable polymers.
- Characterized as one of the low-cost AM processes.
- Suitable for large parts.

Disadvantages:

- Products are of low mechanical strength due to high porosity.
- Limitations of ceramic/binder system.
- Parts are difficult to depowder.
- The process is quite slow.
- Not suited to producing advanced ceramic materials.
- Low resolution.
- Inferior surface quality.
- Green parts require extra work, including sintering, infiltration, and isostatic pressing, for further quality improvements.

6.3.2 Selective Laser Sintering (SLS) of Ceramics

SLS was invented by Deckard and Beaman at the University of Texas at Austin, with the very first patent filed in 1986, and was further developed by the DTM Company, which was acquired by 3D Systems in 2001. In the SLS process, as its name implies, a high-power laser beam is used to selectively irradiate the surface of the target powder bed. The powder then heats up, and sintering (i.e. inter-particle fusion) takes place for bulk joining. After this, a new layer of powder is spread onto the previous surface for the next run of heating and joining. In this way, the process is repeated layer by layer until the designed 3D part is fabricated. No extra support structures have to be intentionally prepared for overhanging regions during an SLS process, as they are surrounded by the loose powder in the bed at all times. A schematic of the SLS process is shown in Figure 6.14.

SLS of ceramics is a powder-based, fusion bed AM process. Compared with metals and polymers, processing of ceramics is hampered by challenges, mainly due to the high melting temperature of ceramics. Accordingly, more laser energy and a longer cooling time are required, which generally lead to non-efficient and less cost-effective performance.

The original intention of SLS was to make wax models for investment casting of metallic prototypes (e.g. aluminum). SLS has been extensively studied to process a broad range of powdered materials, starting with plastic and polymer powders, and later extended to metals. The feasibility of fabricating complex ceramic parts using SLS was first reported by Lakshminarayan and co-workers at the University of Texas at Austin in 1990 using Al_2O_3-based mixed powder systems. As alumina has a melting point as high as 2045°C, secondary components of ammonium phosphate ($NH_4H_2PO_4$) and boron oxide (B_2O_3) with lower melting points (190°C and 460°C, respectively) were introduced as low-temperature binders. 3D ceramic parts with reasonable dimensional accuracy and part definition, such as gears and casting molds, were successfully fabricated. Structural ceramics must be made almost fully dense so that their optimal mechanical performance can be achieved. To maximize the density of the final ceramic parts, infiltration/isostatic pressing can be used along with SLS.

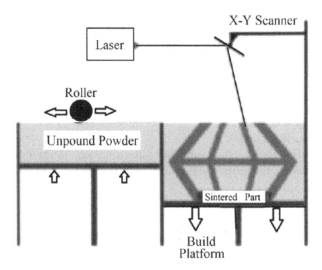

FIGURE 6.14 Diagram illustrating the powder bed fusion technique. (Adapted from Lakhdar, Y. et al., (2021) *Progress in Materials Science*, 116, 100736)

Processing steps of ceramic SLS:

- Preheating the ceramic powder.
- Filling the chamber with nitrogen gas to avoid oxidation.
- Using CO_2 laser to scan.
- Cooling down the part for removing.

This technique allows the processing of 3D products with physical, mechanical, and chemical properties that differ from those of the original powder. However, low packing density in the powder bed is a drawback that may lead to low sintering density and crack formation due to thermal stresses (Shahzad et al. 2014).

Applications and characteristics of SLS of ceramics:
Table 6.3 presents typical ceramic compositions processed by powder-based direct SLS, along with their potential applications.

SLS of ceramics is an advanced AM process that is excellent for parts of complex geometry, since no support structures are needed. The process has been adopted to produce lead zirconate titanate (PZT) ceramics from precursor powders. The properties of the ceramic parts were manipulated to match the requirements of some medical ultrasonic equipment. Bone tissue engineering scaffolds were also produced by SLS. Bioceramics such as HA and TCP were manufactured by SLS with high processing accuracy and biocompatibility, which is excellent for bone generation.

However, there are three main challenges of SLS of ceramics:

1. The part density is usually limited, causing poor mechanical strength.
2. Due to the high processing temperature, an appropriate cooling cycle should be ensured, otherwise failure of the whole part may occur.
3. Ceramic parts of large dimensions are hard to produce by this process.

TABLE 6.3

Typical Ceramic Compositions Processed by Powder-Based Direct SLS with Their Potential Applications

Ceramic Composition	Potential Applications
Alumina (Al_2O_3) and zirconia (ZrO_2)	Automotive, aerospace, and biomedical sector, and implants
Yttria (Y_2O_2) and zirconia (ZrO_2)	Ceramic shell molds and solid oxide fuel cells
Tricalcium phosphate (TCP) / hydroxyapatite (HA)	Implant materials in tissue engineering
Nano-hydroxyapatite (HA)	Bone tissue engineering scaffolds
Cu-Ni and ZrB_2	Electrodischarge machining (EDM) electrodes

Source: Adapted from Bandyopadhyay, A. and Bose, S., (2016) *Additive Manufacturing*, CRC Press, Boca Raton, London, New York. With permission

Future developments will move forward to adjust and optimize the processing parameters to overcome the previously mentioned challenges.

6.3.3 SELECTIVE LASER MELTING OF CERAMICS

SLM is often considered as a variant of SLS. SLM was developed at the Fraunhofer Institute for Laser Technology (ILT) in Germany in 1996, resulting in German patent DE 19649865 (Schleifenbaum et al. 2010). It proceeds in almost the same way as SLS except that it is a time-saving, one-step PBF by full melting, which uses laser sources with much higher energy densities (Figure 6.15) and requires no secondary low-melting binder powders. Thus, SLM can produce nearly fully dense homogeneous parts without post-processing treatments due to its ability to fully melt the powder into the liquid phase, ensuring rapid densification, instead of heating up the powder to a specific point where the particles are partially melted and fused together as in SLS. The technology was originally developed to produce solid parts from metal powders (e.g. aluminum, copper, and stainless steel) and has now been extended to the increasingly used advanced alloys, particularly for the production of lightweight parts for the aerospace industry. Stronger functional and end-use production parts can be produced using SLM with lower porosity and superior control over the crystal structure (Chen et al. 2018).

SLM is one of the most rapidly growing 3D printing techniques. This is mainly attributable to its ability to manufacture strong and durable metal parts in a single-step process, where the final shape and properties of the parts can be obtained simultaneously. The application of SLM to a ceramic powder involves full melting of the powder to form a solid part by high-energy density laser scanning layer by layer, without binders or post-sintering, due to the full melting and fusion of the powder. Thus, complex parts of higher purity, density, and strength are expected to be produced in less time. SLM is considered to be the only 3D printing process that offers the possibility of manufacturing ready-to-use ceramic parts with full density, high strength, and complex net shape from ceramic powders in a single step. The overall quality of an SLM-produced ceramic part is influenced by many factors, such as feedstock properties, fabrication parameters, fabrication position and orientation,

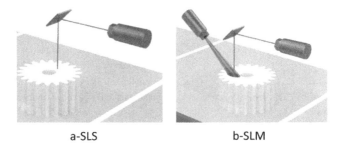

a-SLS   b-SLM

FIGURE 6.15 Comparison between SLS and SLM. (Compiled from: Chen, Z. et al., (2018) *Journal of the European Ceramic Society*, 39, 4, 661–687)

post-processing, and the physical and chemical properties of the interaction in the fabrication process, including the interaction of the energy source and the materials. One of the most important fabrication parameters is the slice thickness, which may influence the production time and the surface roughness of the part. A smaller slice thickness reduces the surface roughness but at the same time leads to longer building time, whereas a larger thickness may cause a significant stair-step effect. The slice thickness relies on the fusion depth, similar to the penetration depth in the SLC process. The fusion depth is directly related to the material properties as well as the laser–material interaction. An optimal combination of processing parameters may exist for balancing the overall quality of the fabricated part.

In general, for the powder bed fusion processes, fabrication of ceramic parts seems to be more challenging compared with other AM processes, and most studies have focused on assessing the feasibility of PBF technology for ceramic materials. With the successful demonstration of high-quality fabrication of ceramic parts using the other AM technologies, PBF has not attracted extensive attention in this area (Yang and Miyanaji 2017a, b).

During the SLM process of ceramics, laser processing parameters are crucial to the quality of the fabricated parts. Direct melting involves an extremely high-temperature interaction between laser and powder, the very short period of interaction leading to large temperature gradients. One of the most significant problems arising from the SLM process is the thermal stress induced by extremely short laser–powder interaction times, namely, the drastic heating and cooling rates upon each laser scan (Mercelis and Kruth 2006). Cracks and distortions are most likely to form in the sintered parts due to such thermal stresses as a result of the limited thermal shock resistance of the ceramic materials. Shishkovsky et al. (2007), from the Ecole Nationale d'Ingenieurs de Saint-Etienne (ENISE) in France, reported the fabrication of ZrO_2 parts using SLM, in which defects such as cracks and large open pores developed (Figure 6.16a and b). The cracks were likely caused by the accumulation of heat during the process and could be alleviated by preheating the powder bed. On the other

FIGURE 6.16 Surface morphology of the sintered ZrO_2 part, produced by SLM, showing: (a) cracks; (b) open pores. (Reprinted from Shishkovsky, I, Yadroitsev, I., Bertrand, P. and Smurov, I., Alumina–zirconium Ceramics Synthesis by Selective Laser Sintering/Melting. *Applied Surface Science*, 254, 4, 966–970, Copyright (2007), with permission from Elsevier)

hand, due to the thermal shocks imposed while spreading the cold ceramic powder, the presence of fine cracks on the surface of fabricated components was initiated.

6.4 BULK SOLID–BASED PROCESSES

6.4.1 LAMINATED OBJECT MANUFACTURING OF CERAMICS

LOM was first introduced in 1991 by Helisys, United States. In this method, multiple laminations are stacked to manufacture a 3D object. A schematic diagram of the LOM process, as shown in Figure 6.17, includes a work table, which can move vertically; a feeder, which is a continuous roll of the material; and an x–y plotter. The feeder sends the laminations/sheets over the build platform on the work table. The bottom side of the sheets is covered with a heat-sensitive adhesive. Using a hot lamination roller, the adhesive melts, and the sheet gets bonded to the layer below. The x–y plotter uses a laser beam to cut the outline of the part at each layer. After the completion of each layer, the build platform moves downward by a depth of the sheet thickness. This process continues layer by layer until the whole part is made. Once the part is manufactured, *decubing* is applied to remove any excess material.

Prior to fabrication of the ceramic roll, a suspension including the ceramic precursor, plasticizer, binder, and dispersant should be prepared. The dispersant content has a significant effect on the viscosity of the suspension. A high amount of dispersant increases the viscosity significantly, which is undesirable due to decrease in green density and increase in firing shrinkage. A suspension with viscosity lower than 20 Pa·s results in a fluid system, while higher viscosity causes the formation of a paste system, which is not helpful in the LOM process.

The next step is the ceramic paper formation. To form a continuous sheet of ceramic paper, prepared suspensions including fibers, binder, and retention agents are transferred to a papermaking machine, and the ceramic paper is prepared through

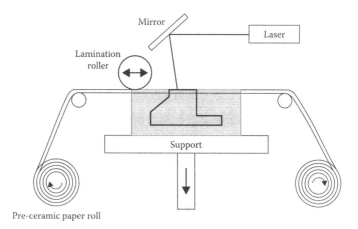

FIGURE 6.17 Schematic of LOM process. (Data from Travitzky, N. et al., (2008) *Journal of the American Ceramic Society*, 91, 11, 3477–3492)

sheet formation, pressing, drying, calendaring, and rolling. Pre-ceramic paper roll is then transferred to the LOM machine to fabricate the 3D structure as discussed earlier. After decubing, post-processing, including resin infiltration or polishing, is applied to enhance the strength and surface quality of the part.

Compared with other rapid prototyping methods, LOM is a high-speed method. In addition, its low operation cost is another advantage. Although LOM is a fast method, the decubing process is time-consuming, and depending on the geometry and complexity of the part, labor work is needed.

Alumina was manufactured using 7 wt.% polyvinyl butyral (PVB) and polyvinyl acetate as binder and adhesive, respectively. Binder was removed at 240–300°C, and samples were sintered at 1580°C. Although binder removal did not cause any damage, the sintering and cooling processes led to distortion and cracking (Zhang et al. 2001).

Weisensel et al. (2004) processed Si–SiC composites using the LOM technique. Though LOM is a simple process, it is only useful for fiber-reinforced composite fabrication. Even then, if there are curved surfaces, it is very difficult to control warpage in LOM processed and cured components. At present, there is no commercial vendor for an LOM-based process for ceramic structures. However, a cheaper version of the same concept using sticky paper and a knife is still available to produce 3D parts from computer-aided design (CAD) files.

6.4.2 Fused Deposition Modeling of Ceramics (FDC)

FDM, also called liquid deposition modeling (LDM), is the most widespread technique among different suppliers. It is a solid-based material extrusion AM process. FDC is a modified FDM process. In 1996, FDC was introduced by researchers at Rutgers University as modified FDM, patented later in 1998.

Figure 6.18 shows the schematic of FDC. In this process, a filament of a semi-solid thermoplastic polymer is fed into liquefier by two rollers, extruded through the liquefier and then a nozzle, and finally deposited on a platform. The heaters in the liquefier heat up the polymer at temperatures above but close to the melting temperature, so that the extruded filament can move easily through the nozzle. After deposition of the first layer, the platform moves downward, and the second layer is deposited on the first one, until the whole part is built. The filament is a composite polymer loaded with high levels of ceramic. In addition to being the most affordable ceramic additive manufacturing technology, FDC is fast, but not as simple as binder jetting, because supports are required under all overhangs of the parts during shaping.

Moreover, due to the deposition of coarse extruded material layer by layer, parts have a real anisotropy in the z direction (vertical direction), the surface has a step-structure, and fine details cannot be realized. This technology is capable of printing parts that are as dense as those printed with SLA or inkjet methods. Zetamix claimed that when debinded and sintered, ceramic parts achieved a density of up to 99%. Moreover, contrary to other technologies like SLA or binder jetting, FDC requires little to no preparation or cleaning. However, it does provide less precision

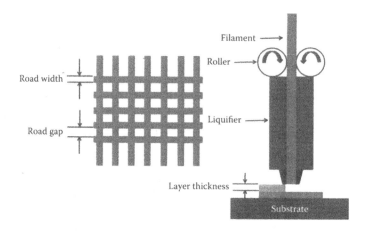

FIGURE 6.18 Schematic of FDC. (From Bandyopadhyay, A. and Bose, S., (2016) *Additive Manufacturing*, CRC Press, Boca Raton, London, New York. With permission)

FIGURE 6.19 Large-size ceramic part produced by FDC. (Courtesy of WASP)

than SLA and cannot print massive parts, unlike binder jetting. Since it is possible to 3D print a wide variety of designs and structures with high density, FDC is a suitable option for many applications, from tooling to cutting-edge objects, depending on the level of precision that is required for those objects. The printing resolution depends on the size of the extruded wire, usually in the range of a millimeter or more. The dimensional accuracy is therefore quite poor compared with other technologies.

This technology inspired many manufacturers to create desktop machines that have found great success, such as the ClayXYZ, launched in 2017, or the LUTM V4, which was introduced to the market in 2018 with a more artistic character. One of the most renowned examples of this technology is the Italian company WASP, which offers ceramic 3D printing on a large scale. Figure 6.19 illustrates a large-size ceramic part produced by WASP FDC technology.

6.5 POTENTIALS OF CERAMIC AM FOR TODAY'S INDUSTRIAL APPLICATIONS

AM of ceramics has great potential for applications in various fields of industry, which will be discussed hereafter.

Engine and propulsion components for aerospace, automobiles, and energy are among the most extensive applications. In these applications, the push for ever-higher operation temperatures in the attempt to improve the performance efficiency has stretched the traditional superalloys to their limit. For example, for the next-generation gas turbine engines with operating temperatures of over 1200°C, advanced ceramics are considered to be the only suitable material options (Clarke et al. 2012). In these applications, ceramics are primarily used in two ways: either as a thermal barrier coating (TBC) or as primary structural materials (Cao et al. 2004). The TBC topcoats usually have a thickness of 100–2000 μm and possess very low thermal conductivity across the entire operational temperature range in order to avoid the softening or melting of the superalloy substrates (Clarke and Phillpot 2005).

The applications of AM in aerospace include fabricating ultra-high-temperature ceramics (UHTCs) (>2000°C), which are not possible to manufacture using conventional methods. The elements of miniature sensors, turbine components, electrode supports, spark plugs, flange insulation, etc. that have complex requirements can be fulfilled using AM processes. UHTCs are used in hypersonic engines that are built with improved thermal conductivity, toughness, oxidation resistance, strength, and emissivity Padture 2016. The scaled-down models of missile nose cones are fabricated with alumina and zirconia boride materials using AMC.

The use of ceramics in electronics applications is widespread. In these applications, ceramic materials are used for their intrinsic dielectric properties and tailorable electrical and electromagnetic properties when doped. Ceramics are used as insulators, conductors, capacitors, piezoelectric sensors and actuators, semiconductors, optical and electromagnetic films, and many other applications. Therefore, the requirements of ceramic materials for these applications are extremely diverse. In many of these applications, a myriad of studies have clarified the importance of microstructure, including phases, grain sizes, grain boundary characteristics, contaminants/impurities, and point defects.

For biomedical applications, bioceramics are used. Alumina and zirconia are the most basic materials that are bioinert in this class. There are also bioactive materials like hydroxyapatite or bioactive glasses used in applications like knees, tendons, ligaments, repair for jawbones, and spinal fusion. Bioglass implants have been successfully implanted for low-load-bearing bone restoration. Using a binder jetting process, cast orthopedic implants, Si_3N_4 ceramics of differing porosity (Rabinskiy, et al. 2017), and many other bioceramic scaffolds can be fabricated at a very economical price.

Over the years, ceramics, especially oxides, have been preferably used for dental applications such as crowns, bridges, and veneers. They are implemented either using all ceramic or with metal restorations. Several researchers have investigated the use of dental ceramic fabrication using AM methods and reported the issues, inaccuracies, and drawbacks of these processes (Silva et al. 2011 and Dehurtevent et al. 2017).

6.6 SELECTION CRITERIA FOR AMC PROCESSES

From the foregoing, it is seen that each AM technology is characterized by its own advantages and limitations. Therefore, there is no best overall ceramic AM process; rather, the selection of the most suitable AM technology to be used to manufacture a particular component directly depends on the selected ceramic material and the end-use application of the intended component. As a result, the following items need to be identified and specified before deciding which AM process should be used (Lakhdar et al. 2021):

1. *Type of ceramic material:* Non-oxide ceramics with high refractive index and absorption for UV and daylight wavelengths (e.g. B_4C and TiC) may prove significantly more difficult to shape using photopolymerization-based processes than more transparent ceramics (e.g. Al_2O_3, ZrO_2, HA) because of insufficient cured depths. The use of direct selective laser sintering (dSLS) and DED with some non-oxide ceramics may also be hindered by the very high melting point, low toughness, and poor thermal shock resistance, resulting in cracking. Alternatively, other AM processes such as ceramic extrusion, ink jetting, binder jetting, and sheet lamination have already been successfully used to process all types of advanced ceramics. Toxic types of ceramics should not be used for biomedical applications.

2. *Final density:* Before choosing an AM process, it is important to consider carefully the requirements in terms of in-service mechanical properties of the end-use component, because this is correlated to the required part density. 3DP and SLS are not suitable for producing high-density parts due to the use of dry coarse powders that have low sintering activity and provide poor packing density, thus resulting in high residual porosity. High sintered densities can be achieved using the slurry-based variants of these processes, namely Slurry-LS and Slurry-BJ, as well as other AM processes that use liquid ceramic feedstock, such as aqueous-based extrusion, RC, ceramic on demand extrusion (CODE), ink jetting, and SLA. SLS and DED have also been used to produce monolithic specimens, although issues with cracking still need to be considered.

3. *Resolution and part dimensions:* The overall dimensions of the component have a deciding influence on the ceramic AM process selection. Some AM processes may be better suited to print small components for which high precision is essential. At the extreme end of the spectrum, TPP enables the production of micron-sized objects with a resolution of a few hundred nanometers. SL, DLP, and RC are also very well suited to manufacturing small components for high-precision applications thanks to their excellent resolution and accuracy in the order of a few tens of microns.

Although all AM processes probably have the technical potential to eventually be scaled up for the production of large parts, processes that tend to have a lower accuracy but a higher throughput, such as powder bed processes, may be easier and more cost-effective to scale up for the production of medium to large parts. In particular, powder-based binder jetting is highly suited to producing large-sized ceramic parts. In this regard, however, direct laser sintering does not (yet) offer the same capability as other powder bed processes due to the extreme thermal gradients, which limit the height that can be manufactured without the formation of pores and cracks.

4. *Surface finish:* Usually strongly correlated to process resolution, surface roughness is mostly a function of two parameters: open porosity at the surface and stair casing effect. While it is commonly known that the best surface finish from AM processes can be achieved using SL and DLP vat photopolymerization technologies, excellent results have also been demonstrated with RC and 3DP. Parts manufactured using these two processes usually do not display any open porosity and can be optimized to be seemingly free of stair casing effect. Slurry-based processes (S-BJ and slurry indirect laser sintering [S-iLS]) also provide a good surface finish compared with their powder-based counterparts. The often poor surface quality obtained with ceramic extrusion technologies is caused by stair casing, especially when using large extrusion nozzles.

5. *Geometrical complexity:* It is important to realize that not all AM processes have the same capabilities and potential regarding geometrical complexity. The process with the most limited potential for improvement in this regard is without doubt DED, due to the fact that neither self-supporting material nor support structures are present to accommodate overhangs. Conversely, LS and 3DP-BJ powder bed processes, in both their dry powder-based and slurry-based variants, are naturally well suited to the manufacture of complex-shaped parts thanks to their self-supporting powder bed, which allows the production of intricate geometries and large overhangs without the need for additional support structures.

6. *Manufacturing costs and time:* Costs and time taken by equipment, materials, pre-processing, processing, and post-processing must be taken into account.
   • Equipment costs include the purchase price of the AM equipment, running costs (electricity, gases), and maintenance charges from the equipment manufacturer.
   • Materials costs include the price of feedstock materials as well as additional pre-processing costs.
   • Processing costs depend mostly on the degree of complexity of the AM process and whether costly equipment is required, such as high-power lasers and inert gases.

- Post-processing costs can amount to a significant share of the total production costs due to high-temperature heat treatments for debinding and sintering, as well as potential consolidation processes such as CIP or HIP, and machining or polishing steps.

Such calculations are difficult to run accurately, even more so for AMC. For instance, while DIW is a relatively low-cost process, it still requires post-processing heat treatment at high temperature to sinter ceramic parts to full density, which can become relatively expensive for some advanced ceramics, such as high-melting-point carbides requiring inert gas flow in furnace chambers and temperatures in excess of 2000°C. On the other hand, SLS is a significantly more expensive process due to the use of high-power laser(s) and argon gas, but post-sintering would typically not be required since parts are already sintered to full density in situ. Therefore, the time saved and lower number of post-processing steps may in fact result in overall manufacturing cost savings.

6.7 COMPARISON AND CONCLUDING REMARKS REGARDING AM OF CERAMICS

AM today offers the ceramics industry a new tool for prototyping and for developing new applications. With LCM, it is possible to achieve the same, or at least very similar, properties as conventional ceramic manufacturing technologies. For additive manufactured ceramic parts, if the necessary tolerances and surface qualities are not reached, grinding and polishing processes can be applied as post-processing processes to achieve the desired values.

AM is not a threat to the ceramics industry but much more an opportunity to move into new fields and develop new applications. AM has huge potential to become a "game changer" for the industry with its new freedom in design. This new way of designing has to be fully understood, and it will take some years to convey the new limits in design to the industry. AM should be seen as a new forming method, which opens up new opportunities. It is not a substitute for traditional forming processes but rather, an addition to the industry to open up new markets. One such promising market is ceramic casting cores for turbines made from superalloys. The cooling channels designed into these products are becoming ever more complex; ceramic injection molding (CIM) is reaching its limits, and AM may be the solution. There is a high probability that some special core designs will be made in the future only by AM. Other products in mechanical or biomedical engineering have been reviewed.

The choice of a ceramic AM technology depends on the application and desired characteristics of the final object: porous or dense, high or medium resolution, high-volume production or prototyping, large or small size, etc. Accordingly, industrial plants are often equipped with two or three different ceramic AM technologies to use one or the other depending on the part at hand.

Table 6.4 compares the ten ceramic AM technologies discussed above regarding equipment price, part density, part size, resolution, surface quality, processing speed, and the fields of application.

TABLE 6.4

Comparison of the Ten Ceramic AMC Technologies Dealt with in the Chapter

AM Technology	SLA	DLP	TTP	IJP	RC/DIW	3DP	SLS	SLM	LOM	FDM
Stock Material	Slurry-Based					Powder-Based			Bulk Solid–Based	
Forming Method[a]	Polymerization	Polymerization	Polymerization	BB	Extrusion	BB	Powder Fusion	Powder Fusion	Sheet Lamination	Extrusion
Power Source	Laser	Laser	Laser	Thermal Energy	Thermal Energy		Laser	Laser		Thermal Energy
Part Size[b]	S/M	S/M	VS	S/M	S/M	M/L	S/M	S/M	S/M	S/M
Resolution	μm	μm	nm/μm	μm	μm/mm	μm/mm	μm/mm	μm/mm	μm/mm	mm
Surface quality[c]	H	H	H	H	L	M	L	L	M	L
Speed[c]	L	M	L	L	M	M	M	M	H	M
Cost[c] Feed Stock	H	H	H	M	L	L	M	L	M	M
Process	M	M	H	H	L	L	M	H	L	M
Support[d]	Y	Y	Y	N	Y	N	N	N	N	N
Post-processing[e]	D/S	D/S	D/S	D/S	D/S	D/S	D/S	N/A	D/S	D/S
Application Fields[f]	Str.	Str.	F	Str./B	F	Str.	B	Str.	Str.	F

[a] Forming Method: BB = binder bonding

[b] Part size: S = small, VS = very small (micro-size), M = medium, L = large

[c] Surface quality, [c]Speed, and [c]Cost: L = low, M = medium, H = high

[d] Support: Yes (Y) or No (N)

[e] Post-processing: D = densification, S = sintering

[f] Application Fields: Str. = structural, B = biomedical, F = functional

6.8 REVIEW QUESTIONS

1. Mention parameters that affect the stereolithography of ceramics (SLC).
2. Suggest only one of the photocurable polymers that you can use in SLC for the following types of ceramic powders: alumina (Al_2O_3), lead zirconate titanate (PZT), titanium oxide (TiO_2), and hydroxyapatite (HA).
3. To achieve a homogeneous dispersion of ceramic powder in polymeric solution with high ceramic powder loading, suggest a suitable dispersant.
4. What are the advantages and limitations of the ceramic SLC process?
5. What are the advantages of SLM over SLS of ceramics?
6. What are the most used additive manufacturing processes for ceramics?
7. How does AMC differ from metal- or polymer-based AM?
8. What are the key advantages of using additive manufacturing for ceramic production?
9. What are the challenges in printing ceramics via additive manufacturing?
10. What post-processing steps are typically involved after printing ceramic parts?
11. Are there specific applications where ceramic additive manufacturing excels?
12. What safety considerations are important when working with ceramic AM?
13. What is the future outlook for ceramic additive manufacturing?

BIBLIOGRAPHY

ASTM. (2009) *International Committee F42 on Additive Manufacturing Technologies (ASTM F2792–10)*. West Conshohocken, PA: Standard Terminology for Additive Manufacturing Technologies.

Bandyopadhyay, A. and Bose, S. (2016) *Additive Manufacturing*, Boca Raton, London, New York: CRC Press.

Bertrand, P., Bayle, F., Combe, C., Goeuriot, P., and Smurov, I. (2007) Ceramic Components Manufacturing by Selective Laser Sintering. *Applied Surface Science*, 254(4), 989–992. https://doi.org/10.1016/j.apsusc.2007.08.085

Blazdell, P., Evans, J., Edirisinghe, M., Shaw, P., and Binstead, M. (1995) The Computer Aided Manufacture of Ceramics Using Multilayer Jet Printing. *Journal of Materials Science Letters*, 14(22), 1562–1565. https://doi.org/10.1007/BF00455415

Bose, S., Roy, M., and Bandyopadhyha, A. (2012) Recent Advances of Bone Tissue Engineering Scaffolds. *Trends in Biotechnology*, 30(10), 546–54. https://doi.org/10.1016/j.tibtech.2012.07.005

Bose, S., Vahabzadeh, S., and Bandyopadhyay, A. (2013) Bone Tissue Engineering Using 3D Printing. *Materials Today*, 16(12), 496–504. https://doi.org/10.1016/j.mattod.2013.11.017

Bourell, D., Kruth, J. P., Leu, M., Levy, G., Rosen, D, Beese, A. M., and Clare, A. (2017) Materials for Additive Manufacturing. *CIRP Annals-Manufacturing Technology*, 66(2), 659–681. https://doi.org/10.1016/j.cirp.2017.05.009

Cai, K. et al. (2012) Geometrically Complex Silicon Carbide Structures Fabricated by Robocasting. *Journal of the American Ceramic Society*, 95(8), 2660–2666. https://doi.org/10.1111/j.1551-2916.2012.05276.x

Cao, X. Q., Vassen, R., and Stoever, D. (2004) Ceramic Materials for Thermal Barrier Coatings. *Journal of the European Ceramic Society*, 24(1), 1–10. https://doi.org/10.1016/S0955-2219(03)00129-8

Castlino, G., Filippis, L., Ludovico, A. D., and Tricarico, L (2002) An Investigation of Rapid Prototyping of Sand Casting Molds by Selective Laser Sintering. *Journal of Laser Applications* 14, 100–106. https://doi.org/10.2351/1.1471561

Cesarano, J. III, Sasaki D. Y, Singh S, and Brinker, C. J. (1997) Oriental Inorganic Thin Film Channel Structures with Uni-directional Monosize Micropores, Sandia National Laboratories, New Mexico and Livermore, California: Technical Report. doi.org/10.2172/548862

Chartier, T, Chaput, C, Doreau, F, and Loiseau, M. (2002) Stereolithography of Structural Complex Ceramic Parts. *Journal of Material Science*, 37(15), 3141–3147. https://doi.org/10.1023/A:1016102210277

Chen, Z., Li, Z., Li, J., Liu, C., Liu, C., Li, Y., and Yuelong, F. (2018) 3D Printing of Ceramics: A Review. *Journal of the European Ceramic Society*, 39(4), 661–687. https://doi.org/10.1016/j.jeurceramsoc.2018.11.013

Clarke, D. R. and Phillpot, S. R. (2005) Thermal Barrier Coating Materials. *Materials Today*, 8(6), 22–29. https://doi.org/10.1016/S1369-7021(05)70934-2

Clarke, D. R., Oechsner, M., and Padture, N. P. (2012) Thermal-barrier Coatings for More Efficient Gas Turbine Engines. *MRS Bulletin*, 37(10), 891–898.

Dehurtevent, M., Robberecht, L., Hornez, J.-C., Thuault, A., Deveaux, E., and Béhin, P. (2017) Stereolithography: A New Method for Processing Dental Ceramics by Additive Computer-aided Manufacturing. *Dent Mater*, 33(5), 477–485. https://doi.org/10.1016/j.dental.2017.01.018

Fielding, G. and Bose, S. (2013) SiO2 and ZnO dopants in 3D printed TCP Scaffolds Enhances Osteogenesis and Angiogenesis in Vivo. *Materials Science, Biology Acta Biomater*, 9, 9137–9148. https://doi.org/10.1016/j.actbio.2013.07.009

Gmeiner, R. Mitteramskogler, G., Stampfl, J., and Boccaccini, A. R. (2015) Stereolithographic Ceramic Manufacturing of High Strength Bioactive Glass. *International Journal of Applied Ceramic Technology*, 12(1), 38–45. https://doi.org/10.1111/ijac.12325

Griffith, M. L. and Halloran, J. W. (1994) Ultraviolet Curable Ceramic Suspensions for Stereolithography of Ceramics. *Materials Science and Engineering*, 68–2, 529–534.

Lakhdar, Y, Tuck, C, Binner, J, Terry, A, and Goodridge, R. (2021) Additive Manufacturing of Advanced Ceramic Materials. *Progress in Materials Science*, 116, 100736. https://doi.org/10.1016/J.Pmatsci.2020.100736

Lakshminarayan, U., Ogrydiziak, S., and Marcus, H. (1990) *Selective Laser Sintering of Ceramic Materials, International Solid Freeform Fabrication Symposium*. Austin, TX, USA: The University of Texas.

Leukers, B., Gulkan, H., Irsen, S. H., Milz, S., Tille, C., Seitz, H., and Schieker, M. (2005) Biocompatibility of Ceramic Scaffolds for Bone Replacement Made by 3D Printing. *Materialwissenschaft und Werkstofftechnik*, 36(12), 781 –787. https://doi.org/10.1002/mawe.200500968

Mercelis, P. and Kruth, J.-P. (2006) Residual Stresses in Selective Laser Sintering and Selective Laser Melting. *Rapid Prototyping Journal*, 12(5), 254–265. https://doi.org/10.1108/13552540610707013

Padture, N. P. (2016) Advanced Structural Ceramics in Aerospace Propulsion. *Nature Materials*, 15(8), 804–809. https://doi.org/10.1038/nmat4687

Pham, T. A., Kim, D. P., Lim, T. W., Park, S. H., Yang, D. Y., and Lee, K. S. (2006) Three-dimensional SiCN Ceramic Microstructures via Nano-stereolithography of Inorganic Polymer Photoresists. *Advanced Functional Materials*, 16(9), 1235–1241. https://doi.org/10.1002/adfm.200600009

Rabinskiy, L. N., Sitnikov, S. A., Pogodin, V. A., Ripetskiy, A. A., and Solyaev, Y. O. (2017) Binder Jetting of Si3N4 Ceramics with Different Porosity. *Solid State Phenomena*, 269, 37–50. https://doi.org/10.4028/www.scientific.net/SSP.269.37

Reis, N., Seerden, K. A. M., Derby, B., Halloran, J. W., and Evans, J. R. G. (1998) Direct Inkjet Deposition of Ceramic Green Bodies: II - Jet Behaviour and Deposit Formation. *MRS Online Proceedings Library (OPL)*, 542, 147–152. https://doi.org/10.1557/PROC -542-147.

Sachs, E. M., Haggerty, J. S., Cima, M. J., and Williams, P. A. (1993) Three-dimensional Printing Techniques. Google Patents, US5204055A

Schleifenbaum, H. Meiners, W., Wissenbach, K., and Hinke, C. (2010) Individualized Production by Means of High Power Selective Laser Melting. *CIRP Journal of Manufacturing Science and Technology*, 2(3), 161–169. https://doi.org/10.1016/j.cirpj .2010.03.005

Schwentenwein, M. and Homa, J. (2014) Additive Manufacturing of Dense Alumina Ceramics. *International Journal of Applied Ceramic Technology*, 12(1), 1–7. https:// doi.org/10.1111/ijac.12319

Seerden, K. A, Reis, N, Evans, J. R., Grant, P. S., Halloran, J. W., and Derby, B. (2001) Ink-jet Printing of Wax-based Alumina Suspensions. *Journal of the American Ceramic Society*, 84(11), 2514–2520. https://doi.org/10.1111/j.1151-2916.2001.tb01045.x

Shahzad, K., Deckers, J., Zhang, Z., Kruth, J.-P., and Vleugels, J. (2014) Additive Manufacturing of Zirconia Parts by Indirect Selective Laser Sintering. *Journal of the European Ceramic Society*, 34(1), 81–89. https://doi.org/10.1016/j.jeurceramsoc.2013 .07.023

Shishkovsky, I, Yadroitsev, I., Bertrand, P., and Smurov, I. (2007) Alumina–zirconium Ceramics Synthesis by Selective Laser Sintering/Melting. *Applied Surface Science*, 254(4), 966–970. https://doi.org/10.1016/j.apsusc.2007.09.001

Silva, N. R. F. A., Witek, L., Coelho, P. G., Thompson, V. P., Rekow, E. D., and Smay, J. (2011) Additive CAD/CAM Process for Dental Prostheses. *Journal of Prosthodontics*, 20(2), 93–96. https://doi.org/10.1111/j.1532-849X.2010.00623.x

Stuecker, J. N., Cesarano, J., and Hirschfeld, D. A. (2003) Control of the Viscous Behavior of Highly Concentrated Mullite Suspensions for Robocasting. *Journal of Materials Processing Technology*, 142(2), 318–325. https://doi.org/10.1016/S0924 -0136(03)00586-7

Tang, Y., Fuh, J. Y. H., Loh, H. T., Wong, Y. S., and Lu, L. (2003) Direct Laser Sintering of a Silica Sand. *Materials & Design*, 24(8), 623–629. https://doi.org/10.1016/S0261 -3069(03)00126-2

Tarafder, S., Balla, V. K., Davies, N. M., Bandyopadhyay, A., and Bose, S. (2013) Microwave-sintered 3D Printed Tricalcium Phosphate Scaffolds for Bone Tissue Engineering. *Journal of Tissue Engineering and Regenerative Medicine*, 7, 631–641. https://doi.org /10.1002/term.555

Tarafder, S. and Bose, S. (2014) Polycaprolactone-coated 3D Printed Tricalcium Phosphate Scaffolds for Bone Tissue Engineering: In Vitro Alendronate Release Behavior and Local Delivery Effect on in Vivo Osteogenesis. *ACS Applied Materials & Interfaces*, 6(13), 9955–9965. https://doi.org/10.1021/am501048n

Travitzky, N., Windsheimer, H., Fey, T., and Greil, P. (2008) Preceramic Paper-derived Ceramics. *Journal of the American Ceramic Society*, 91(11), 3477–3492. https://doi.org /10.1111/j.1551-2916.2008.02752.x

Vorndran, E. (2008) 3D Powder Printing of β-tricalcium Phosphate Ceramics Using Different Strategies. *Advanced Engineering Materials*, 10(12), B67 –B71. https://doi.org/10.1002 /adem.200800179

Wang, Jia-Chang, Dommati, Hitesh, and Hsieh, Sheng-Jen. (2019) Review of Additive Manufacturing Methods for High-performance Ceramic Materials. *The International Journal of Advanced Manufacturing Technology*, 103(1–4). 2627–2647. https://doi.org /10.1007/s00170-019-03669-3

Weisensel, L., Travitzky, N., Sieber, H., and Greil, P. (2004) Laminated Object Manufacturing (LOM) of SiSiC Composites. *Advanced Engineering Materials*, 6(11), 899–903. http://doi.org/10.1002/adem.200400112

Yang, Li and Miyanaji, Hadi. (2017a) Ceramic Additive Manufacturing: A Review of Current Status and Challenges. In *Solid Freeform Fabrication: Proceedings of the 28th Annual International Solid Freeform Fabrication Symposium – An Additive Manufacturing Conference*. Louisville, KY: Department of Industrial Engineering, University of Louisville, 652–672.

Yang Li, and Miyanaji Hadi (2017b) Solid Freeform Fabrication. In *Proceedings of the 28th Annual International. Solid Freeform Fabrication Symposium - an Additive Manufacturing* Conference (Volume 2). Austin, Texas: The University of Texas at Austin.

Zhang, Y, He, X, Du, S, and Zhang, J. (2001) Al_2O_3 Ceramics Preparation by LOM (Laminated Object Manufacturing). *The International Journal of Advanced Manufacturing Technology*, 17(7), 531–534. https://doi.org/10.1007/s001700170154

Zhao, X., Evans, J., Edirisinghe, M., and Song, J. (2003) Formulation of a Ceramic Ink for a Wide Array Drop-on-demand Ink-jet Printer, *Ceramics International*, 29(8), 887–892. https://doi.org/10.1016/S0272-8842(03)00032-4

7 Feedstock Materials for Additive Manufacturing Processes

7.1 INTRODUCTION TO ADDITIVE MANUFACTURING (AM) MATERIALS

Materials are key to determine the application scope of AM. The rapid prototyping parts should be close to the final requirements and meet certain requirements of strength, stiffness, moisture resistance, and thermal stability. Costs, material properties (such as mechanical properties and chemical stability), finished product details after post-treatment, and application environments are all important factors that need to be considered in the selection of AM materials.

Materials have a vital part to play in the complete understanding of AM processes. A wide spectrum of raw materials is currently in use for different processes, and an appreciable quantum of research is in progress toward the development of newer materials meant for specific applications. Polymers are another important group of materials that still constitute important AM raw materials. With due course and advancements, a variety of other materials, like metals, composites, ceramics, and so on, have found utilization for various applications. Consequently, an impressive material spectrum is available these days for processing via the AM route.

A comparison of different type of raw materials (metal, polymer, and ceramic) along with their state of fusion, material feedstock, and AM processing strategies, is presented in Table 7.1.

7.2 POLYMERS

Polymers are formed by addition or condensation polymerization.

a. **Addition polymerization**: an initiator (or catalyst) reacts with a starting monomer. The result of this initiation reaction is a monomer attached to the initiator with an unsatisfied bond. The unsatisfied bond is free to react with another monomer, thus adding to the chain. The process repeats over and over again until two chains combine or another initiator binds to the end of the chain, either of which will terminate the chain.

DOI: 10.1201/9781003451440-7

TABLE 7.1

AM Processing for Different AM Materials (Metal, Polymer and Ceramic)

Type of material	State of fusion	Material feedstock	Material distribution	Basic AM principle	AM process category
Metallic	Molten state	Filament/wire	Deposition nozzle	Selective deposition of material	DED
	Solid + molten state	Powder	Powder bed	Selective fusion of material on a bed	PBF
	Solid state	Sheet	Sheet stack	Fusion of stacked sheets	SL
Polymer	Thermal reaction bonding	Filament/wire	Deposition nozzle	Extrusion of melted material	ME
		Melted material–liquid	Printhead	Multi-jet material printing	MJ
	Chemical reaction bonding	Powder–printhead	Powder bed	Selective fusion of material on a bed	PBF
		Liquid material	Printhead	Curing (reactive)	BJ
			Vat	Curing by photopolymer by light	MJ, VP
		Sheet	Sheet stack	Fusion of stacked sheets	SL
Ceramic	Solid state	Powder and liquid suspension	High-density green compact	Selective fusion of particles in a high-density green compact	PBF
	Solid + liquid state	Powder material	Powder bed	Selective fusion of particles on a bed	

Source: Reprinted from *Journal of Materials Research and Technology*, 21, Srivastava, M., Rathee, S., Patel, V., Kumar, A. and Koppad, P, A review of various materials for additive manufacturing: Recent Trends and Processing Issues, 2612–2641, Copyright (2022), with permission from Elsevier.

b. **Condensation polymerization**: a monomer with an exposed H (hydrogen) atom binds with a monomer with exposed OH (oxygen–hydrogen) atoms. During the reaction, water is released (compensated) as the H and OH combine to form H_2O (water).

7.2.1 PHOTOPOLYMERIZATION

Polymerization is the process of linking small molecules (known as monomers) into chain-like larger molecules (known as polymers). When the chain-like polymers are linked further to one another, a crosslinked polymer is formed. Photopolymerization is polymerization initiated by a photochemical process, whereby the starting point is usually the induction of energy by a radiation source. The polymerization of photopolymers is normally an energetically favorable or exothermic reaction. However, in most cases, the formulation of a photopolymer can be stabilized to remain unreacted at ambient temperature. A catalyst is required for polymerization to take place at a reasonable rate. This catalyst is usually a free radical, which may be generated either thermally or photochemically. The source of a photochemically generated radical is a photoinitiator, which reacts with an actinic photon to produce the radicals that catalyze the polymerization process (Chua et al. 2003).

7.2.2 VAT POLYMERIZATION

Vat photopolymerization (VP) is a category of additive manufacturing (AM) processes that create 3D objects by selectively curing liquid resin through targeted light-activated polymerization. Stereolithography, the first AM process to be patented and commercialized, is a VP technique. The most common processes that use vat polymerization technology are:

- Stereolithography (SLA)
- Digital light processing (DLP)
- Continuous liquid interface production (CLIP)
- Day light polymer printing (DLPP)
- Drop-on-demand (DOD)
- PolyJet (PJ)

These processes use photosensitive monomer resins that react to the radiation of ultraviolet (UV) light. Upon irradiation, these materials undergo a chemical reaction (photopolymerization) to become solid. A solid 3D part can be produced by curing one layer over a previous layer. This method of 3D printing has some advantages over other methods, including achieving a higher amount of detail in objects and creating less waste.

7.2.3 POLYMERIC MATERIALS

Polymers used in AM can be supplied in the form of liquid, filament, or powder depending on the AM technology.

Natural polymers derived from plants and animals include wood, cotton, leather, rubber, wool, and silk. Plastics are human-made solid polymers derived from petroleum products. They are a perfect example of designing better, cheaper, and completely human-made polymers. Plastic polymers can be divided into:

1. Hydrocarbons, plastic polymers that contain hydrogen and carbon atoms.
2. Saturated hydrocarbons, when the carbon atoms in a polymer are bound to four other atoms.
3. Unsaturated hydrocarbons, when the carbon atom is not bound to four other atoms and forms double or triple bonds with another carbon atom.

Around 1850, billiards was becoming increasingly popular. The balls were made of ivory, which is in very limited supply, is thus very expensive, and requires the killing of elephants to obtain. In 1856, the first human-made plastic (Parkesine) was patented by Alexander Parkes from Birmingham, England. Often called synthetic ivory, it was composed of nitrocellulose—cellulose treated with nitric acid and a solvent. It was the first thermoplastic, but it failed as a commercial product due to poor product quality control (www.e-education.psu.edu/).

There are several different types of polymer printing processes out there. For instance, fused filament fabrication (FFF) is one of the most popular forms. This involves melting a filament of polymer and extruding it through a nozzle onto a build platform, where it cools and hardens into shape. Another process is called selective laser sintering (SLS), which is a powder bed fusion system that uses laser heat to fuse small particles of polymer material into a solid object. SLA, DLP, and continuous digital light processing (CDLP) are three other examples of polymer printing processes, each leveraging light or laser to systematically cure liquid polymer and build an object.

7.2.3.1 Resin-Based Polymers

Resin-based polymer is a type of synthetic material composed of an organic component and a resin, or monomer. Monomers can be combined in various ways to create polymers with specific properties and characteristics. SLA and DLP are two of the most common AM processes that produce components from liquid polymers (resins). Both processes polymerize synthetic resins with the help of light.

Table 7.2 provides the name of each resin, the main components and weight percent of each main component, viscosity, and the marketed or visual qualities of resins. In this respect, the photosensitive resins are comprised of four main components: monomers, solvents, photoinitiators, and additives. The solvent is used to adjust the monomer concentration, which affects the reaction rate. Additives like pigments or fillers are added to the resin to give the cured product a better appearance or improved mechanical properties. For faster printing, it is recommended to use a low- to medium-viscosity resin and avoid high-viscosity resins. The most durable results used hexamethylene diacrylate and isooctyl acrylate monomers, which are therefore recommended in cases where durability is prioritized.

TABLE 7.2

Name of Each Resin, Main Components and Weight Percent of each Main Component, Viscosity, and Marketed or Visual Qualities of the Resin

Resin	Monomer (wt%)	Plyimide (PI) (wt%)	Solvent/Epoxy (wt%):	Additives (wt%)	Viscosity (mPa·s)	Comments
Any cubic resin provided	Isooctyl acrylate, 1,3- propanediyl diacrylate (45%)	Phosphine oxide (TPO) (5%)	Propylidynetri-methanol, esters with acrylic acid (45%	Unspecified pigments/ fillers (5%)	190	Translucent green color. Resin provided with printer.
ELEGOO ABS-like resin	Hexamethylene diacrylate (40%)	Hydroxycyclohexyl phenyl ketone (5%)	Epoxy resin (50%)	Unspecified pigments/ fillers (5%)	210	Opaque gray resin, medium viscosity. Marketed as comparable to ABS polymer in FDM printing.
ELEGOO Waterwashing resin	Hexamethylene diacrylate (40%)	Hydroxycyclohexyl phenyl ketone (5%)	Epoxy resin (50%)	Unspecified pigments/ fillers (5%)	140	Transparent blue resin, very low viscosity, almost clear in appearance.
F69 Flexible TPU-like resin	4-Acryloylmorpholine (15%)	Phosphine oxide (TPO) (5%)	Acrylated aliphatic urethane (80%)	Unspecified pigments/ fillers (<1%)	980	Opaque black resin, very thick and had a bad odor.
iFun Toughness resin	E03TMPTA (30%)	Phosphine oxide (TPO) (5%)	Polyester acrylate (60%)	Unspecified pigments/ fillers (5%)	350	Opaque white resin, marketed as durable and slightly flexible.
Wanhao flexible resin	Four different monomers (crosslinking) (50%)	Phosphine oxide (TPO) (5%)	Bisphenol A epoxy diacrylate (45%)	Unspecified pigments/ fillers (<1%)	1080	Opaque black resin, very thick and had a bad odor.
Any cubic plant-based resin	Isooctyl acrylate, 1,3- propanediyl diacrylate (45%)	2-Methylpropan-1-one (5%)	Fatty acids, Soya, epoxidized, Bu esters (45%)	Unspecified pigments/ fillers (5%)	190	Clear resin, same viscosity as provided by any cubic resin. Marketed as plant based, more sustainable and environmentally friendly.

Source: Wilson, T., Application of SLA 3D Printing for Polymers, 1490, 2022. Williams Honors College, Honors Research Projects. With permission from Taylor and Francis.

7.2.3.2 Filament-Based Polymer

Filament-based polymer is arguably the most popular kind, used widely by fused deposition modeling (FDM) machines. It is fed into 3D printers as a spool of tube and then melted to form designs layer by layer. There are many subtypes of filament-based polymer, including polylactic acid (PLA), polycarbonate (PC), and polyether ether ketone (PEEK). Filament-based polymers are used for a wide variety of applications, such as consumer products, medical devices, and tools. Higher-grade varieties like PC and PEEK are common in the fields of engineering and aerospace.

A. Polycarbonate (PC)

Polycarbonate (PC) is a versatile *thermoplastic polymer*, which is easy to process and has many applications within different industries. It is a high-performance, synthetic *filament* used for FFF 3D printing. It is a recyclable material that can be engineered to meet specific application or processing requirements, such as low warping. It is mainly used for medical devices and parts, as well as being a stronger substitute for glass. Polycarbonate is not UV or scratch resistant as standard. However, there are versions that are UV resistant and have an extra scratch-resistant layer. An electrostatic discharge (ESD)–safe coating is also available on a PC, which avoids the risk of electrostatic discharge. The main characteristics of PC plastic are its high impact resistance and light transmittance. Additionally, it is resistant to high temperatures and easy to reshape. The properties of polycarbonate include the following (https://www.bkbprecision.com/)

- a. High impact resistance—at least 250 times stronger than glass
- b. Light transmittance
- c. Easy to shape
- d. Suitable for intensive use
- e. Resistant to high temperatures
- f. Bendable
- g. Easy to process
- h. Option of applying UV- and scratch-resistant layer
- i. Option of ESD-safe coating

PC advantages

- a. Outstanding toughness
- b. Ability to reduce wall thickness up to 25% without sacrificing any properties
- c. Good stiffness relative to other plastic materials
- d. Transparency
- e. Temperature resistance up to 250°F
- f. A high degree of elasticity; resistant to denting and kinking when bent upon itself

The application fields of PC engineering plastic include the glass assembly industry, the automobile industry, electronics, the electrical industry, industrial machinery parts, office equipment, medical and health care, leisure and protective equipment, etc.

B. Polyether ether ketone (PEEK)

PEEK polymers are obtained by step-growth polymerization by the dialkylation of bisphenolate salts. A typical example is the reaction of 4,4′-difluorobenzophenone with the disodium salt of hydroquinone, which is generated in situ by deprotonation with sodium carbonate. It is a semi-crystalline *thermoplastic* with excellent mechanical and chemical resistance properties that are retained to high temperatures. It is regarded as one of the highest-performing thermoplastics in the world, together with other polymers of the poly ether ether ketone (PEEK) family, such as poly ether ketone ketone (PEKK). PEEK is used to fabricate items used in the aerospace, automotive, oil and gas, and medical industries. PEEK filament has unique properties because it does not come into contact with water during the production process and is directly packaged in a vacuum packaging. These properties make the PEEK filament particularly suitable for usage in FDM and FFF 3D printers. The material has an excellent adhesion between layers, which results in improvement of the impact resistance, strength, durability, and printing process (www.3d4makers.com/).

C. Acrylonitrile butadiene styrene (ABS)

ABS is an amorphous thermoplastic polymer comprised of three monomers: acrylonitrile, butadiene, and styrene. It is most commonly polymerized through the emulsification process or the expert art of combining multiple products that don't typically combine into a single product. ABS material is the preferred engineering plastic for the melting deposition molding process, which is mainly prefabricated into silk and powder for use at present. Its scope of application covers almost all daily necessities, engineering supplies, and some mechanical supplies.

The acrylonitrile in ABS plastic provides chemical and thermal stability, while the butadiene adds toughness and strength. Moreover, the styrene gives the finished polymer a nice, glossy finish. ABS has a low melting point, which enables its easy use in 3D printing. It also has high tensile strength and resistance to physical impacts and chemical corrosion. This, in turn, allows the finished plastic to withstand heavy use and adverse environmental conditions.

ABS plastic takes color easily, allowing finished products to be dyed in exact shades to meet precise project specifications. More recently, ABS has been playing a key role in the rise of 3D printing (https://adrecoplastics.co.uk/). It has the following advantages:

a. Reasonable production costs.
b. Able to withstand heating and cooling several times, which makes it suitable for recycling.
c. Versatile in the range of color and surface texture options, such that it is manufactured to a high-quality finish.

d. Lightweight and suitable for a vast range of applications.
e. Has low heat and electrical conductivity, making it especially helpful for products requiring electrical insulation.
f. Offers excellent impact resistance and can absorb shock effectively and reliably.

ABS disadvantages include:

a. Low melting point, making it inappropriate for high-temperature applications and medical implants.
b. Has poor solvent and fatigue resistance.
c. Doesn't stand UV exposure and weathering unless it is properly protected.
d. Low conductivity. so it cannot be used in situations where this prove a hindrance to the overall design.
e. High smoke generation, which could cause air pollution.

D. **PPSF/PPSU (Polyphenylsulfone)**

PPSF (polyphenylsulfone) is an engineering thermoplastic with a wide range of benefits that include chemical, temperature, and moisture resistance. It has the highest strength ratings of the FDM plastics. This material lends itself well to sterilization via steam and chemical sterilization processes. It is an advanced thermoplastic that is well worth the cost when used in corrosive and high-temperature applications. This advanced material should not be used for visual prototypes or parts that will not be exposed to harsh environments; it is best to consider cheaper alternatives for parts with less extreme operating conditions. PPSF material possesses the highest strength, the best heat resistance, and the highest corrosion resistance among all thermoplastic materials. It can be used for 3D printing using FDM to produce high load–bearing products and has become the preferred material to replace metal and ceramic (Hayden Thompson Black et al. 2016).

a. Has the greatest heat and chemical resistance of all FDM materials.
b. Ideal for aerospace, automotive, and medical applications, or any application where prototypes are exposed to extreme conditions.
c. Produced parts are not only mechanically superior but also dimensionally accurate. This allows truly accurate future performance predictions from the prototype.

E. **ABS-PC blend**

ABS-PC stands for acrylonitrile butadiene styrene polycarbonate. It is a thermoplastic blend that combines the desirable properties of both ABS plastic and polycarbonate. PC-ABS filament is a 3D printing material of increased impact strength (comparable to PC), increased heat resistance, and decreased moisture absorption. PC-ABS filament is a V-0 flame retardant blend of polycarbonate and ABS—two of the most used thermoplastics for engineering and electrical applications. It exhibits a balance of strength, impact resistance, and heat resistance, making it highly suitable for a wide range of applications in automotives, electronics, and telecommunications, where durability and performance are crucial. ABS-PC material possesses a remarkable set of properties that make it highly desirable for a wide range of applications (https://europlas.com.vn/).

a. The combination of ABS and PC results in superior impact resistance. It can withstand high-stress situations and absorb energy, making it suitable for applications where durability and resistance to impact are essential.
b. Exhibits excellent heat resistance, allowing it to withstand elevated temperatures without significant deformation or degradation. This property makes it suitable for applications that require exposure to heat, such as automotive components and electronic enclosures.
c. Maintains its shape and dimensions even under varying temperature conditions due to its low coefficient of thermal expansion. This property ensures that components remain stable and perform consistently over a wide temperature range.
d. Offers good resistance to a variety of chemicals, including oils, greases, and some solvents. This property makes it suitable for applications where exposure to chemicals is expected, such as automotive parts, electrical connectors, and housings for industrial equipment.
e. Combines the clarity and transparency of polycarbonate with the versatility of ABS. This results in a material that can be easily molded and colored, allowing aesthetically pleasing and visually appealing finished products.
f. Exhibits excellent electrical insulation properties, making it suitable for applications in the electrical and electronics industries. It provides reliable insulation and protection for electrical components, ensuring safe and efficient operation.
g. Can be easily processed using common methods such as injection molding, extrusion, and thermoforming. This ease of processing makes it cost-effective and allows the efficient production of complex shapes and designs.

Acrylonitrile styrene acrylate (ASA)

ASA, also called acrylic styrene acrylonitrile, is an amorphous thermoplastic developed as an alternative to ABS, but with improved weather resistance. It is an acrylate rubber-modified styrene acrylonitrile copolymer used for general prototyping in 3D printing, where its UV resistance and mechanical properties make it an excellent material for use in fused filament fabrication printers, particularly for outdoor applications like weather station radiation shields for environmental sensors. It is also widely used in the automotive industry. ASA is a synthetic, amorphous thermoplastic that works best in material extrusion (ME) printing. It is a high-impact material that can be used to produce functional parts for automotive applications, electronics, tooling, and more. ASA offers a range of features that make it suitable for prototyping, tooling, and even end-use parts. The material's exceptional resistance to weathering, UV rays, impact, and wear makes it a reliable choice for producing components used in everyday items. From gutters, drains, and electrical panels to outdoor furniture, it can fulfill diverse needs. When reinforced with additives like fiberglass, it can even contribute to the construction of pedestrian bridges. The automotive industry also benefits from ASA, using it for designing car bumpers and side mirrors. Furthermore, the wide array of colors and finishes available in ASA make it ideal for creating visually appealing prototypes across various industries (www.3dnatives.com/).

G. ULTEM/PEI

ULTEM, the branded name for polyetherimide (PEI), is one of the few commercially available amorphous thermoplastic resins that retain their mechanical integrity at high temperatures. High-performance plastic polymers fall into two categories: thermosets and thermoplastics. Thermosetting polymers are those that solidify to an irreversible hardness after curing due to chemical bonds within the plastic. The polymer hardens in a crosslink pattern that prevents it from re-melting even under extreme heat. ULTEM is easy to thermoform and can be manufactured using FDM. This material is strong, chemical and flame resistant, easy to use, and able to withstand extremely high temperatures while retaining a set of stable electrical properties. It is often used in the production of circuit boards, eyeglasses, food preparation and sterilization equipment, and aircraft parts. ULTEM is known for its strength and durability (https://sybridge.com/).

H. Polyethylene terephthalate glycol (PETG)

PETG or polyethylene terephthalate glycol is a thermoplastic polyester commonly used in manufacturing plastic beverage bottles and food products. The G stands for glycol, which adds durability and strength, and contributes to the compound's impact resistance and ability to withstand high temperatures. This clear, amorphous thermoplastic can also be colored during processing. The PETG filament is quickly becoming a popular choice among 3D printing for several reasons. Its high impact resistance and durability make it an ideal option. PETG is also BPA-free (does not contain any traces of Bisphenol A, which may leach into foods and beverages like baby food containers and water bottles). These elements, combined with PETG's natural transparency (even more transparent than PET), make it clear to see how it has become a popular choice in the food and beverage industry. Its transparency and durability also make it perfect for medical and pharmaceutical packaging. Additionally, PETG is incorporated into medical implants and prostheses. The transparent plastic can be easily colored to create eye-catching storefront signage. It's also used in creating display stands of various color and size (jewelry stands). Due to its tough resistance to heat and chemical stressors, PETG is also used to create parts such as protective guards, testing components, and manufacturing tools and aids. PETG can be extruded as a filament for 3D printing. 3D printing with PETG filament delivers a result that gives the materials a clear look, which is less brittle and simpler to use than the base form of PET. Figure 7.1 shows a polymer-based filament feedstock fabrication process.

Benefits of PETG filament include the following:

- Excellent layer adhesion.
- Warp resistance.
- Reduced shrinkage.
- Higher density.
- Chemical resistance to both acidic and alkaline compounds.
- Flexible printing on glass, acrylic, glass, blue tape, and polyimide tape.
- Odorlessness during printing.

I. **Polystyrene (PS)**

Polystyrene (PS) is a synthetic polymer made from monomers of the aromatic hydrocarbon styrene. General-purpose polystyrene is clear, hard, and brittle. It is a poor barrier to air and water vapor and has a relatively low melting point. PS is one of the most widely used plastics, with the scale of its production being several million tonnes per year. Polystyrene is naturally transparent but can be colored with colorants. Uses include protective packaging (such as packing peanuts and in the jewel cases used for storage of optical discs such as CDs and occasionally DVDs, containers, lids, bottles, trays, tumblers, and disposable cutlery, in the making of models, and as an alternative material for phonograph records.

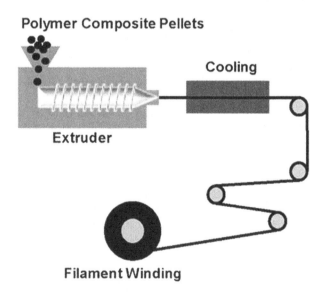

FIGURE 7.1 Polymer-based filament feedstock fabrication. (Adapted from Park, S. et al., *Matter*, 5, 1, 43–76, 2021)

As a thermoplastic polymer, polystyrene is in a solid (glassy) state at room temperature but flows if heated above about 100°C, its glass transition temperature. It becomes rigid again when cooled. This temperature behavior is exploited for extrusion (as in Styrofoam) and also for molding and vacuum forming, since it can be cast into molds with fine detail. The temperature behavior can be controlled by photocrosslinking. Under ASTM standards, polystyrene is regarded as not biodegradable.

J. Polylactic acid (PLA)

PLA is a fully biodegradable thermoplastic polymer consisting of renewable raw materials. Among all 3D printing materials, PLA is one of the most popular materials used for AM for filament fabrication. PLA is a bioplastic, used in 3D printing employing the FDM technology, and along with ABS, this material is one of the standard materials for this technology. There is often a tendency to compare these plastic materials, as they are the two most common alternatives available for consumer printers. PLA plastic material is easy to use and offers some interesting mechanical properties. Table 7.3 shows the glass transition and melting temperatures of thermoplastics widely used in structural applications.

7.2.3.3 Powder-Based Polymers

Powder-based polymer is a class of material that starts out in dry granular form and is combined with a controlled emission of energy to form a polymer. This type of material can be used in the production of parts, tools, or products that have specific characteristics depending on the ingredients used in its formulation. The main

TABLE 7.3

Physical Properties of Thermoplastics Widely Used in Extrusion-Based 3D Printing

Polymer	Glass transition temperature (°C)	Melting temperature (°C)
Polyether ether ketone (PEEK)	135 to152	335– to 343
Low-density polyethylene (LDPE)	<−100	100 to 110
Polypropylene(PP)	−30 to −20	160 to 165
Polystyrene (PS)	90 to 105	–
High-density polyethylene (HDPE)	<−100	125 to 135
Polylactic acid (PLA)	53 to 64	145 to 186
Polyethylene terephthalate glycol (PETG)	75 to 80	–
Acrylonitrile butadiene styrene (ABS)	−63 to 127	–
Polycarbonate (PC)	145	–
Nylon 6	50 to 80	225 to 235
Nylon 66	70 to 90	225 to 265

Source: Modified from Grigorescu, R.M. et al., *Recycling*, 4, 3, 32, 2019.

advantages of powder-based polymers are their low cost, wide range of colors and textures available, ease of use (especially when compared with liquid resin systems), and versatile performance. Figure 7.2 shows how polybutylene terephthalate (PBT) powder is produced by mixing with polyethylene glycol (PEG).

A. Polyamide (PA)

Polyamide (PA), commonly called nylon, is a semi-crystalline thermoplastic with low density and high thermal stability. Polyamides are typically made by combining two monomers, namely, adipic acid and 1,6-diaminohexane. Once these two monomers have reacted together, they form water as a by-product of each polymer chain linkage. This linking of the two monomers is known as polymerization. This creates a nylon salt, which is then heated to evaporate the water. This heating is done inside an autoclave at 280°C and 18 Bar. After the polymerization process, various additives and pigments are added. These additives can change the physical properties of the polymer. After the additives are added, the molten polyamide nylon is extruded through holes to form long laces of nylon. These laces are extruded into a water bath, which allows the laces to cool and solidify (https://matmatch.com/). It is available in the form of powder for the SLS process.

Nylon plastic is one of the most important construction plastics due to its excellent wear resistance, good coefficient of friction, and good temperature resistance and impact strength. PA material can be directly used to manufacture equipment parts; PA–carbon fiber composite plastic parts manufactured by FDM technology possess high toughness and can be used for mechanical tools instead of metal. The most common nylons used for SLS 3D printing are PA 11 and PA 12, but there are many more. PA 11 powders are used for parts that require UV and impact resistance, while PA 12 is preferred for enhanced part strength and stiffness. Table 7.4 shows the properties and applications of different types of nylon (PA).

B. High-density polyethylene (HDPE)

HDPE is a pure hydrocarbon, which makes it non-polar, water-repellent, and highly resistant to chemicals. At room temperature, HDPE is therefore not attacked by many solvents, alkalis, or acids. It is a high-density polyethylene (0.94–0.97 g/cm^3)

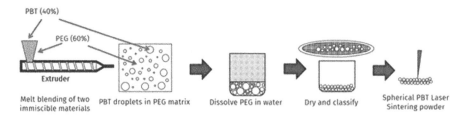

FIGURE 7.2 Production of polybutylene terephthalate (PBT) powder for SLS. (Adapted from Kleijnen, R.G. et al., *Appl. Sci.*, 2019, 9, 2019)

TABLE 7.4

Characteristics of Different Types of Nylon (PA)

SLS Material	Properties	Applications
PA 12 (nylon 12)	abrasion resistant, chemical resistance to oils, greases, aliphatic hydrocarbons, and alkalis; ductile (can meet biocompatibility standards for skin contact)	connectors, hinges, housings, complex assemblies, enclosures, watertight applications, prosthetics
PA 11 (nylon 11)	chemically resistant, ductile, impact-resistant	insoles, snap fits, hinges, prosthetics, sports goods
PA 6 (nylon 6)	durable, flexible, excellent surface appearance, low viscosity, electrical insulating properties, high water absorption	electronics and electrical, consumer goods, packaging
PA CF (carbon fiber-filled nylon)	high stiffness and thermal resistance, less shrinkage than other PAs, ductile, impact-resistant, oil and grease resistant	rugged industrial applications, snap fits, connectors, hinges, housings, high-performance racing applications
PA ESD (electrostatic discharging nylon)	PA that dissipates electrostatic charges	electronic housings; jigs, fixtures, and manufacturing aids for the electronics industry
PA FR (flame retardant)	all the qualities of PA plus flame retardancy	air plenums, housings
PA GB (nylon with glass beads)	high stiffness and strength plus a better surface finish and finer feature details than standard PAs, high heat deflection temperature (HDT) values	functional parts, enclosures, housing, tooling

Source: https://all3dp.com/

and is particularly characterized by its very good resistance to chemicals and greases as well as its water-repellent effect. At room temperature, HDPE has a hard flexible appearance and in addition to its very good mechanical properties, has good sliding behavior and increased wear resistance. HDPE is, therefore, employed for products used for food and packaging for the chemical industry. Thus, containers, bottles, and pipes for chemicals, fuels, water, gas, or oil are manufactured from HDPE as standard. HDPE offers a significant price advantage, as HDPE is a widely used mass plastic, which is much cheaper to produce than PA12 or PA11.

HDPE is used for processing by SLS, where a high laser power is required to melt the powder particles. The mechanical properties of the HDPE are negatively influenced by the high thermal load due to the line-by-line exposure to the laser, which results in embrittlement of the material. In the case of high-speed sintering (HSS), the energy is applied to the powder bed surface by means of an infrared lamp, which significantly lowers the maximum temperatures that can be realized and consequently, reduces the thermal stresses on the material and preserves the proven mechanical properties of HDPE (www.voxeljet.com/).

7.2.3.4 Sheet-Based Polymers

Urea formaldehyde resin is used for laminated sheets. The monomer of this resin is urea (NH_2CONH_2) and formaldehyde (HCHO).

7.2.4 SUMMARY

Table 7.5 shows the AM technologies and materials of polymers.

a. Stereolithography (SLA), digital light processing (DLP), continuous liquid interface production (CLIP), day light polymer printing (DLPP, Drop-on-demand (DOD), and PolyJet (PJ) use resin-based polymers, photocured polymers that do not exhibit great strength or toughness. In the solid-form fabrication (SFF) family, the SLA process is the most accurate and has emerged as the industry standard for creating a master pattern that might then be used as the basis for a casting or injection mold.

b. Fused deposition modeling (FDM) and fused filament fabrication (FFF) use thermoplastic polymer-based filaments including acrylonitrile butadiene styrene (ABS), polystyrenes (PS), polylactic acid or polylactide (PLA), and polyamides (PA).

TABLE 7.5
AM Technologies and Polymeric Materials

Technologies	Materials
Stereolithography (SLA)Digital light processing (DLP)Continuous liquid interface production (CLIP)Day light polymer printing (DLPP)Drop on-demand (DOD)PolyJet (PJ)	Photosensitive resins
Fused deposition modeling (FDM) Fused filament fabrication (FFF)	ThermoplasticsAcrylonitrile butadiene styrene (ABS)*Polystyrenes (PS)*Polylactic acid or polylactide (PLA)Polyamides (PA)*
Binder jetting (BJ)	Polymer powderAcrylonitrile butadiene styrene (ABS)*Polyamides (PA)*Polycarbonate (PC)
Selective laser sintering (SLS)	Polymer powderPolyamides (PA)*Polystyrenes (PS)*Thermoplastic elastomers (TPE)Polyaryl ether ketones (PAEK).
Laminated object manufacturing (LOM)	Plastic foil

*Possibility of being in either filament or powder form.

c. Binder jetting (BJ) uses powder-based polymers such as acrylonitrile buta-diene styrene (ABS), polyamides (PA), and polycarbonate (PC). On the other hand, SLS uses polymer-based powders such as polyamides (PA), polystyrenes (PS), thermoplastic elastomers (TPE), and polyaryl ether ketones (PAEK).

d. Laminated object manufacturing (LOM) uses plastic foil made from most of the thermoplastic polymers.

e. Some thermoplastics can be supplied in the form of filament or powder.

7.3 COMPOSITE MATERIALS

Composite materials are commonly used for lightweight components and structures in various industries. They are often used in transportation, construction, automotive, and aerospace, since they offer good mechanical properties, flexibility in design, relatively low cost, and high performance. In the aerospace industry, they have been used for fabricating several aircraft components. Currently, 50% of the airframe of aircraft is fabricated from composite materials. Composite materials can be classified into three common types:

1. Fiber-reinforced composites that consist of fibers or whiskers of one material, embedded in a matrix material.
2. Particulate-reinforced composites that consist of particles of one material in a matrix material.
3. Laminated composites that consist of layers of two or more materials.

Youssef et al. (2023) present the classification, and detailed definitions and structure, of the constituents of composite materials. Composites, typically, comprise a core polymer material and a reinforcing material, like chopped or continuous fiber. The composite material offers higher strength and stiffness compared with non-reinforced polymers. In some cases, it can even replace metals like aluminum.

AM is an effective fabrication method for composite materials where carbon fiber (CF) is commonly mixed with thermoplastic polymers such as nylon (PA), polycarbonate (PC), and even PEEK (polyether ether ketone). Such composite AM materials result in parts that are both stronger and more lightweight.

Composites are usually classified by the type of material used for the matrix. The four primary categories of composites are polymer matrix composites (PMCs), metal matrix composites (MMCs), ceramic matrix composites (CMCs), and carbon matrix composites (CAMCs).

Producing polymer-based composites through AM enhances their properties and expands their applications across various industries. Natural fibers such as wool, hemp, flax, kenaf fiber, and vegetable fibers have been successfully utilized as replacements for artificial fibers in composite manufacturing using AM.

Different types of polymers, including thermoplastics, liquid polymers, and reactive polymers, are used in AM. Recent advancements have focused on incorporating fillers such as nanotubes, CFs, nanofibers, nanoparticles, and synthetic fibers into the polymeric products. Fiberglass is the most common and familiar composite, in which small glass fibers are embedded within a polymeric material (normally an epoxy or polyester). The glass fiber is relatively strong and stiff (but also brittle), whereas the polymer is ductile (but also weak and flexible).

7.3.1 POLYMER-BASED COMPOSITES

PMCs are usually fabricated using directed energy deposition (DED) and SLA processes. These will also be classified later.

7.3.1.1 Resin-Based Polymer Composites

SLA is used to fabricate PMCs but with fewer mechanical properties. Shape-memory polymer composites are made using fiber or fabric reinforcements and shape-memory polymer resin as the matrix. These composites have the ability to be manipulated into various configurations when they are heated above their activation temperatures and exhibit high strength and stiffness at lower temperatures. They can also be reheated and reshaped repeatedly without losing their material properties. High strain composites are designed to perform in a high-deformation setting. The high strain composite is generally dependent on the fiber layout as opposed to the resin content of the matrix. Castalite ceramic resin is a UV curable that can print ceramic molds used to cast metal parts after firing. It is suitable for SLA and DLP 3D printers.

As can be seen in Table 7.5, DOD material jetting (MJ) with UV solidification and Drop-on-demand BJ uses liquid polymers.

7.3.1.2 Filament-Based Polymer Composites

Fiber-reinforced thermoplastics polymers use short fiber, long fiber, or long-fiber-reinforcement.

FDM is commonly used for fiber-reinforced polymers because it allows the addition of the reinforcement by adding CF into a plastic matrix. Fiber-reinforced PMCs fabricated by FDM increased the tensile strength and decreased the toughness and ductility (Yakout and Elbestawi 2017). 3D printed polylactic acid (PLA) composite parts with CF processed using the FDM technique have enhanced tensile, flexural, and interlaminar shear. The low strain in PLA was increased by adding CF, which led to a high-performance PLA-based polymeric composite. The strength of composites fabricated through AM is negatively affected by the agglomeration of fiber reinforcements at high volume fractions (Table 7.6, Table 7.7).

Ionic polymer–metal composites (IPMCs) are composed of an ionic polymer like Nafion or Flemion that is chemically plated or physically coated with conductors such as platinum or gold. The unique actuation and sensing properties of IPMCs are exploited in 3D printing to create electroactive polymer structures for application in soft robotics and bio-inspired systems. The extrusion of a Nafion filament (Figure 7.4 was conducted at between 280 and 300°C. The extrusion speed varied between 25 and 125 mm/s depending on temperature and the amount of material in the auger. The extrusion was conducted in a fume hood due to the potential for the production of hydrogen fluoride (HF) and other toxic gases when Nafion precursor is heated.

TABLE 7.6

Characteristics of 3D Printing Technologies for Polymers and Composites

	Printing principle	Polymer/ composite state	Typical polymer materials
ME (FFF/ DIW)	ExtrusionPressurized extrusion	Solid filamentLiquid polymer	Thermoplastics, such as polycarbonates, ABS,PLA, and nylonLiquid polymer, hydrogel, and colloidal suspension
VP	UV-induced curing	Liquid photopolymer	Photocurable resin (epoxy or acrylate-based resin)
PBF	Heat-induced sintering	Solid powder	Polyamide (PA), polystyrene (PS), and polycaprolactone (PCL) powder
MJ	DOD material jetting with UV solidification	Liquid photopolymer	Photocurable resin
BJ	Drop on demand binder jetting	Liquid polymer	Bonding agents + acrylate-based powder (metal and sand)
SL	Layer-by-layer adhesive	Polymer sheet	Bonding agents + polymer composites

Source: Adapted from Park, S. et al., *Matter*, 5, 1, 43–76, 2021.

TABLE 7.7

Ranges of Powder Size, Sphericity, and Range of Size Distribution in Commonly Used Metal Powder–Based AD Processes

Additive manufacturing process	Powder feeding method	Particle size (μm)	Sphericity requirements	Range of particle size distribution	Utilization of powder	Production efficiency
SLM	Prespread powder	15-33	High	Small	Low	Low
EBM	Prespread powder	53-150	High	Small	Low	Low
BJ	Synchronous powder feeding	53-150	Middle	Middle	High	High

Source: Shanthar, R. et al., *Advanced Engineering Materials*, 25, 2300375, 2023.

Examples of the Nafion filament obtained are also shown in Figure 7.3. The discoloration in the filament on the right is attributed to contamination, possibly caused by accelerated oxidation of the extruder components (Carrico et al. 2015).

7.3.1.3 Powder-Based Polymer Composites

SLS is used for fabricating particulate-reinforced polymers by mixing the powder and then sintering the mixture using a laser source. One of the challenges in SLS

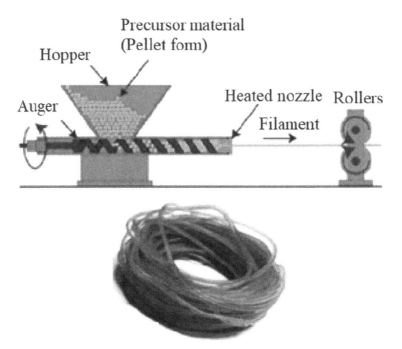

Melt processed (extruded) Nafion filament

FIGURE 7.3 Nafion precursor filament extruder showing the main components such as an auger, a hopper, a heated nozzle, and rollers, and a sample of Nafion precursor filament obtained by extruding. (Adapted from Carrico, J.D. et al, *Smart Materials and Structures*, 24, 12, 125021, 2015)

of PMCs is to have a uniform mixture between the matrix and the reinforcement. Moreover, the percentage of reinforcement weight affects the amount of energy density required for melting. DED is a powder-fed process, while SLS is a powder-bed process. The use of DED in PMC fabrication allows better flexibility in reinforcement distribution. Fiber-reinforced thermosetting polymer composites are made by 3D printing of chopped fiber–reinforced composite material. These composites include paper composite panels, while advanced thermoset polymer matrix composites incorporate aramid fiber and carbon fiber in an epoxy resin matrix. Figure 7.4 shows a massive production of powder compound of ASA 20 CF using a twin screw (see Table 7.4). MMCs (i.e. aluminum–matrix, titanium–matrix, and TiAl–matrix composites) have been fabricated by powder-based processes such as SLM and DED processes. The reinforcement may be particulates in the form of powder. On the other hand, a fiber in the form of a filament is used in FDM. Graphite, silicon carbide, titanium carbide, and tungsten carbide are additives commonly used in AM of MMCs (Yakout and Elbestawi 2017). Table 7.4 shows the characteristics of 3D printing technologies for polymers and composites.

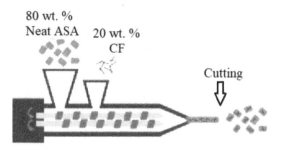

FIGURE 7.4   Massive production of compound of ASA 20 CF using a twin screw. (Adapted from Sanchez, D.M. et al., *Materials & Design*, 191, 108577, 2020)

7.4.   METALLIC MATERIALS FOR AM

A metal is a material (an element, compound, or alloy) that is typically hard when in solid state, opaque, shiny, and has high mechanical properties (tensile strength, fracture toughness, etc.) and electrical and thermal conductivity. Metals are ductile (able to be drawn out into a thin wire), malleable (can be hammered or pressed permanently out of shape without breaking or cracking), and fusible (able to be fused or melted). Metals have a crystalline structure with atoms closely packed, which renders high density for most of them. Alloys are combinations of metals, or of a metal and another element, which grant higher properties than their constituents. These unique fundamental properties make metals ideal materials for use in a diverse range of applications.

Metallic materials used as feedstock for AM processes take the form of powders, wires, and sheets, as described in Chapter 5. In this section, the special characteristics and applications of powders and wires used for specific AM processes are discussed.

7.4.1   Powder Metallurgy as a Basis for Metal AM Powders

Metals were manufactured in the powder form a long time ago for the purpose of the well-known processing technology, powder metallurgy (PM). This technique involves preparation of the powder, blending and mixing of constituents, and then consolidating it into a solid near net form by the application of pressure (compaction) and heat (sintering) at a temperature below the melting point of the main constituent. Therefore, this technology is adopted for refractory metals such as molybdenum, tantalum, and tungsten with high melting temperatures, as well as alloys with immiscible phases, which will not form solutions under normal melting conditions. (These can be made to form alloy systems by the intimate mixture of particles prior to further processing, known as mechanical alloying.) Metal powders can be mixed with ceramic powders to manufacture MMCs. The technique provides controlled porosity and permeability for self-lubrication or filtration. Material and energy utilization efficiency is higher than with casting or metal forming processes, and above all, it is the most environmentally friendly technology for reducing energy consumption and allowing recycling of the powder. The development of PM technology diversified with time to include other superior processes based on powder, such as hot isostatic pressing (HIP) and metal injection molding (MIM). The tremendous developments in this field represent one of the major contributors to AM technology, where most of its operations are used during AM processing or post-processing. Accordingly, all the advantages of PM technology are conveyed to AM processes based on metallic powders in addition to its unique privileges.

7.4.2 POWDER PREPARATION TECHNIQUES FOR ADDITIVE MANUFACTURING

There are various techniques to produce metal powders, but they can generally be divided into three categories: thermal/mechanical, chemical, and electrolytic methods, as presented in Figure 7.5. The thermal/mechanical methods include atomizing the powder from the molten metal, comminution (crushing, milling in a ball mill, or grinding brittle or less ductile metals and oxide powders into small particles), and mechanical alloying. The electrolytic method is based on electrochemical deposition of metallic powders. Chemical methods include reduction of metal oxides, precipitation from solution, and decomposition of metal salts such as carbonyls. From the 1990s to the present time, powder preparation has prospered at an escalating pace to match the feedstock powder requirements for the innovative emerging technology of AM. Therefore, only the powder preparation techniques suitable for AM processes are presented here.

7.4.2.1 Atomization of Metal Powder

This method is based on injecting molten metal through an orifice to form a stream, which is broken down (atomized) into droplets by a jet of water, air, or gas, or through centrifugal force. Metal powders for AM are most commonly produced using the gas atomization process, where the molten metal stream is atomized under the effect of

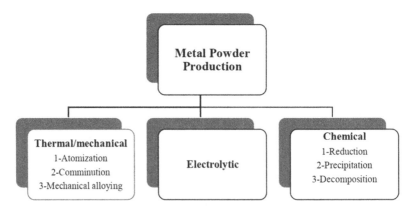

FIGURE. 7.5 Metal powder production techniques. (From Youssef, H. et al., *Manufacturing Technology; Materials, Processes and Equipment*, CRC Press, California, United States, 2023. With permission from CRC Press)

a high-pressure neutral gas jet into small metal droplets, thus forming metal powder particles after rapid solidification. Powders produced by gas atomization have a spherical shape, which is very beneficial for powder flowability, while powders produced by water atomization will have an irregular shape. Spherical-shaped particles ensure good powder density and particle size distribution. This technique leads to good reproducibility of particle size distribution and enables a very wide range of alloys.

Gas atomization: The principle of the process is shown in Figure 7.6. The melting of metal takes place at the top of the atomization tower. For a small volume of metal (a few liters), melting is carried out by inductive heating in a ceramic crucible, such as in a vacuum induction melting (VIM) furnace; the process is thus known

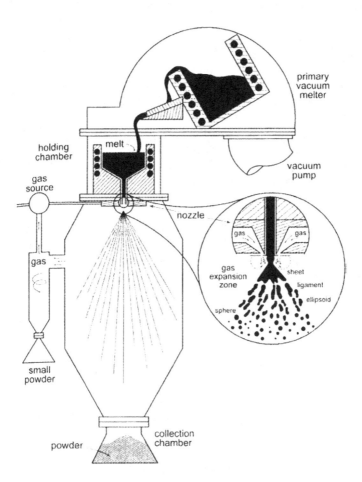

FIGURE 7.6 Principle of gas atomization. (From Peyre, P. and Charkaluk, E. *Additive Manufacturing of Metal Alloys 1, Processes, Raw Materials and Numerical Simulations.* 2022 ISTE Ltd and John Wiley & Sons, Inc. Copyright Wiley-VCH Verlag GmbH & Co. KGaA. Reproduced with permission)

as vacuum induction melting inert gas atomization (VIGA). This process is recommended for superalloys so as to avoid, in particular, oxygen pick-up when working with alloys with reactive elements such as Ti and Al. The crucible has a channel in its lower part, the orifice of which is closed by a vertical bar (stopper rod, not shown in the figure) during the heating phase; the bar is then raised to allow the liquid metal to flow. For larger volumes, more than 200 L (1.6 tons of steel, for example), the melting takes place in an additional module; the melt is gradually poured from a primary furnace into an intermediate tank, which feeds, via a channel, the interaction zone with the atomization gas. The actual atomization occurs in the *atomization chamber* at a pressure close to atmospheric pressure. Metal droplets, resulting from the small sheets and ligaments created by the shearing of the liquid metal by the gas (see insert in Figure 7.6), are formed in the metal/gas interaction zone. They are shapeless at the beginning, but under the action of the liquid/gas surface tension, they spheroidize during their drop, while they are still liquid, and finally solidify and reach the walls or the bottom of the chamber, which is relatively cold. Several configurations are possible. In Figure 7.6, the heaviest (largest) particles fall by gravity into a container at the bottom of the installation. The gas, carrying the lightest (finest) particles, exits through a side duct connected to a cyclone, which has the function of separating the particles from the gas.

The choice of the atomizing gas affects the properties of the powder and the process cost. Three types of gas are generally used: noble gases (most often argon and sometimes helium), nitrogen, and air. The latter two can react with some alloys in the liquid state, leading to nitriding or oxidation of the particle surface. Fine particles are more contaminated with O and N than coarse particles because of their larger specific surface area.

Plasma atomization (PA): This was introduced in 1996 as a unique metal powder preparation technology by AP&C in Canada. The process is capable of producing extremely pure batches of fine metal powders with near-perfect spherical-shaped particles through melting droplets at the tip of a wire. Three plasma torches are symmetrically positioned at the top of the melting chamber to create a high-temperature plasma focus, which leads to superheating of the material, allowing particles to spend a longer duration of time in the melted phase to complete the spheroidization process. The feeding apparatus feeds the metal wire into the plasma focus, where it is rapidly melted or vaporized and scattered by the plasma's high-speed impact. The flying deposition process in the atomization tower exchanges heat with the cooled argon gas introduced into the atomization tower and cools and solidifies into an ultrafine powder. The principal diagram of the equipment is shown in Figure 7.7. Because of the high temperature of the plasma torch, the PA method can produce all high-temperature metal alloy powders. However, as it uses wire feed atomization to make powder, it limits the usage of more difficult-to-deform alloy powders, such as titanium and aluminum, and increases the cost of powder production.

Plasma Rotating Electrode Method (PREP): This is a centrifugal spherical powder preparation process developed in Russia. The principle is shown in Figure 7.8. The metal or alloy is processed into a bar, and the end of the bar is heated by plasma.

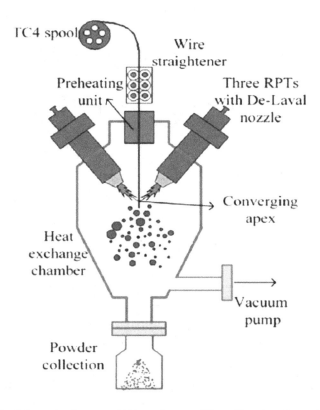

FIGURE 7.7 Schematic diagram of PA equipment. (From Shanthar, R. et al., *Advanced Engineering Materials*, 25, 2300375, 2023)

At the same time, the bar is rotated at high speed, relying on centrifugal force to refine the molten droplets and solidify them in an inert gas environment under the action of spheroidization to form a powder. The PREP method is suitable for the preparation of alloy powders such as titanium alloys and high-temperature alloys. The metal powder prepared by this method has high sphericity and good fluidity, but the powder particle size is relatively coarse. The yield of fine-grained (0–45-μm) powder used in the SLM process is low, and the cost of fine powder is relatively high. Increasing fineness of the powder (the droplet size) mainly depends on increasing the speed of the bar or its diameter, which inevitably puts higher requirements on equipment sealing and vibration, leading to higher costs.

7.4.2.2 Comminution (Solid State Reduction)

This was for a long time the most widely used method for the production of iron powder. Selected ore is crushed, mixed with carbon, and passed through a continuous furnace, where reaction takes place, leaving a cake of sponge iron, which is then further treated by crushing, separation of non-metallic material, and sieving to produce powder. Since no refining operation is involved, the purity of the powder is

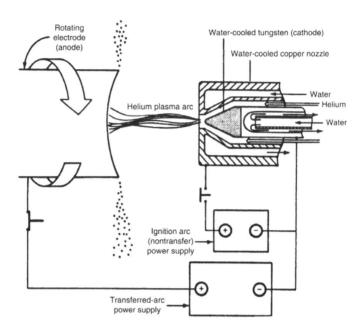

FIGURE 7.8 PREP device schematic diagram. (From Shanthar, R. et al., *Advanced Engineering Materials*, 25, 2300375, 2023)

dependent on that of the raw materials. The irregular sponge-like particles are soft and readily compressible, and give compacts of good green strength. Refractory metals are normally made by hydrogen reduction of oxides, and the same process can be used for copper. There are several plants producing powder by the reduction of iron oxide (mill scale) by means of hydrogen or carbonaceous material such as coke (EPMA, 2008, www.epma.com).

7.4.2.3 Mechanical Alloying

This is a relatively recent technique, developed in the 1960s by crushing mixed powders of two or more pure metals in a ball mill. Under the impact of the hard balls of the mill, the metal particles are distorted and fragmented, and the generated heat agglomerates fine particles, achieving atomic-level alloying in solid state. This technique has now been proved to be capable of synthesizing a variety of equilibrium and non-equilibrium alloys starting from blended elemental or pre-alloyed powders. It can mix brittle and ductile powders of different materials to produce entirely new alloys and compounds, which do not normally form at room temperature or may be immiscible by conventional alloying methods. Notably, mechanical alloying does not undergo the cooling process from liquid to solid phase, which can avoid the solute separation. Immiscible metallic systems provide new opportunities for tailoring material properties due to their unique microstructures, which enables obtaining desirable combinations of properties and opens up new opportunities for the customization of material performance. Typical examples of ductile–ductile powder materials mechanically alloyed include Fe–Cu, Cu–Ag, and Cu–Cr immiscible systems (Shuai, et. al. 2021).

7.4.2.4 Electrodeposition of Metal Powders

An electrolytic cell is set up, in which the source of the required metal is the anode, which is slowly dissolved under the applied current, transferred through the electrolyte, and deposited on the cathode (Figure 7.9). The deposit takes the form of a loosely adhering powdery or spongy layer that can be easily disintegrated into fine powder, or a dense, smooth, brittle layer is deposited, which should be removed by crushing, then washed and dried, giving a very high-purity metallic powder.

Copper is the main metal to be produced using this technique, but chromium and manganese powders are also produced by electrolysis. Electrolytic iron powders can also be prepared, but currently, atomization is used to get the required powder characteristics.

7.4.3 Powder Characterization

Metal powders are characterized by the particles' chemical composition, geometric features, including particle size and distribution, shape (morphology), and surface area, as well as inter-particle friction and flow characteristics, density and porosity (packing), and surface films.

1. *Chemical composition*

Regarding the chemical composition of the powder, main alloying elements are very important, but it is also essential to take into account interstitials, such as oxygen, nitrogen, carbon, and sulfur, as well as trace elements and impurities. All these constituents may significantly affect material properties. Measurement techniques for the chemical composition are ICP (inductively coupled plasma) spectroscopy or spectrometry. With the gas atomization process, all powder particles have the same chemical composition, but finer particles tend to have a higher oxygen content due to the higher specific surface. It should be noted that chemical composition can change slightly after multiple uses in AM machines.

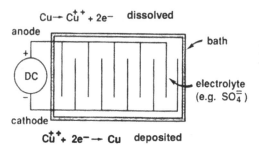

FIGURE 7.9 Electrolytic cell for manufacturing copper powder. The resulting powder usually takes the shape of a dendritic structure. (From EPMA, 2008, www.epma.com)

2. *Particle size and size distribution*

The most common method to define the particle size is to screen the powder through a series of sieves with different mesh sizes. The higher mesh size indicates a smaller size of particles (mesh size 200 represents a screen with 200 openings in an inch). The screening method has a practical upper limit of 500 (corresponding to particle size of 25 μm) due to the difficulty of making a finer screen and because of tendency for agglomeration of finer powders. Another method for examining finer sizes is microscopy. Optical microscopes can be used for measuring particles of a small diameter of 0.5 μm, and electron microscopes can measure particle diameters of 0.01 μm.

The particle size distribution (PSD) of metal powder is essential to ensure there is uniform spread within the machine bed and the surface of the final product is not rough. Particle size analyzers are used for measuring size distribution. Common analyzers use a laser diode and a charge-coupled device (CCD) detector for high sensitivity, resolution, and reproducibility of the particle size measurement. They can cover a size range from 0.3 nm to >3 mm for different applications. These analyzers give an index of the PSD, indicating what sizes of particles are present in what proportions (i.e. the relative particle amount as a percentage of volume where the total amount of particles is 100%) in the sample particle group to be measured.

There are two main types of PSD, depending on the AM technology and equipment:

- Powders usually below 50 μm for most powder bed systems. In this case, finer powder particles below 10 or 20 μm should be avoided, as they are detrimental to the powder flowability.
- Powder between 50 and 100 to 150 μm for EBM and LMD technologies.

The PSD is a major point in AM, as it can influence many aspects, such as powder flowability and ability to spread evenly, powder bed density, energy input needed to melt the powder grains, and surface roughness of the build. Table 7.5 represents the particle sizes and ranges of size distribution in the commonly used powder AD processes.

3. *Particle shape (morphology)*

Metal powder shapes can be categorized into various types, as shown in Figure 7.10. Spherical and rounded types are represented by the diameter. Other shapes are described in terms of the aspect ratio or shape factor (index). Aspect ratio is the ratio of the largest to the smallest dimension of the particle, from a minimum ratio of 1 for spherical particles to a maximum of 10 for flaky or acicular particles. The shape factor (index), K_{sf}, is a measure of the surface area to the volume of the particle, with reference to a spherical particle of equivalent diameter.

$$K_{sf} = \frac{Asp}{vp}D \tag{7.1}$$

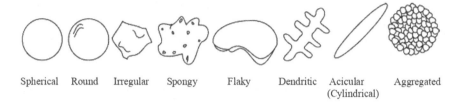

Spherical   Round   Irregular   Spongy   Flaky   Dendritic   Acicular   Aggregated
(Cylindrical)

FIGURE 7.10 Possible particle shapes in PM. (From Youssef, H. et al., *Manufacturing Technology; Materials, Processes and Equipment*, CRC Press, California, United States, 2023. With permission from CRC Press)

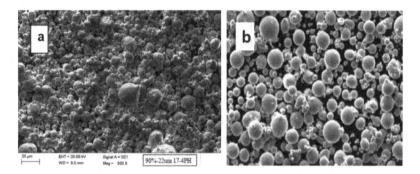

FIGURE 7.11 (a) SEM picture of gas atomized 17-4PH (Stainless Steel) powder <20 μm, (b) SEM picture of VIM gas atomized Pearl® Micro Ni718 powder. (Panel (a) courtesy of Sandvik Osprey Ltd; panel (b) courtesy of Erasteel)

For a sphere with surface area $As = \pi D^2$ and volume $v = \pi D^3/6$, where D is the diameter of the particle, $K_{sf} = 6$, and for other particle shapes, K_{sf} is larger than 6. A higher shape factor means higher surface area for the same total volume or weight of metal powders, and correspondingly, a greater area for surface oxidation to occur. Small powder size also leads to more agglomeration of particles, which is a disadvantage for automatic feeding of the powder through nozzles. Spherical particles are generally preferred, as they pack together more efficiently than other shapes for uniform powder bed density and a better-quality final product.

The recommended particle morphology for AM is the spherical shape, because it is beneficial for powder flowability and also to help in forming uniform powder layers in powder bed systems. The powder morphology can be observed by scanning electron microscope (SEM) (Figure 7.11).

4. ***Inter-particle friction and flow characteristics:***

Friction between particles restricts the ability of powder to flow willingly, thus leading to a higher pile when the powder is poured freely from a narrow nozzle. A practical measure of inter-particle friction is the ***angle of repose*** (side-to-base angle) for this pile. A smaller particle size indicates greater friction and thus, a larger angle of repose. Spherical shapes lead to the lowest friction and smallest angle. Higher shape factors lead to higher friction. A common measure of flow is the time required for a certain weight of powder to flow through a standard-sized nozzle. Larger flow times demonstrate more difficult flow and higher inter-particle friction. Lubricants are usually added to reduce inter-particle friction and facilitate flow. Dynamic image analyzers are able to record images of the particles in motion. Both particle size and up to 28 different shape parameters can be examined in real time.

5. *Density and packing*

There are two measures of density in powder-based manufacturing. *The bulk density* is the density of the powders in the loose state. This includes the effect of the pores between particles. *The true density*, on the other hand, is the density of the consolidated product. The true density is larger than the bulk density. The ratio of the bulk density to the true density is known as the *packing factor*, and thus,

$$\text{Packing Factor} = \text{Bulk Density} - \text{True Density} \qquad (7.2)$$

Typical values of the bulk density for loose powders range between 0.5 and 0.7. This packing factor depends on the particle shape and the distribution of particle sizes. The packing factor can be increased by large variation of powder sizes.

Porosity represents an alternative means of assessment of the packing factor of a powder. It is defined as the ratio of the volume of pores (empty spaces) in the powder to the bulk volume. Therefore,

$$\text{Porosity} = 1 - \text{Packing Factor} \qquad (7.3)$$

Another common name used in AD is the TAP density, which is a measure of how well the powder particles pack together. There are TAP density analyzers that are non-destructive and are also able to show the total porosity of the metal powders when used together. Porosity content can be evaluated either by SEM observation or by helium pycnometry. The presence of excessive amounts of large pores or pores with entrapped gas can affect material properties.

The porosity of a metal powder bed has a large influence on the mechanical strength of the final product. Some applications, such as artificial bone implants, should be light and match the high porosity of the surrounding bone area while still maintaining mechanical strength.

6. *Surface films*

Metallic powders are classified as either elemental, consisting of pure metal, or pre-alloyed, where each particle is an alloy. Surface films represent an obstacle in AM technology because of the large area of powders per unit weight of metals. Possible surface films include oxides, silica, absorbed organic materials, and moisture. Such films necessitate extra processes for removal prior to processing.

7. *Surface topography*

Agglomeration of particles, surface roughness, and irregularly shaped particles are some of the factors that impede efficient powder flow during sintering. Surface roughness also has a strong correlation with fatigue strength in the final metal object. SEM and X-ray analysis can both examine the surface topography and provide elemental composition analysis.

8. *Other powder physical properties*

Rheological properties are very important for metal powders used in AM equipment, both for powder handling from powder container to working area and in the case of powder bed systems to form uniform layers of powders.

Rheology is a complex matter, but some standard test methods are available, though not always fully appropriate for the particle sizes typical of AM systems: density (apparent or tap), flow rate, and angle of repose.

Additional points are important to consider when selecting metal powders for AM processes:

- Storage, handling, and aging of powders
- Reusability of powder after AM cycles
- Health, safety, and environmental issues.

The ASTM F3049-21 Standard Guide for Characterizing Properties of Metal Powders Used for Additive Manufacturing Processes is published jointly by the international standards organizations ASTM and ISO (ASTM 2014).

7.4.4 Common Metal Powders Used in AM

Powders of metallic materials such as titanium alloys, steels, some grades of lightweight metal alloys (Al and Mg), Ni-based alloys, etc. are highly compatible with AM systems. In this section, the most commonly used powders are presented.

Ferrous alloys
Different types of steels (austenitic, precipitation hardened, martensitic, duplex, etc.) are widely used ferrous alloys and are processed via Powder bed fusion (PBF)–laser and DED–laser AM techniques. Grades of austenitic stainless steels including 304-, 316-, 304L, and 316L AISI types are most commonly used. EBM technology uses powdered stainless steel to produce dense, super-strong, waterproof parts for extreme environments like jet engines, rockets, and even nuclear facilities. In biomedical applications, Maraging Steel (Fe–Mn alloy) has been used extensively to produce bone scaffolds by SLM. Mn is added to control the high degradation of Fe. Moreover, an Fe–HA (iron–hydroxyapatite) composite has been manufactured using different particle sizes to achieve better corrosion rates and closer mechanical properties to those of bone. For example, the tensile strength of pure iron is 215 MPa,

whereas for Fe + 2.5 wt% HA (1–10 μm), it is 117 MPa, which is close to that of the human femur bone (135 MPa, longitudinal tension).

AM produces fine-grained steel components compared with conventional manufacturing techniques due to rapid solidification along with non-equilibrium conditions. Heat treatments are generally applied to AM-produced steels to achieve desirable properties.

Titanium alloys

Titanium alloys are among the most commonly researched AM materials. These alloys have excellent properties in terms of high strength-to-weight ratio, good fracture and fatigue resistance, good corrosion resistance, and formability, due to which they are widely utilized in the aerospace, automobile, and biomedical sectors. One of the most popular titanium alloys used for part fabrication via PBF and DED routes is Ti6Al4V due to its compatibility with numerous biomedical applications. During AM fabrication of parts using this alloy, the printing environment should be carefully selected, especially the alpha and beta phases of Ti, to achieve the required properties. Titanium alloys are also used in AM to produce a wide range of industrial components, including blades, fasteners, rings, discs, and vessels. Titanium alloys are also used to produce high-performance race engine parts like gearboxes and connecting rods.

Aluminum alloys

Aluminum (Al) alloys are widely utilized in various engineering sectors due to their good strength-to-weight ratio and corrosion resistance. Aluminum is sintered in the DMLS process or melted in the SLM process. Fine detail down to 25 microns and wall thicknesses of as little as 50 microns are possible when aluminum is used. Parts typically have a textured, matte surface, which distinguishes them from traditional manufactured aluminum parts. Due to the geometrically complex structures possible with AM, further weight reduction is often possible with little or no compromise in strength and overall performance. A typical Al alloy, AlSi10Mg Silumin, is used in the DMLS technology applied for thin-walled parts with complex geometry. The alloy offers good strength, hardness, and dynamic properties, and consequently, is used in the production of parts subjected to high loads.

On the other hand, the AM of Al alloys is still limited due to the poor weldability and low laser absorption of Al alloys. Another reason is that Al alloys are melted during the fusion-based AM process, risking the solubility of hydrogen, which could be entrapped, leading to the formation of pores. These solidification-related defects weaken the mechanical properties of the manufactured part. To avoid these issues, the process zone should be shielded using additional shielding gas.

Magnesium alloys

Magnesium alloys are promising materials for use as degradable biomaterials, having similar stiffness to bone, which can minimize the stress-shielding effects. The applications of Mg alloys are increasing at a rapid rate, including orthopedics and other medical specialties. AM of Mg alloys is attracting interest due to their ease of design as compared with traditional manufacturing techniques. AM has the capability to develop biodegradable implants. Some AM processes, such as PBF, SLM, electron beam melting [EBM]), face the problem of oxidation and evaporation of Mg during processing. However, this difficulty can be overcome by printing Mg alloy in an inert atmosphere with optimized process parameters. In such cases, indirect AM processes are playing an important role in developing biodegradable Mg alloys.

Cobalt–chrome alloys

AM parts are fabricated from cobalt–chrome alloys like ASTM F75 CoCr when excellent resistance to high temperatures, corrosion, and wear is critical (typically in the aerospace industry). It is an appropriate selection where nickel-free components are required, such as in orthopedic and dental applications. Medical implants produced from cobalt–chrome metal powder possess the hardness and biocompatibility necessary for long-term performance. Cobalt–chrome alloys are used in AM to print parts that often benefit from hot isostatic pressing (HIP), which combines high temperatures and pressures to induce a complex diffusion process that strengthens grain structures, producing fully dense metal parts. The CoCr alloys most often employed in medical applications are Co–Cr–Mo, Co–Ni–Cr–Mo, and Co–Cr–W–Ni.

Nickel-based alloys

Nickel chromium superalloys like Inconel 718 and Inconel 625 produce strong, corrosion-resistant metal parts. These alloys are often used in high-stress, high-temperature aeronautical, petrochemical, and auto racing environments. The mechanical properties of nickel-based alloys used in AM processes, such as Inconel 625, are considerably enhanced by the use of significant amounts of nickel, chromium, and molybdenum in the metal. It resists pitting and cracking when exposed to chlorides. Inconel 718 is an age-hardened version of 625. The hardening process generates precipitates that better secure metal grains in place. Inconel 718 is a metal that is also highly resistant to the corrosive effects of hydrochloric acid and sulfuric acid. It also demonstrates excellent tensile strength and good weldability. Although nickel is very toxic, a titanium oxide layer is formed that prevents nickel oxidation. An example of Ni–Ti alloy is Nitinol, which contains approximately 50% of Ni and 50% of Ti. Nitinol is a shape-memory alloy, which retains its original shape after severe deformations. It is used for hard tissue implants and in dentistry as well as many other industries.

Precious metals

It is possible to sinter powdered gold, silver, platinum, and palladium for AM in SLM processes. Extremely fine metal powder is partially melted to create jewelry. Unique and beautiful pieces of jewelry feature interlocking or interwoven designs only possible with additive manufacturing. Forbes profiled a jeweler creating one-of-a-kind items with six-figure valuations, customized to customer preferences. To commemorate a Parisian honeymoon, one client requested and received a charm featuring a 3D-printed Eiffel Tower perched atop a pearl.

TABLE 7.8

Types and Main Uses of Metal Materials for AM Applications

Metal type	Main alloy and number	Main use
Iron and steel materials	Stainless steel (304 L, 316 L, 630, 44°C), hemp aging steel (18NI), tool steel, mold steel (SKD-11, M2, H13)	Medical equipment, precision tools, forming molds, industrial parts, art products
Nickel-based alloy	Superalloy (IN625, IN718)	Oxygen turbine, aerospace parts, chemical parts
Titanium and titanium-based alloys	Titanium metal (CPT), titanium alloy (Ti–6Al–4V alloy), Ti–Al, Ti–Ni alloy	Heat exchangers, medical implants, chemical parts, aerospace parts
Cobalt-based alloy	F75 (C0–Gr, Co–Cr–Mo alloy), Super alloy (HS188)	Dental crowns, orthopedic implants, aerospace parts
Aluminum alloy	Al–Si–Mg alloy (6061)	Bicycle and aerospace parts
Copper alloy	Bronze (Cu–Sn alloy), Cu–Mg–Ni alloy	Forming molds, marine parts
Precious metals	18k gold, 14k gold, Au–Ag–Cu alloy	Jewelry, art products
Other special metals	Nonquality materials (Ti–Zr–B alloy), liquid crystal alloy (Al–Cu–Fe alloy), multielement high-entropy alloy, biodegradable alloy (Mg–Zn–Ca alloy)	Still in the development and research stage, mainly used for industrial parts, precision molds, auto parts, medical equipment, etc.
Conductive ink	Ag, etc.	Used in inkjet printing electronic equipment

Table 7.8 presents different powder metallic materials, specific alloys, and their main applications

7.4.5 Wire Feedstock Materials

Metal wires for AM wire feed processes are manufactured by wire drawing processes to reduce the diameter of a rod or larger-diameter wire by pulling it through a converging die. These are usually cold forming processes, which require high ductility of the material. To avoid internal defects in the wire, the limiting drawing ratio of one drawing process makes it inevitable that a large number of successive drawing processes will be required with intermediate annealing in between to reach the small diameter required for AM. Therefore, only highly ductile materials and alloys can be used as feedstock wires.

Nowadays, wire-fed DED AM is a promising process for the fabrication of metal components such as nickel alloys, aluminum, steel, and chiefly, titanium. Particularly, wire and arc additive manufacture (WAAM) processes can result in reduced fabrication and post-processing time in comparison with traditional processes for bulk production of large components. For instance, WAAM technology enabled nearly 90% of raw material saving in a Ti–6Al–4V external landing gear assembly. The unit wire material cost is nearly half that of the powder material. Common wire diameters in the range of 1 to 4 mm are available wound on spools, as shown in Figure 7.12.

FIGURE 7.12   Spooled pure titanium wire for use in wire-fed machines. (From www.perrymanco.com/)

7.5 FEEDSTOCK MATERIALS FOR AMC

7.5.1 Feedstock Forms for AMC

Like other printed materials, ceramics use the same as stock material forms. The main differences between the seven additive manufacturing technologies (AMTs) discussed in Chapter 3 are the feedstock material forms and how they are used to form layers. Briefly, vat photopolymerization (VP) uses liquid photopolymers that are selectively cured by light. With ME, filament materials are extruded from a nozzle or orifice and form a solid structure. As the name implies, PBF uses a material powder, which is selectively fused together. In MJ, ceramic-loaded liquid material is selectively jetted from a printhead to build parts. In contrast to MJ, BJ utilizes a liquid bonding agent to selectively deposit to bind powders together. With DED, powder or filaments are thermally fused together when they are being deposited. In LOM, material sheets are bonded together to form a layered structure.

According to their physical states, AM materials can be divided into liquid, powder, filiform, and sheet/plate states, etc., and the corresponding manufacturing technologies that they are adapted to are shown in Table 7.9.

7.5.2 Ceramic Powders and Their Processing

Commonly, ceramic materials can be divided into two main categories: traditional and advanced (or technical). Traditional ceramics are those made from natural materials, such as clay and sand. In contrast, advanced ceramics are typically synthesized using advanced manufacturing techniques, and examples include silicon carbide (SiC), silicon nitride (Si_3N_4), boron nitride (BN), aluminum oxide (Al_2O_3), zirconium oxide (ZrO_2), composites, and many others.

TABLE 7.9

Classification of Materials for AMC According to Physical States and Their Applied Technologies

Physical state	Representative materials	Applied technologies
Liquid: photosensitive resin	Epoxy resin, acrylic resin, liquid silica gel	VP
Ceramic powder	Alumina	3DP, SLS, SLM
Filiform	ABS, PLA, PA, PC	FDM
Sheet/plate	PC	LOM

Ceramic powders are made by crushing or grinding (followed by sieving)—the ceramic particles are mostly very irregular.

Particle shape is an important issue because:

- Irregular particles "key" together well and so, when pressed together, tend to adhere, leading to stronger "green" components than rounded particles.
- Rounded particles flow more easily than irregular particles and can pack together more uniformly and densely than irregular particles. For the highest packing density, use mainly rounded particles with a range of particle sizes (small ones to fill up the gaps between big ones). With a single size, highly controlled porosity can be achieved.

7.5.3 NECESSITY OF BINDERS FOR AMC

Most current AM processes focus on plastics and metals. Another type of engineering material, ceramics and ceramics-based composites, is severely limited for current AM processes. To overcome such material limitations, researchers have studied AM processes using high-performance ceramic powders. A developed method based on SLS is to use a high-energy laser beam to locally sinter ceramic particles such that they will bond with each other. However, due to the high sintering temperature of ceramics, a low-melting-point medium such as a polymeric binder is usually added with ceramic powders. As shown in Figure 7.13a, the low-melting-point particles serve to bind ceramic particles together in the SLS process. Finally, some post-processing procedures are performed to burn out the added binder and to fuse the ceramic particles.

Due to the good properties of SLA, such as high accuracy and good surface finish, efforts to apply SLA in ceramics fabrication have been made. In such an AM process, a photosensitive resin is used to bind ceramic particles together (Figure 7.13b). That is, liquid resin is mixed in ceramic suspension and selectively solidified by controlled light-induced photopolymerization. Consequently, green parts with different shapes can be fabricated using a ceramic suspension that is a mixture of ceramic powders and photosensitive resin. Post-processing of the fabricated green parts is needed to burn out the organics in the photosensitive resin and to fuse ceramic particles together in order to get dense ceramic components (Song et al. 2015).

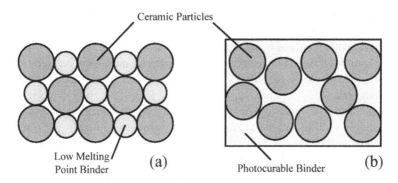

FIGURE 7.13 Different methods of bonding ceramic powders in AMC. (a) Low-melting point particles serve to bind ceramic particles together in the SLS process, (b) Photosensitive resin is used to bind ceramic particles together. (Reprinted from *Journal of Manufacturing Processes*, 20, 3, Song, X., Chen, Y., Lee, T.W., Wu, S. and Cheng, L. Ceramic Fabrication Using Mask-Image-Projection-based Stereolithography Integrated with Tape-casting, 456–464, Copyright (2015), with permission from Elsevier)

7.5.4 Feedstock Preparation for AMC

Unfortunately, every AM process has its own unique disadvantages in building ceramic or ceramic–metal parts. For example, SLA exhibits the best part surface finish among currently available techniques; however, it makes use of very expensive and a limited range of raw materials (photosensitive resins), while additional care must be taken because of the environmentally hazardous solvents used to clean up parts. In contrast, 3D printing achieves a limited surface finish of the manufactured parts yet allows the use of a wide range of feedstock materials (in principle, any material available in powder form). Moreover, 3D printing offers the possibility to control not only the shape of the ceramic parts but also the composition, microstructure, and properties throughout a component. Thus, there are still obstacles for these technologies to overcome before they become "everyday manufacturing methods" in place of "conventional" techniques. In spite of the fact that significant progress has been made in the development of metallic and ceramic feedstock materials, the most difficult limitations for AM technologies are the restrictions set by material selection for each AM method and aspects considering the inner architectural design of the manufactured parts. Hence, any future progress in the field of AM should be based on the improvement of the existing technologies or alternatively, the development of new approaches with an emphasis on parts allowing the near net formation of ceramic structures while optimizing the design of new materials and of the part architecture.

7.5.4.1 Feedstock Filament Preparation for FDM and RC

Feedstocks for extrusion free forming (EFF) should have the following common requirements:

- A high solid loading to counteract shrinkage and cracking due to sintering or binder burnout.

- A homogeneous particle distribution to ensure a constant flow and to avoid flaws due to agglomerates.
- Entrapped air should be minimized.
- Solvent migration and sedimentation of particles must be prevented.
- Suitable rheology to ensure shape retention and good welding of the filaments.
- Suitable particle size and distribution for the applied nozzle dimensions.
- Suitable solidification kinetics (vapor pressure, thixotropy, thermal conductivity).
- Suitable interface properties for adhesion and fusion of filaments.

1. Fused Deposition Modeling (FDM)

Figure 7.14 shows the filament-type FDM operation, which uses a ram extruder, with the filament pushing the softened material out of the nozzle. This technology is capable of producing dense structural ceramic parts. However, the surface finish is limited, and the staircase effect is an issue, as previously described. Also, there is a limitation on the maximum wall thickness that can be produced without the formation of cracks during the subsequent heat treatment.

In FDM, mostly, a dried and surfactant pre-coated powder (~1 μm) is incorporated into the organic binder vehicle at high temperatures (just above the melting

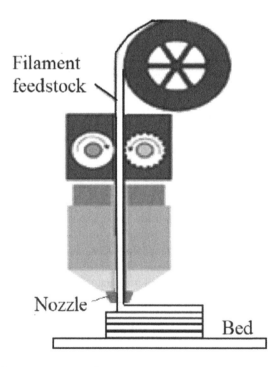

FIGURE 7.14 Schematics of the filament type fused deposition modeling of ceramics. (From Gonzalez-Gutierrez, J. et al., *Materials*, 11, 5, 840, 2018)

point of the polymer) with a tackifier to add tackiness and flexibility, wax to lower the viscosity and add stiffness, and a plasticizer to add flexibility. When homogeneity is achieved, the feedstock is extruded to form a filament, which fits into the liquefier of the FDM machine. Furthermore, shape retention by crosslinking the binder molecules can be achieved by mixing two components continuously in a mixing chamber shortly before exiting the nozzle while deposition is carried out (Table 7.10).

Figure 7.15 illustrates the extrusion-based AM process consisting of raw mixing, vacuum mixing, 3D printing, debinding, and sintering.

Special attention has to be paid to the mechanical properties of the feedstock. The wire-like extrudate must be wrappable on spools and rigid enough to act as a piston for the liquefier (buckling must be prevented). Most preparation routes for RC and FDC feedstocks follow the same principle. Most recently, RC feedstock recipes were improved by using a single additive (carboxymethyl-cellulose) as a dispersant and rheology additive. In FDM of ceramics, the nozzle diameter usually varies between 0.4 and 0.6 mm and the layer thickness from 100 to 300 μm. Also, nozzle temperature and printing speed are important parameters.

2. Robocasting (RC)

Robocasting, also called direct ink writing (DIW), is an AM based on ME in which a ceramic slurry (paste) is selectively dispensed through a nozzle (Figure 7.16). DIW is a low-cost and fast AM process, which produces parts with limited geometrical complexity, and poor resolution and surface finish.

RC necessitates developing suspensions of ceramic powder, for example zirconia, and aqueous solution with dispersant, binder, and coagulant agent. Another alternative is to use a pre-ceramic silicone as the binder to create scaffolds made of

TABLE 7.10
Feedstock Ingredients for Several FDC and RC Techniques

AM Process	FDM	RC
Particle size	~1 μm	30 nm to 2.2 μm
Particle vol fraction (%)	<55	<61
Solvent	–	Water/Low molecular weight (LMW) organic solvents (water, 2-propanol)
Dispersant	Oleates (steric), oleyl-alcohol, stearic acid	Polyelectrolytes (Polyacrylic acid [PAA]), steric dispersants
Binder/viscosifier	High-molecular-weight amorphous polyolefin, ethylvinylacetate	Cellulose derivatives, Polyvinylbutyral (PVB)
Other additives	High molecular weight partly crystalline wax, hydrocarbon resin (tackifier), elastomer, low-molecular-weight polyolefin (plasticizer)	Coagulant (polyelectrolyte/salt, pH modifiers); defoamer (octanol); crosslinkers

Source: Adapted from Travitzky, N. et al., *Advanced Engineering Materials*, 16, 6, 729–754, 2014.

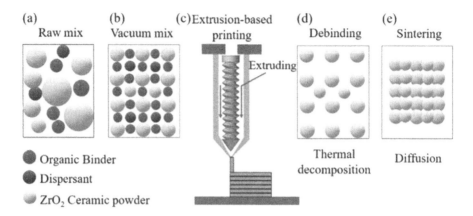

FIGURE 7.15 Schematic of mixing, extrusion-based AM and post-processing process: (a) raw powder mixture, (b)vacuum mixture, (c) extrusion process, (d) debinding, and (e) sintering. (Reprinted from *Ceramics International*, 46, 4, Yu, T., Zhang, Z., Liu, Q., Kuliiev, R., Orlovskaya, N. and Wu, D., Extrusion-based additive manufacturing of yttria-partially-stabilized zirconia ceramics, 5020–5027, Copyright (2019), with permission from Elsevier)

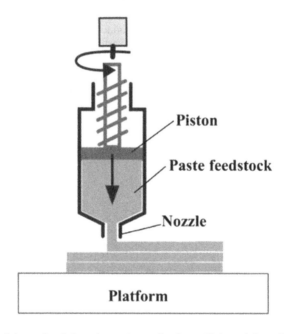

FIGURE 7.16 Schematic of the robocasting technology. (Adapted from Ruscitti, A. et al., *Cerâmica*, 66, 380, 354, 2020)

a variety of ceramic and glass materials with potential applications in bone tissue engineering. RC has been shown to be suitable for the fabrication of highly porous ceramic scaffolds and foams, with porosity that can exceed 80%. Gaddam et al. (2021) produced zirconia scaffolds using pre-ceramic polymer with microporosity of about 70%, an average compressive strength of ~236 MPa, and percentage shrinkage varying between 17% and 24%. The surface quality is a major issue, and roughness (Ra) above 20 μm has been reported.

The high-loaded ceramic slurry (typically 40–50 vol%) must have adequate rheological behavior and should be smoothly extruded through a narrow nozzle without clogging. Besides, it should be self-supporting (to avoid collapsing) and be able to retain its shape. Thus, suitable additives must be chosen. Hydroxypropyl methylcellulose and polyethyleneimine have been extensively used as viscosifying and coagulant agents, respectively. Also, a variety of commercial dispersants have been employed, depending on the selected ceramic powder.

In addition, the choice of nozzle diameter must be properly selected to avoid choking or sudden release of the slurry. The nozzle should be at least 15 times the size of the largest particle, and the optimum diameter is typically between 400 and 800 μm. In addition, the printing speed is a parameter that should be selected carefully. The printing speed should be adequate to form a continuous strut and to achieve interlayer bonding, and small modification of this parameter may result in vastly different outcomes.

Lynxter is the only manufacturer that provides a commercially available solution for advanced ceramics.

Lynxter 2022, and Rapidia 2022 are developing feedstocks of aluminum oxide and zirconium oxide. Lastly, systems focused on building and structures have emerged. WASP has systems able to deal with concrete mortar (Wasp 2022).

Rheology of FDM and RC: The most important property of a ME feedstock is the rheological behavior, which influences shape retention and extrudability. All ME feedstocks show a shear thinning behavior during deposition, mostly accompanied by a significant yield stress <1000 Pa. They are viscoelastic. Furthermore, these materials are thixotropic, which promotes filament fusion and requires additional considerations regarding delayed deposition and handling of extrudates. Viscosities range from 10 to 100 Pa·s in the extrusion process, while the shear elastic modulus ranges between 0.1 and 1 MPa.

Regarding FDC, temperature has a pronounced influence on rheology. The temperature must be optimized in order to promote welding of the filaments, which is also influenced by heat capacity and thickness of the filaments; thus, filaments thinner than 100 μm are difficult to handle.

In RC, the flow profile in the extrusion nozzle proceeds in a three-zone profile. According to Pfeiffer et al. (2021), a slip layer forms adjacent to the nozzle wall on which the filament glides. Thus, the filament itself is hardly sheared. A yielded zone forms next to the slip layer, which promotes filament fusion. At lower shear stresses present in the middle of the filament, a solid core is assumed to exist, which is beneficial for shape retention.

7.5.4.2 Feedstock Filament Preparation for SL, DLP, and 2PP

The photosensitive slurry (feedstock) has several requirements concerning ceramic loading (>40 vol%), proper rheological behavior (<3 Pa·s), stability, etc. Investigations have been carried out to study the formulation of photosensitive ceramic suspensions for VP. Such a slurry is composed of monomers, photoinitiators, ceramic powders, and additives such as dispersants, diluents, defoamers, plasticizers, and light absorbers. VP of ceramics can produce dense parts (>99%) with mechanical strength comparable to those of conventional methods. Using these slurries, zirconia with flexural strength greater than 700 MPa can be printed by VP. On the other hand, printing carbides and borides with this technology is challenging due to the high refractive index (RI) of these materials, which would cause light scattering due to the RI mismatch with the usual photosensitive materials, leading to poor resolution and reducing the photopolymerization reaction. Such materials are not yet commercially available. One alternative approach to deal with this issue is the use of polymer-derived ceramics (PDCs), which are converted into ceramics without the addition of ceramic particles. Adopting PDCs proved that SiOC, SiC, SiCN, and SiBCN can be produced by ceramic VP. This AM technology is relatively expensive. These feedstocks are costly because they depend on high-priced photosensitive materials.

Stability and rheological behavior of ceramic suspension:
Ceramic powder has negligible solubility in polymer solutions. To achieve a homogeneous dispersion of ceramic powder in polymeric solution with high ceramic powder loading, dispersants are needed. Many dispersants, such as quaternary ammonium acetate and Triton X-100, have been introduced to enhance these characteristics. Using quaternary ammonium acetate as dispersant allows 50 vol.% of alumina powder loading in Hexanediol diacrylate (HDDA) resin; However, without dispersant, 50 vol.% of alumina powder loading in HDDA resin results in a stiff, paste-like colloidal gel (Brady and Halloran 1997).

The viscosity of an SLA suspension is another crucial factor that affects the rheological behavior of the suspension, whose viscosity is usually larger than that of the photocurable resin. The viscosity of the suspension should be in the range of 2–5 Pa·s to guarantee satisfactory layer recoating. Therefore, decreasing the suspension viscosity is vital for successful ceramic SLA. Suspension with low viscosity is achievable by using the appropriate polymer, dispersant, and diluent (Hinczewski et al. 1998).

Cured depth and cured width:
Both are important parameters that control accuracy and processing time. Cured depth Cd can be theoretically calculated according to the Beer–Lambert law:

$$Cd = Dp \ln (E/Ec) \tag{7.4}$$

Where
Dp = Penetration depth, which depends on vol. fraction of ceramic powder, particle size, RI difference between ceramic and polymer solution.
Ec = Minimum energy of polymerization of the monomer.
E = Provided energy.

However, cured width Wc should be low enough to ensure high resolution and quality. SLA of ceramics is much more complicated than polymeric SLA due to the scattering phenomenon. A linear relationship was found between mean particle diameter and cured width Wc, with rates depending on the energy density (Jacobs 1992).

Polymer derived ceramic (PDC) and preceramic polymers (PCPs):

The polymer derived ceramic (PDC) technique employs liquid pre-ceramic polymers (PCPs) without the need for adding ceramic particles to the solution, making possible the decomposition of organic contents and their conversion into ceramic materials via pyrolysis (Colombo et al. 2010). The advantages of the PDC technique lie in the control of composition and microstructure through design at the molecular level. In this way, high ceramic yields can be achieved following sintering with the crosslinked active groups present in PCPs. Moreover, both the liquid and solid polymer PDC forms possess simple shaping and machinability, and their ceramic forms possess stability. This breaks the limitations of traditional ceramic materials and revolutionizes the ceramic preparation process.

At present, research into the SL and DLP-based 3D printing of PDCs is still in its infancy. There is still space for improvements in the mechanical properties and high-temperature stability of manufactured porous ceramics. Brigo et al. (2018) fabricated micro/nano SiOC structures by means of two photopolymerizations. More recently, Wang et al. (2019) demonstrated the DLP 3D printing of Si_3N_4 components using polysilazane-based polymers.

Figure 7.17 depicts the fabrication process of lattice-structured SiOC ceramic components, including the preparation of precursor resins, the design of 3D models with lattice structures, and DLP 3D printing and pyrolysis treatment.

Well-known pre-ceramic polymers (PCPs) contain a primary Si backbone and usually consist of C, O, N, B, and H atoms, such as polysiloxanes, polysilazanes, and polycarbosilanes (Colombo et al. 2010). Figure 7.18 shows the common Si-based polymers. These kinds of silicon-based pre-ceramic polymers can be converted to various types of ceramics, such as SiC, silicon oxide (SiO_2), silicon oxycarbide (SiOC), and silicon carbonitride (SiCN), after pyrolysis. The schematic chemical structure of the most used Si-polymer precursor for ceramics, the polysiloxane (PSO), is shown in Figure 7.19.

Compared with ceramic powders, pre-ceramic polymers offer much more flexibility for effectively fabricating ceramic components with complex geometric structures and shapes using a wide range of processes, such as casting, molding, and AMC. Pre-ceramic polymers can have different configurations/microstructures, which can affect the composition, microstructure, porosity, yield, and properties of fabricated ceramics. Common silicon-based pre-ceramic polymers, such as polycarbosilanes, polysiloxanes, polysilazanes, and polyborosilazanes, have already been adopted for AM processes to produce ceramic components.

AM processes have been adopted for fabricating advanced ceramic materials, such as CMCs, particularly AM processes using pre-ceramic polymers, because of their excellent processibility for forming complex structures and shapes. The CMCs are composites with reinforcement material embedded in a ceramic matrix. CMCs

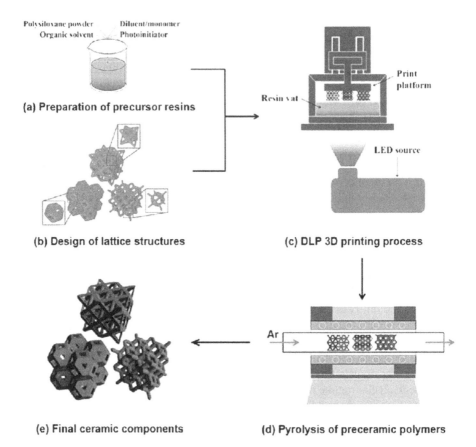

FIGURE 7.17 Additive manufacturing of PDCs: (a) preparation of photosensitive resins for the ceramic precursor; (b) design of CAD models with varied lattice structures; (c) DLP 3D printing of the as-prepared resin; (d) pyrolysis of the as-printed resin components in a tube furnace under an argon atmosphere; (e) as-pyrolyzed SiOC ceramic components with lattice structures. (From Ziyong, L. et al., *Additive Manufacturing of Lightweight and High-strength Polymer Derived SiOC Ceramics*, Taylor & Francis, Virtual and Physical Prototyping Taylor and Francis, 1–15, 2020. With permission from Taylor and Francis, Florida, USA.)

can overcome the known limitations of printed ceramics and offer improved properties because of the synergistic combination of properties from the reinforcement and matrix materials. Therefore, CMCs have many critical applications and are in high demand in various fields, including aerospace, defense, energy and power, electrical and electronics, and more. A recent report indicated that the global CMCs market size was expected to grow at a rate of 12.8% from 2023 to 2030 (Region and Segment Forecasts Report 2023). Typically, reinforcement materials (also known as fillers) in CMCs have one or more properties superior to the matrix. Combining a ceramic matrix with fillers achieves enhanced properties that exceed those of the individual constituents alone.

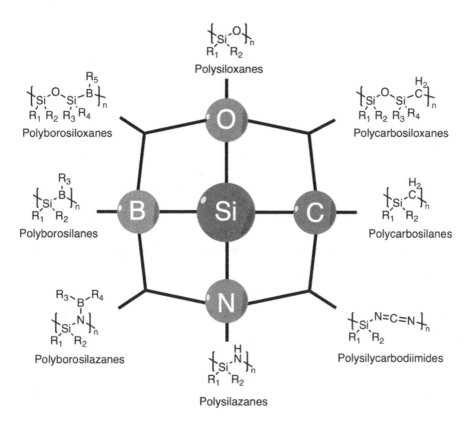

FIGURE 7.18 Main classes of Si-polymer precursors for ceramics. (Adapted from Colombo, P. et al., *J. Am. Ceram. Soc.*, 93, 1805–1837, 2013)

$$-\!\left[\underset{R}{\overset{R}{Si}}\!-\!O\right]_{x}\!-\!\left[\underset{OH}{\overset{OH}{Si}}\!-\!O\right]_{1\text{-}x}\!-$$

FIGURE 7.19 General chemical formula of polysiloxane. (From Ziyong, L. et al., *Additive Manufacturing of Lightweight and High-strength Polymer Derived SiOC Ceramics*, Taylor & Francis, Virtual and Physical Prototyping, Taylor and Francis, 1–15, 2020. With permission from Taylor and Francis, Florida, USA.)

Composites, including reinforcement materials at the micro size or larger, are usually called CMCs. For example, CFs, SiC fibers, or microscale particles have been extensively used to fabricate CMCs for aerospace applications. Composites using zero-, one-, or two-dimensional nanomaterials as reinforcements are well known as ceramic matrix nanocomposites (CMNCs). For example, nanoparticles (such as Y_2O_3, Al_2O_3, SiC, Si_3N_4, and nano diamonds); nanotubes and nanofibers (such as carbon nanotubes, nanofibers, and SiC nanofibers); and nanosheets (such as graphene and BN) have been used as reinforcement materials to fabricate CMNCs with improved mechanical, thermal, or electrical properties.

7.5.4.3 Feedstock Preparation for Binder Jetting (BJ)

Binder jetting, or the 3D method, is an indirect printing technique. It is the most widely used method for commercial applications. This technology is the most suitable for large parts. Moreover, it is capable of producing complex parts with overhanging structures due to its self-supporting powder bed. On the other hand, the ceramic particles should be large (>30 μm) to ensure flowability in the layer spreading, which decreases the density of the final part and provides a poor surface finish. Thus, BJ is best suited to porous parts; it is not adequate for structural parts. Maleksaeedi et al. (2014) showed that infiltration can also improve surface quality; the surface roughness (Ra) was reduced from 13.2 to 0.9 μm.

The binder must be compatible with the process, with viscosity around 10 mPa·s and surface tension from 23 to 30 mN/m. It is usually water-based with some additives for surface tension and viscosity adjustment, such as isopropyl alcohol, diethylene glycol, polyvinyl alcohol, and glycerol (Bui et al. 2022).

The remarkably high speed of 3DP is based on the inkjet principle and on the fact that no phase change of material is involved during the building process. Consequently, the combination of numerous potential materials and low-cost equipment together with outstanding speed means that 3DP offers great prospects for the future.

Since dimensional accuracy is directly influenced by material properties (particle size, pourability, and wetting behavior) as well as process parameters (nominal dimensions and build orientation), the 3DP technique still has some precision range compared with other AM methods. Fine powders produce a smoother surface yet become difficult to spread thoroughly on the working area. Powder bed stability can be optimized by moistening the powder during printing, resulting in enhanced surface roughness and geometric accuracy of the specimens (Butscher et al. 2013).

Desired characteristics of feedstock for BJ (3DP):
Desired characteristics, such as appropriate rheological properties, have to be satisfied for the binder solution to be successfully ejected through the printheads. Sintering is generally required to remove the organic binder so that the desired mechanical properties can be obtained. Typically, this post-treatment process also causes shrinkage of the part depending on the percentage of binder present. The building plate size ranges from a few centimeters to several meters. Print resolution is in the range from 50 to 150 μm. The dimensional accuracy of the whole processing chain (printing + debinding–sintering or infiltration) does not exceed 0.2 mm. The 3DP method was originally designed to rapidly produce components from a larger variety of materials than was possible by the other existing AM techniques available at that time, such as SLA, SLS, and LOM. The feedstock materials mentioned

included ceramics, metals, and plastics in powder form as well as their combinations. Known manufacturers of 3DP BJ machines compatible with ceramic powders are Voxeljet, Exone, 3DSystems, Johnson Matthey, and Desamanera. Tethon 3D and Additive Elements are providers of binders and ceramic materials especially dedicated to BJ technologies.

The most critical limitation of 3DP is that it works better with coarse powders, allowing the particles to be easily spread and allowing the liquid binder to quickly percolate down into the powder bed in order to glue the particles of the printed layer. Unfortunately, these coarse powders are more difficult to sinter afterwards, thus limiting potential applications for porous parts or infiltrated structures. However, lightweight and complex parts with high performance can be considered using this technology. Post-processing, such as depowdering and sintering, is another challenge of this method. Due to the low green density of the printed ceramic, depowdering can cause cracking of the part (Khalyfa et al. 2007).

7.5.4.4 Feedstock Preparation for Material Jetting (MJ)

Figure 7.20 presents a schematic of the MJ technology. It is a direct ink printing technique (DIP). The build plate (substrate) may be heated to begin to evaporate the solvent from the printed ink. In order to complete the solvent evaporation, in this process, a layer thickness of around 10 μm has been used. This technology is best suited for compact parts, as it has excellent resolution. In contrast, MJ has low productivity (~1 mm height per hour). 3D ceramic parts printed by MJ may form "coffee stains" in the drying step, in which solid particles segregate from the center to the edge of the printed patterns.

The accuracy of the dimensional resolution of this process is affected not only by the ink parameters (solid content, particle size, and viscosity) but also by the extrusion parameters, such as the extrusion rate, nozzle travel speed, and the distance between the nozzle and the previously deposited layers. Often, an ink additive is

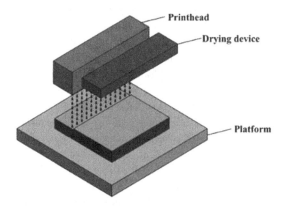

FIGURE 7.20 Schematic of the material jetting technology. (From Camargo I.L. de et al., *Cerâmica*, 68, 329–347, 2022. With permission from Associacao Brasileiro De Ceramica, Brazil)

used to prevent drying and clogging of the nozzles. This method is considered the only method among all additive processing techniques that is able to produce dense ceramic bodies without a post-processing treatment.

This technique has also been applied to create parts from wax-based alumina inks, demonstrating the suitability of low-melting-point waxes for hot-melt inkjet printing. A functionally graded composite was printed from zirconia and alumina inks by Mott and Evans (1999). Alcohol-based zirconia inks were also applied to manufacture complex-shaped geometries, which may be used as miniature heat exchangers. As well as creating dense parts, this technique can be used as a cost-effective and flexible way to manufacture coatings and devices from functional ceramics.

For this process, the feedstock is a suspension with well-dispersed ceramic particles in the liquid solvent, and with suitable stability, viscosity, and surface tension. Unlike the pastes used in RC technology, MJ feedstock is ink-based, ejecting a low-viscosity (~20 mPa·s) fluid with surface tension not exceeding 60 mN/m. To avoid clogging and blockage, the diameter of the nozzles should be 100 times bigger than the particle size. Consequently, nanoparticles are more desirable in formulating inks, since the nozzle diameters usually range between 20 and 30 μm. On the other hand, smaller particles are more likely to agglomerate. In addition to deionized water, some additives such as dispersants (triethanolamine, PEG, polyacrylic acid, glycerol, and ethanol) are added to the suspension to adjust the ink properties and performance (Qu et al. 2021).

Characteristics of ceramic inks for IJP:
The performance of IJP of ceramics very much hinges on the critical factors of the ceramic powder and ink formulation as well as their properties, in particular the rheological characteristics such as dispersivity, stability, viscosity, and surface tension. In addition, a moderate pH value must be maintained to prevent possible corrosion of the jetting system by the ink. A uniform PSD with particles less than 1/100 of the nozzle diameter (at the micron scale) can prevent the clogging and blockage of nozzles and capillaries, as required by the printer manufacturers (Kosmala et al. 2012). Thus, adequate filtering is generally applied to remove larger particles. In general, nanoparticles are more desirable in formulating inks. However, smaller particles are more likely to undergo agglomeration within the inks. Therefore, homogeneous dispersion of the ceramic powder within the ink is a crucial prerequisite for the ink to pass smoothly through the printhead nozzles.

The ejection behaviors of ceramic inks are largely determined by their viscosity. Insufficient jetting or too high a velocity may result from too large or too small a viscosity, respectively (Peymannia et al. 2015). Ceramic inks often possess low viscosities, down to a few mPa·s, as a result of low solid loading, causing a lengthy drying time and considerable shrinkage, which could adversely affect the final accuracy of the printed part. An increase in solid loading might be beneficial, but this can lead to changes in the ink's rheological behavior. Therefore, various optimizations and compromises need to be made to ensure that proper solid loading and rheological properties are achieved. Seerden et al. (1989) and Reis et al. (1998) showed that the use of alumina wax ink could minimize drying shrinkage, while good sintering density remained difficult to achieve.

A quality ink that is compatible with a DOD inkjet printer should exhibit proper printing performance, which can then be described as the "printability" of the ink. A quantitative characterization based on the ink's physical properties in terms of printability was developed by Fromm (1984).

Another major concern related to ceramic inks used in IJP is the coffee stain effect that can occur in the course of drying as-printed patterns (Deegan et al. 1997). This effect appears as a segregation of solid particles from the center to the edge of the printed patterns on the substrate. This problem can lead to defects in the printed structure and is mostly observed for inks with low solid volume fractions. Measures have been taken by many researchers to minimize such effects (Friederich et al. 2013).

7.5.4.5   Feedstock Preparation for Selective Laser Sintering (SLS)

In SLS, the model is sliced into layers with thicknesses typically in the 100 µm range, depending on the material used and its interaction with the laser beam, and the part is built up layer by layer. This enables fabrication of very complex-shaped parts, which are not reproducible using other shaping processes. A well-known challenge is the induced thermal stresses during SLS due to high heating and cooling rates. Because of the limited thermal shock resistance of ceramic materials, thermal stresses during SLS can lead to crack formation in sintered parts. However, powder bed preheating can reduce thermal stresses and thus, crack formation in ceramic parts produced using SLS.

Binder and powder characteristics for SLS:
Binder and powder characteristics such as powder packing density, particle size, powder flowability, powder wettability, layer thickness, binder drop volume, and binder saturation, as well as drying time and heating rate, play crucial roles in the success of the printing process. Powder packing density is the relative density of powder after spreading into the build bed. Binder drop volume is the volume of binder drop as released from the nozzle. Binder saturation S is calculated according to an equation in Zocca et al. (2013). Tarafder et al. (2013) reported that high binder saturation can cause binder spreading over multiple layers of powder as well as bumping appearance in build layers, whereas low binder saturation leads to layer displacement and/or unhardenable parts.

Powder flowability is another critical parameter that depends on powder particle size, PSD, particle shape, and surface roughness. Flowability is enhanced by using large particle size, while the densification of the part and the sinterability of ceramic particles are compromised due to the smaller surface area of the particles. On the other hand, using fine particles causes severe agglomeration (Hausner 1981). Similarly, powder wettability is related to particle characteristics such as chemistry and surface energy. Low wettability causes weak integration between the ceramic particles and the binder; however; high wettability results in binder spreading among different layers (Uhland 2001).

To date, extensive studies have been conducted on a number of aspects, including ceramic powder and binder properties, binder–powder bed interaction, and process parameters. A study focusing on the investigation of binder performance (Moon et al. 2002) reported that rheological properties, such as the surface tension and viscosity of the binder, should be optimized to appropriate levels for smooth jetting and high dimensional accuracy.

There are two types of SLS:

1. **Direct SLS**

   Bertrand et al. (2007) fabricated ceramic parts from pure zirconia (YSZ) powder by direct SLS. They found that powder bed density, which is a function of powder size and shape, has a significant influence on the SLS process. The higher the density of the powder bed, the higher the density of the ceramic that can be achieved. The achieved densities were in the range of 56%.

2. **Indirect SLS**

   With this method, primary particles are bonded through the binder phase and not directly, so the process is called indirect selective laser sintering. After SLS, a post heat treatment (e.g. in a furnace) is needed for full densification of the produced parts. A distinction can be made between two groups of binders:

   • *Organic binders.* Thermoplastics are used as organic binders. Deckers et al. (2012) used polyamide (PA) as the binder phase for indirect SLS of alumina parts. Alumina and PA powder (22 wt%) were mixed together in a ball mill. The powder bed was pre-heated to temperatures slightly below the melting point of PA (T_m = 179°C). SLS was performed under an N_2 atmosphere to prevent oxidation of PA. It was found that specific laser energy is required to achieve manageable green parts. Energy that is too low leads to fragile parts, and energy that is too high leads to polymer degradation. Due to low powder bed density, the sintered part density that can be achieved is limited. In order to achieve higher green densities, the laser-sintered samples were cold isostatic pressed for 1 min at 200 MPa. This made it possible to double the green density. After cold isostatic pressing, the samples were debinded and finally sintered in a furnace at 1600°C.

   • *Inorganic binders.* Inorganic binders for indirect SLS are typically metal phosphates, metal silicates, and metal borides. The working principle is the same as for organic binders, but one must consider that inorganic binders cannot be burnt out and thus remain as a secondary phase in the final product.

A lot of work has been done on ceramic composite materials. Examples include ceramic–glass composites and ceramic–polymer composites. Here, the low-melting glass/polymer provides densification due to the liquid phase formation. Most of these studied material systems are biocompatible and/or bioactive and designated for bio-medical applications that require biocompatibility of materials and complex shapes of products as key mechanical properties.

The latest developments in the field of SLS were adjusted to enhance the resolu-tion of produced parts from about one hundred microns to a few tens of micrometers. This was the development of selective laser micro-sintering. To reach the intended resolution, sub-micron ceramic powder and near infrared lasers with a wavelength of about 1 μm were used. In addition to the known problems of SLS of ceramics, poor absorption of ceramics in the near infrared wavelength spectrum was also observed.

7.5.4.6   Feedstock Preparation for Laminated Object Manufacturing

LOM is a process whereby sheets of material are laid on top of each other with a layer of adhesive material between the layers to help their interconnection. The bonding of subsequent layers is ensured by mechanical compression or thermally. This layered structure is cut with the help of a laser beam to achieve the desired final shape. In the case of ceramics, tape casting is used to prepare sheets of material.

The low-temperature lamination process prevents distortion and deformation due to low thermal stresses. The material, machine, and process costs are low compared with other SFF techniques. In relation to part sizes up to a volume of $500 \times 800 \times 500$ mm^3, the accuracy over the work envelope is 0.25 mm.

A disadvantage of LOM is the decubing of the parts, which prevents the cre-ation of very detailed surfaces. Additionally, in post-processing, the organic binder material has to be removed between and within the single layers, which results in anisotropic mechanical properties. LOM of ceramics began with Al_2O_3 tapes, which exhibited mechanical properties comparable to pressed samples. Bending strength was also discovered to be isotropic.

LOM has focused on using ceramic tapes as building materials. TheAl_2O_3 tapes, consisting of powder, a styrene–acrylic latex binder, water, and some additives, show good ability for LOM processing due to their high green strength, which facilitates the coiling of the tape, and their lamination behavior. Figure 7.21 shows the top view of a laser cut LOM-processed Si–SiC gear wheel manufactured from SiC-filled pre-ceramic paper and a representative SEM image showing the polished cross section of the fabricated Si–SiC composite.

7.6   REVIEW QUESTIONS

1. Explain, using a schematic diagram, a simplified free-radical photopolymerization.
2. Explain what is meant by vat photopolymerization.
3. List the processes that adopt the vat photopolymerization technique.
4. What are the main types of polymeric materials?
5. Mention AM processes that use resin-based polymers.
6. What are the main materials supplied in the form of filament-based polymer?
7. Mention AM processes that use filament-based polymer.

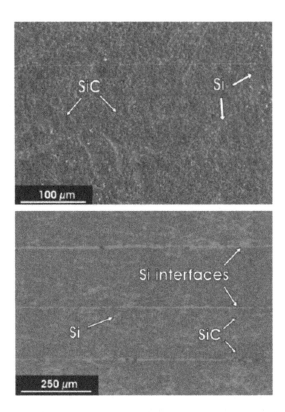

FIGURE 7.21 Top view of laser cut LOM-processed Si–SiC gear wheel manufactured from SiC-filled pre-ceramic paper and representative SEM image showing the polished cross section of the fabricated Si–SiC composite (dark phase is SiC, bright phase is Si). (From Windsheimer, H., Travitzky, N., Hofenauer, A. and Greil, P: Laminated Object Manufacturing of Preceramic-Paper-Derived Si-SiC Composites. *Advanced Materials*. 2007. 19, 24. 4515–4519. Wiley Online Library. Copyright Wiley-VCH Verlag GmbH & Co. KGaA. Reproduced with permission.

8. List the polymeric materials that can be supplied as a filament or powder.
9. What are the main materials supplied in the form of powder-based polymer?
10. What are the processes that adopt powder-based polymers?
11. Mention some processes that use resin-based polymer composites
12. What processes adopt filament-based polymer composites?
13. What process is suitable for powder-based polymer composites?
14. What is the process that uses sheet-based polymers?
15. What is the role of powder metallurgy in developing AM technology?
16. Give examples of metal powders prepared by thermal, mechanical, electrolytic, and chemical techniques.
17. Why is the atomization technique the most popular for AM powder preparation?
18. What are the differences between gas atomization and VIGA?

19. Differentiate between the two plasma atomization techniques PA and PREP, showing the advantages and limitations of each.
20. What are the distinguishing applications of mechanical alloying as compared with conventional metallurgical alloying?
21. Why is electrodeposition suitable for preparing copper powder?
22. Discuss how to characterize the particle size and size distribution.
23. Aluminum particles prepared by electrolysis passed the US Standard mesh size 400. What are the particle size, aspect ratio, and shape factor? And, what are the relative density and packing factor if the bulk density is 1.5 g/cm^3?
24. How is the inter-particle friction of particles measured, and how does it affect the quality of PBF products?
25. Is it recommended to reuse the remaining powder after a LBM process? And, does this affect the powder characteristics and the quality of further products?
26. What ceramics are used in additive manufacturing?
27. What ceramics can be 3D printed?
28. How is ceramic powder made?
29. What types of feedstock materials are used in VP AM processes?
30. Enumerate some types of advanced ceramics that are typically used in AMC.
31. For the highest packing density, what is your recommendation when selecting a ceramic powder for a ceramic AM process?
32. Propose feedstock materials for FDM and RC processes to produce ceramics by AM.
33. Enumerate the main factors that influence the rheology of the feedstock for FDC.
34. What is a pre-ceramic polymer? Discuss its importance for the different ceramic AMTs.
35. Differentiate between CMCs and CMNCs, and suggest some AM processes to produce them.

BIBLIOGRAPHY

Bertrand Ph.; Bayle F.; Combe C.; Goeuriot P.; Smurov I. (2007). Ceramic Components Manufacturing by Selective Laser Sintering., Applied Surface Science 254(4), 989–992. doi:10.1016/j.apsusc.2007.08.085

Brady, G. Allen and Halloran John W. (1997). Stereolithography of Ceramic Suspensions, Rapid Prototyping Journal,3(2), 61–65. doi:10.1108/13552549710176680

Brigo, Laura, Schmidt, Johanna Eva Maria, Gandin, Alessandro, Michieli, Niccolò; Colombo Paolo, Brusatin, Giovanna (2018). 3D Nanofabrication of SiOC Ceramic Structures. Advanced Science, volume 5 issue 12, 1800937–. doi:10.1002/advs.201800937

Bui H M, Fischer R, Szesni N, Tonigold M, Achterhold K, Pfeiffer F, Hinrichsen O, (2022) Development of a Manufacturing Process for Binder Jet 3D Printed Porous Al_2O_3 Supports used in Heterogeneous Catalysis. Additive Manufacturing Volume 50 102498. doi.org/10.1016/j.addma.2021.102498

Butscher, A.; Bohner, M.; Doebelin, N.; Galea, L.; Loeffel, O.; Müller, R. (2013) Moisture Based Three-Dimensional Printing of Calcium Phosphate Structures for Scaffold Engineering. Acta Biomaterialia, 9(2), 5369–5378. doi:10.1016/j.actbio.2012.10.009

Camargo, I. L. de, Fortulan, C. A., Colorado H. A. (2022) A Review on the Ceramic Additive Manufacturing Technologies and Availability of Equipment and Materials. Cerâmica 68, 329–347.Brazil, doi.org/10.1590/0366-69132022683873331

Carrico, James D; Traeden, Nicklaus W; Aureli, Matteo; Leang, Kam K (2015). Fused Filament 3D Printing of Ionic Polymer-Metal Composites (IPMCs). Smart Materials and Structures, 24(12), 125021–. doi:10.1088/0964-1726/24/12/125021

Chua C. K, Leong, K f. and Lim, C. S. (2003) Rapid Prototyping: Principles and Applications, 2nd Edition. World Scientific Publishing Co. Pte. Ltd.

Colombo, P.; Mera, G.; Riedel, R.; Sorarù, G.D. (2013) Polymer-Derived Ceramics: 40 Years of Research and Innovation in Advanced Ceramics. J. Am. Ceram. Soc. 2010, 93, 1805–1837. doi: 10.1002/9783527631940.ch57

Colombo, Paolo et al. (2010). Polymer Derived Ceramics: From Nano-structure to Applications. Philadelphia: DEStech Publications, Inc.

Deckers, J.; Shahzad, K; Vleugels, J.; Kruth, J.P. (2012). Isostatic Pressing Assisted Indirect Selective Laser Sintering of Alumina Components. Rapid Prototyping Journal, 18(5), 409–419. doi:10.1108/13552541211250409

Deegan R.D., Bakajin O., Dupont T.F., Huber G., Nagel S.R., Witten T.A (1997) Capillary Flow as the Cause of Ring Stains From Dried Liquid Drops, Nature 389 (6653) 827.doi .org/10.1016/j.addma.2021.102394

EPMA (2008), Introduction to Powder Metallurgy - The Process and Its Products, European Powder Metallurgy Association, www.epma.com

Friederich, A.; Binder, J. R.; Bauer, W.; Derby, B. (2013). Rheological Control of the Coffee Stain Effect for Inkjet Printing of Ceramics. Journal of the American Ceramic Society, 96(7), 2093–2099. doi:10.1111/jace.12385

Fromm, J. E. (1984). Numerical Calculation of the Fluid Dynamics of Drop-on-Demand Jets. IBM Journal of Research and Development, 28(3), 322–333. doi:10.1147/rd.283.0322

Gaddam A; Daniela S. Brazete;Ana S. Neto;Bo Nan;Hugo R. Fernandes;José M. F. Ferreira; (2021). Robocasting and Surface Functionalization with Highly Bioactive Glass of ZrO_2 Scaffolds for Load Bearing Applications. Journal of the American Ceramic Society, 105- 1753. doi:10.1111/jace.17869.

Gonzalez-Gutierrez, Joamin; Cano, Santiago; Schuschnigg, Stephan; Kukla, Christian; Sapkota, Janak; Holzer, Clemens (2018). Additive Manufacturing of Metallic and Ceramic Components by the Material Extrusion of Highly-Filled Polymers: A Review and Future Perspectives. Materials, 11(5), 840–.doi:10.3390/ma11050840

Grigorescu, Ramona; Marina Grigore; Iancu, Lorena; Ghioca, Paul; Ion, Rodica (2019). Waste Electrical and Electronic Equipment: A Review on the Identification Methods for Polymeric Materials. Recycling, 4(3), 32–. doi:10.3390/recycling4030032

Hausner Henry H. (1981). Powder Characteristics and Their Effect on Powder Processing, Powder Technology 30(1), 3–8. doi:10.1016/0032-5910(81)85021-8

Hayden Thompson Black, Mathias C. Celina, James R. McElhanon (2016) Additive Manufacturing of Polymers: Materials Opportunities and Emerging Applications, Sandia Report, SAND2016-6644. Sandia National Laboratories.

Hinczewski, C.; Corbel, S.; Chartier, T. (1998). Stereolithography for the Fabrication of Ceramic Three- Dimensional Parts. Rapid Prototyping Journal, 4(3), 104–111. doi:10.1108/13552549810222867

https://matmatch.com/

https://www.3d4makers.com/

https://www.3dnatives.com/

https://www.e-education.psu.edu/

https:// www.perrymanco.com/

Jacobs P F (1992) Rapid Prototyping & Manufacturing: Fundamentals of Stereolithography. SME.Dearborn, MI.

Jacobs, P.F. (1996) *Stereolithography and other RP&M Technologies*, Society of Manufacturing Engineers (SME), Chapter 2: 29–35, Derburn, Michigan, USA.

Khalyfa A; Vogt S; Weisser J; Grimm G; Rechtenbach A; Meyer W; Schnabelrauch M (2007). Development of a New Calcium Phosphate Powder-binder System for the 3D Printing of Patient Specific Implants, 18(5), 909–916. doi:10.1007/s10856-006-0073-2

Kleijnen Rob G., Schmid Manfred and Wegener Konrad (2019) Production and Processing of a Spherical Polybutylene Terephthalate Powder for Laser Sintering. Appl. Sci., 9, 1308; doi: 10.3390/app9071308

Kosmala A, Zhang Q, Wright R, Kirby P, (2012) Development of High Concentrated Aqueous Silver Nanofluid and Inkjet Printing on Ceramic Substrates, Mater Chem Phys 132 (2–3) 788–795. doi:10.1016/j.matchemphys.2011.12.013

Lynxter, "PAS11, toolhead for 3d printing of ceramics" https://lynxter.fr, acc. 21/01/2022.

Maleksaeedi, S.; Eng, H.; Wiria, F.E.; Ha, T.M.H.; He, Z. (2014). Property Enhancement of 3D-Printed Alumina Ceramics Using Vacuum Infiltration. Journal of Materials Processing Technology, 214(7), 1301–1306. doi:10.1016/j.jmatprotec.2014.01.019

Moon Jooho; Grau Jason E.; Knezevic Vedran; Cima Michael J.; Emanuel M. Sachs (2002). Ink-Jet Printing of Binders for Ceramic Components, Journal of the American Ceramic Society 85(4), 755–762. doi:10.1111/j.1151-2916.2002.tb00168.x

Mott M; Evans J R G (1999). Zirconia/Alumina Functionally Graded Material Made by Ceramic Ink Jet Printing. Materials Science and Engineering A271 344–352. doi:10.1016/s0921-5093(99)00266-x

Park Soyeon, Shou Wan, Makatura Liane, Matusik Wojciech, and Fu Kun (Kelvin) (2021) 3D Printing of Polymer Composites: Materials, Processes, and Applications. Matter Review. Cell Press. doi.org/10.1016/j.matt.2021.10.018

Peymannia, Masoud; Soleimani-Gorgani, Atasheh; Ghahari, Mehdi; Jalili, Mojtaba (2015). The Effect of Different Dispersants on the Physical Properties of Nano CoAl$_2$O$_4$ Ceramic Ink-jet Ink. Ceramics International, 41(7), 9115–9121. doi:10.1016/j. ceramint.2015.03.311

Peyre Patrice and Charkaluk Eric (2022) Additive Manufacturing of Metal Alloys 1, Processes, Raw Materials and Numerical Simulations, ISTE Ltd and John Wiley & Sons, Inc.

Pfeiffer S; Florio K; Puccio D; Grasso M; Colosimo B M; Aneziris C G; Wegener K; Graule T; (2021). Direct Laser Additive Manufacturing of High Performance Oxide Ceramics: A State-of-the-art Review. Journal of the European Ceramic Society, (), –. doi:10.1016/j. jeurceramsoc.2021.05.035

Qu Piao, Xiong Dingyu, Zhu Zhongqi, Gong Zhiyuan, Li Yanpu, Li Yihang, Fan Liangdong, Liu Zhiyuan, Wang Pei, Liu Changyong, Chen Zhangwei (2021) SOFCs Using High Quality Ceramic Inks for Performance Enhancement. Additive Manufacturing, Volume 48, Part A, December 2021, 102394

Rapidia, "Materials", www.rapidia.com, acc. 21/01/2022

Region, and Segment Forecasts (2023) Ceramic Matrix Composites Market Size, Share & Trends Analysis Report by Product (Oxide, Silicon Carbide, Carbon), by Application (Aerospace, Defense, Energy & Power, Electrical & Electronics). Available online:https://www.grandviewresearch.com/industry-analysis/global-ceramic-matrix -composites-market (accessed on 12 May 2023).

Reis N, Seerden K, Derby B, Halloran J, Evans J(1998) Direct Inkjet Deposition of Ceramic Green Bodies: II–Jet Behaviour and Deposit Formation, MRS Online Proceedings Library Archive 542.

Ruscitti A, Tapia C, N.M. Rendtorff N. M. (2020) A Review on Additive Manufacturing of Ceramic Materials Based on Extrusion Processes of Clay Pastes. Cerâmica 66, 380, 354. doi: 10.1590/0366-69132020663802918

Sanchez, Daniel Moreno; de la Mata, Maria; Delgado, Francisco Javier; Casal, Victor; Molina, Sergio Ignacio (2020). Development of Carbon Fiber Acrylonitrile Styrene Acrylate Composite for Large Format Additive Manufacturing. Materials & Design, 191(), 108577–.doi:10.1016/j.matdes.2020.108577

Seerden K, Reis N, Derby B, Grant, P Halloran J, Evans J, (1998) Direct Ink-jet Deposition of Ceramic Green Bodies: I-Formulation of Build Materials, MRS Online Proceedings Library Archive 542.

Shanthar Rajinth, Chen Kun, and Abeykoon Chamil (2023) Powder-Based Additive Manufacturing: A Critical Review of Materials, Methods, Opportunities, and Challenges, Advanced Engineering Materials, 25, 2300375, doi.org/10.1002/adem.202 300375

Shuai, Cijun, He, Chongxian, Peng, Shuping, Qi, Fangwei, Wang, Guoyong, Min, Anjie, Yang, Wenjing, Wang, Weiguo (2021). Mechanical Alloying of Immiscible Metallic Systems: Process, Microstructure, and Mechanism. Advanced Engineering Materials, 23, 2001098, doi:10.1002/adem.202001098

Song, Xuan; Chen, Yong; Lee, Tae Woo; Wu, Shanghua; Cheng, Lixia (2015). Ceramic Fabrication Using Mask-Image-Projection-based Stereolithography Integrated with Tape-casting. Journal of Manufacturing Processes, Volume 20, Part 3, Pages 456–464), S1526612515000675–. doi:10.1016/j.jmapro.2015.06.022

Srivastava Manu, Rathee Sandeep, Patel Vivek, Kumar Atul, Koppad Praveennath (2022), A review of various materials for additive manufacturing: Recent Trends and Processing Issues . Journal of Materials Research and Technology, 21, 2612–2641, Elsevierdoi.org /10.1016/j.jmrt.2022.10.015

Tarafder Solaiman; Balla Vamsi Krishna; Davies Neal M; Bandyopadhyay Amit; Bose Susmita (2012). Microwave-sintered 3D Printed Tricalcium Phosphate Scaffolds for Bone Tissue Engineering. J Tissue Eng Regen Med 2013; 7: 631 – 41. doi:10.1002/term.555

Travitzky, Nahum; Bonet, Alexander; Dermeik, Benjamin; Fey, Tobias; Filbert-Demut, Ina; Schlier, Lorenz; Schlordt, Tobias; Greil, Peter (2014). Additive Manufacturing of Ceramic-Based Materials. Advanced Engineering Materials, 16(6), 729–754. doi:10.1002/adem.201400097

Uhland Scott A.; Holman Richard K.; Morissette Sherry; Cima Michael J.; Sachs Emanuel M. (2001). Strength of Green Ceramics with Low Binder Content, 84(12), 2809–2818. doi:10.1111/j.1151-2916.2001.tb01098.x

Wang, M., Xie, C., He, R., Ding, G., Zhang, K., Wang, G., & Fang, D. (2019). Polymer-derived Silicon Nitride Ceramics by Digital Light Processing Based Additive Manufacturing. Journal of the American Ceramic Society. doi:10.1111/jace.16389

Wasp, "Crane WASP", www.3dwasp.com, acc. 21/01/2022.

Wilson, Taylor, "Application of SLA 3D Printing for Polymers" (2022). Williams Honors College, Honors Research Projects. 1490

Windsheimer H.; Travitzky N.; Hofenauer A.; Greil P. (2007). Laminated Object Manufacturing of Preceramic-Paper-Derived Si-SiC Composites, Advanced Materials 19(24), 4515–4519. Wiley Online Library. doi:10.1002/adma.200700789

Xiong Shuling (2020) Materials, Application Status and Development Trends of Additive Manufacturing Technology. Institute of Scientific and Technical Information of China, No. 15 Fuxing Road, Beijing 100038, China Materials Transactions. The Japan Institute of Metals and Material

Yakout, Mostafa and Elbestawi M. A. (2017) Additive Manufacturing of Composite Materials: An Overview. 6th International Conference on Virtual Machining Process Technology (VMPT), Montréal, Canada.

Youssef H., El-Hofy, H. and Hamed, M. (2023) Manufacturing Technology; Materials, Processes and Equipment. CRC Press,, California, USA. doi.org/10.1201/97810033 73209

Yu, Tianyu; Zhang, Ziyang; Liu, Qingyang; Kuliiev, Ruslan; Orlovskaya, Nina; Wu, Dazhong (2019). Extrusion-based additive manufacturing of yttria-partially-stabilized zirconia ceramics. Ceramics International, Volume 46, Issue 4, Pages 5020–5027),. doi:10.1016/j.ceramint.2019.10.245

Yusheng Shi, Chunze Yan, Yan Zhou, Jiamin Wu, Yan Wang, Shengfu Yu, Chen Ying (2021) Materials for Additive Manufacturing, Elsevier, Academic Press.

Ziyong Li, Zhangwei Chen, Jian Liu, Yuelong Fu, Changyong Liu, Pei Wang, Mingguang Jiang & Changshi Lao, (2020) Additive Manufacturing of Lightweight and High-strength Polymer Derived SiOC Ceramics, Talor & Francis, Virtual and Physical Prototyping Taylor and Francis, 1–15. doi.org/10.1080/17452759.2019.1710919

Zocca, Andrea; Gomes, Cynthia M.; Bernardo, Enrico; Müller, Ralf; Günster, Jens; Colombo, Paolo (2013). LAS glass–ceramic scaffolds by three-dimensional printing. Journal of the European Ceramic Society, 33(9), 1525–1533. doi:10.1016/j.jeurceramsoc.2012.12.012

8 Post-Processing Techniques in Additive Manufacturing Processes

8.1 IMPORTANCE OF POST-PROCESSING AND FINISHING

As previously discussed in Chapter 2, additive manufacturing (AM) includes three steps. These are the pre-processing (computer-aided design [CAD], converting to STL file, and slicing), the processing (3D printing), and the post-processing steps. According to Wohlers Associates (2021), post-processing accounts for nearly 27% of the production cost of AM (www.materialstoday.com/).The chart shown in Figure 8.1 presents the cost segmentation between pre-processing, printing, and post-processing among companies that offer both metal and polymer AM.

Post-processing is an often overlooked part of the AM process. It is a critical and final step, where parts receive adjustments, such as smoothing and strengthening. Despite all the advantages of AM over traditional manufacturing methods, it is still imperfect, and some post-processing techniques are generally needed to enhance components or to overcome AM's limitations. Depending upon the AM technique and starting material used, the reason and the type of post-processing vary.

Additively manufactured parts can still have poor surface quality and high porosity, which can affect the mechanical properties of the component. The main objective of post-processing is to eliminate these potentially dangerous defects. These processes can include heat treatment, which is needed to reduce the stress on components before their removal from the build plate; separating the components from the support structure; and surface finishing procedures such as computer numerically controlled (CNC) machining, blasting, and polishing. The post-processing of additively manufactured parts can be just as vital as the fabrication itself. Some post-processing processes may involve thermal treatment to eliminate residual and thermally induced stresses in the part to enhance the final part properties. The residual stresses are the stresses that persist in a part even in the absence of external loading. In contrast, thermal stresses are usually induced by any change in a material's temperature. Different AM processes have different results in terms of accuracy, and thus, machining to final dimensions may be required. Some AM processes produce relatively fragile components that may require the use of infiltration and/or surface coatings to strengthen the final part.

Finishing may include media blasting, peening, sanding, abrasive flow machining, or grinding to smooth surface features and allow visual inspection. Washing

DOI: 10.1201/9781003451440-8

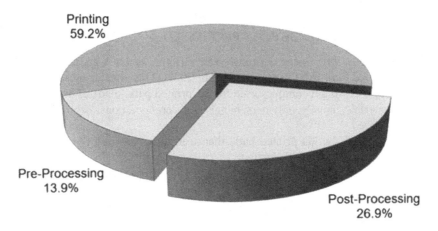

FIGURE 8.1 Cost segmentation between pre-processing, printing, and post-processing steps in AM. (From Wohlers Associates, www.materialstoday.com/)

may be used to help remove powders from internal features. In cases such as medical applications, sterilization may be specified. Coating and painting, as used in plastic prototype finishing, may improve surface finish or appearance. These operations may require specialized equipment and processes provided by a dedicated service provider.

Post-processing of printed parts may require a large amount of skilled manual work in terms of machining, surface preparation, and coating. Post-processing operations may be manual, semi-automated, or automated, and they can be performed through either batch or serial processing. Automation of the post-processing step will be necessary to minimize labor and time cost for production. Companies that can efficiently and accurately post-process parts to a customer's expectations can often charge a premium for their services, whereas those that compete primarily on price may sacrifice post-processing quality in order to reduce costs.

In conclusion, post-processing procedures vary by printing techniques and materials. Some AM processes may not require post-processing, but some processes require careful and lengthy post-processing. For example, photosensitive resins printed by stereolithography (SLA) require the component to cure under the UV environment before the final application. If support structures are used during printing, the removal of such support is required. The support can be removed mechanically by cutting it from the printed object, or chemically by dissolving it in a solvent that will not damage the target part.

Finally, post-processing of 3D printed components offers many advantages:

- Higher cost-efficiency when automated post-processing operations are used
- Improved component performance due to an improved surface finish
- New fields of application due to improved component performance
- Faster implementation of the 3D printing technology thanks to cost savings during the post-processing phase

8.2 DEFECTS IN PRODUCTS PROCESSED BY AM THAT NEED TO BE POST-PROCESSED

The need for post-processing depends on the printing technology, materials used, defects initiated, and print settings (Albright 2021). This chapter presents the different post-processing techniques used for 3D printed parts made of different engineering materials, namely, polymers, metals, and ceramics, considering the printing technology used.

The defects of the 3D printed parts that need to be treated by post-processing techniques may include the following:

- pores
- cracks
- anisotropy
- residual stresses
- thermal stresses
- laser spattering
- poor surface roughness
- part distortion

These defects have serious influences on the internal microstructure and mechanical behavior of the final parts. Therefore, after the parts are manufactured, post-processing operations are usually required to improve the mechanical properties and the surface quality, thus achieving their intended utilization. There are many post-processing technologies, such as the thermal post-processing method to release thermally induced residual stress and laser peening to reduce micro-defects and improve surface quality.

Post-processing includes, therefore, all of the processes that are performed after the parts are removed from a 3D printer. It has two main categories:

1. Primary post-processing includes the necessary steps that must be performed on all 3D printed parts to make them suitable to be used in any particular application. Primary post-processing activities generally include the removal of support as well as cleaning.
2. Secondary post-processing includes optional finishing processes that improve the properties, performance, functions, or even aesthetics of the part. Secondary post-processing activities are those that are not necessary or essential but would depend on the needs and purposes of the user, such as painting and vapor smoothing. Secondary post-processing plays a significant role in improving mechanical, chemical, and aesthetic properties, meeting tight tolerances, providing near–injection molded finishes, and achieving added durability. Additionally, it improves the surface properties of prints beyond the initial print, such that it controls wetting behavior, gloss, scratch resistance, and other properties. These essential surface characteristics allow post-processing to significantly extend the range of applications (Ryan et al. 2021).

8.3 CONSIDERATIONS FOR SETTING UP A POST-PROCESSING SYSTEM

Most AM companies aim to increase the printing speed, printing resolution, size, and other aspects. For AM to be adopted in the industry mainstream, several factors related to post-processing must be considered, including part quality, cost, time, uniformity, repeatability, workforce skills, safety, sustainability, and automation.

The following items provide some important considerations for the selection of post-processing techniques:

- *Methods:* Automation would address most, if not all, of these factors. Automation would also increase efficiency and productivity. Moreover, an interconnected hardware and software solution that can trace parts (with vision systems) throughout the workflow is ideal for manufacturing activities.
- *Type of material:* The type of material is the first determining factor when designing post-processing operations. The development of new polymer 3D printing materials will warrant new post-processing methods for these materials. Materials (metals, polymers, or ceramics) for different applications would need different post-processing techniques, such as in electronics, biomedical applications, oil and gas, aerospace, and many others.
- *Part features:* In addition, the size, geometry, structure, complexity, and purpose of the part, dimensional accuracy/tolerance, and the number of parts to post-process should also be considered, as these factors are decisive for the selection of the post-processing operation and consequently, the properties of the printed part. Knowing these would aid in the optimization of the post-processing operations. After analyzing these factors, the next step is to identify the technologies/machines and materials/solutions that are needed to set up a post-processing system.
- *Targeted properties:* Further, different post-processing techniques are needed to enhance different properties (e.g. mechanical, electrical, surface roughness, etc.) to cope with the intended applications. For example, 3D printed sorbents for oil–water separation (for the oil and gas industry) will need to enhance their hydrophobic/oleophilic properties (Tijing et al. 2020).
- *Quality:* For newly developed materials, where no post-processing devices have yet been developed, manual processing results in varying properties, as the quality depends on the skills of the worker. For such cases, the workforce should be trained in this industry, or dedicated devices and facilities should be developed to ensure part quality, especially in case of mass production.
- *Safety:* Post-processing operations are accompanied by corresponding hazards in all aspects of operation, handling, storage, and disposal. Considering safety in all these aspects entails additional costs not usually previously considered during planning.

As previously mentioned, the type of engineering material to be printed is the first determining factor when designing post-processing operations. Accordingly, it is wise to treat the post-processing techniques as based on the material of the printed part.

8.4 POST-PROCESSING TECHNIQUES IN AM OF POLYMERS

This section presents the most commonly used post-processing techniques applied to 3D printed polymers, arranged according to the different types of printing technologies.

8.4.1 POST-PROCESSING IN FUSED DEPOSITION MODELING (FDM)

FDM 3D printing uses a layer-by-layer technique that generates stair stepping, commonly known as the staircase effect. This effect is more noticeable on curved and oblique surfaces. The intensity of the staircase effect can be decreased by reducing the layer thickness. However, this does not completely eliminate the problem. Layer thickness typically ranges from 15 to 500 µm. When the layer thickness is below 50 µm, the naked eye will in most cases not be able to recognize the stair steps associated with a layered manufacturing approach. For thicker layers, post-processing may be used to remove support structures or to improve surface properties. Additional manual post-processing is often required. For this reason, FDM 3D prints have a rough surface finish, and post-processing is usually done in order to achieve the desired texture as well as to improve its mechanical properties. Figure 8.2 shows the staircase effect on a 3D printed part (Pandey et al. 2003).

The common FDM polymers that exist in the market today are polylactic acid (PLA), acrylonitrile butadiene styrene (ABS), polyethylene terephthalate (PET), nylon, flexible thermoplastic polyurethane (TPU), and polycarbonate (PC). These polymers, especially the engineering grade materials, enable the production of parts

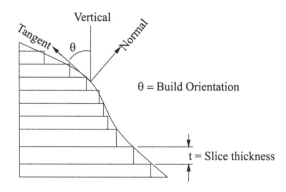

FIGURE 8.2 Staircase effect on 3D printed part. (From Pandey, P.M. et al., (2003) *Rapid Prototyping Journal*, 9, 5, 274–288.With permission from Emerald)

and prototypes with outstanding chemical and thermal resistance as well as excellent strength-to-weight ratios.

Post-processing techniques that can be applied in FDM of polymers include (Ryan et al. 2021):

1. *Material support removal:* Material support removal is considered the most basic form of post-processing. This process does not require any specialized tool beyond using simple needle-nose pliers and flush cutters. It is simply done by cutting, shaving, or deburring excess material. Although it can be time-consuming and labor-intensive, it does not alter the overall geometry of the part. This process, however, sometimes leaves marks on the surface of the part. As an alternative to this process, some 3D printers have both AM and a combined machining/polishing facility. Some 3D printers use a second extruder head, where a water-soluble support like polyvinyl alcohol (PVA) is extruded. Material support removal, as described in FDM, is also done for parts produced with other 3D printing technologies.

2. *Sanding or surface polishing:* Sanding is done after the supports have been removed. In this process, a 3D printed part is smoothened using a sander/sandpaper to remove any obvious imperfections. It is recommended to use finer grades of sandpaper of Mesh 400, 600, 1200, 1500, and 2000 after sanding the print with a coarse grade paper first to remove bumps and scratches.

3. *Gap filling:* This uses inexpensive filling materials on the market to protect and give an aesthetic effect on an FDM printed surface. Gap filling is simply filling the voids and gaps with epoxy/fillers. A filler may be used for larger gaps, which will require additional sanding to remove excess material.

4. *Coating:* This is the use of paint or resin that would improve the aesthetic of the print. Painting may be done manually by using an air spray or brush. As with the other post-processing methods listed above, this method is very simple to apply. Two types of coating polymers are used, namely, polyurethane elastomer and liquid silicone. Both products are commonly used in outdoor waterproofing.

5. *Polymer coating:* While coating is usually done using spray paints and other formulations for aesthetic purposes, polymer coating, on the other hand, is found to increase the adhesion of 3D printing materials on textile fabrics by initially coating the latter with a soluble polymer layer. One of its advantages is that adhesion can be substantially enhanced without significantly changing the haptic properties and the bending stiffness of a fabric. Plastisols are solutions of polymers that require solvent evaporation after coating. Epoxy resins are thermoset polymer coatings that involve curing.

While these post-processing techniques can be applied to all FDM materials, it is still advisable to use a two-component solution (such as epoxy) or a one-component spray to smoothen the surface of rigid plastics such as PC, PET, or PLA. On the other hand, softer materials such as PS, ABS, or acrylonitrile styrene acrylate (ASA) may be dissolved with a solvent or sanded by hand for a streak-free or shiny effect.

It should be noted that there are other post-processing techniques that can only be applied on specific materials such as ABS, PLA, etc. Here are some of the suggested post-processing techniques used for both materials (Ryan et al. 2021):

1. *Acrylonitrile butadiene styrene (ABS):*

This polymer is commonly used due to its low manufacturing cost and high machinability. Post-processing applied for ABS includes:

- **Cold welding:** Solvent and other chemical substances can also be used to weld broken parts of ABS materials. For instance, the use of acetone solvent will not change the color of the print surface as much as other glues. Once dried, the joint will exhibit the properties of bulk ABS, making further finishing more uniform and simple.
- **Vapor smoothing:** The usual method of post-processing an ABS printed object involves the use of solvents to dissolve its surface layer. Acetone is usually used as a solvent for ABS vapor smoothing. However, this method affects the object's dimensional accuracy, as the amount of material being removed cannot be controlled. It must be taken into consideration that some solvents are more toxic, and therefore, proper precautions must be taken when using this post-processing technique.
- *Polylactic acid (PLA):* PLA is another commonly used material in FDM 3D printing. It comes in many colors, making it ideal for diverse applications. The most common post-processing method for PLA is annealing. This is a post-processing technique performed by heating the PLA to a temperature below its melting temperature (173–178°C) and above its glass transition temperature (60–65°C). Annealing PLA results in a stronger and stiffer piece when done correctly. On average, the stiffness could increase by 25%, while strength could increase by 40% (Jorgenson 2021 and Bhandari et al. 2019).

Post-processing of food-safe polymers:
Furthermore, food-safe materials (PLA, nylon-6, PP, PET, co-polyester, high-impact polystyrene (HIPS), and polyethylene terephthalate (PET), and some brands of polyetherimide (PEI), ASA, and ABS) are specially post-processed. These are all fused deposition modeling (FDM) materials that are labeled food safe, which meets the Food and Drug Administration (FDA)'s food code for the intended use. FDA and European Food Safety Authority (European Union) regulations should be met for food-safe coatings and sealants used as a post-processing coating technique.

Dip coating: This is a post-processing method done by dip coating the FDM 3D printed parts into food-grade polyurethane or epoxy resin, or an FDA-approved polytetrafluoroethylene (PTFE), acting as food-safe coatings and sealants to reduce the risk of bacterial build-up and particle migration (Ryan et al. 2021).

8.4.2 POST-PROCESSING IN STEREOLITHOGRAPHY (SLA) AND DIGITAL LIGHT PROCESSING (DLP)

The next step after printing polymers by VAT process is the post-processing, which has a high impact on the final part shape. The post-processing of the manufactured features consists of two main steps: cleaning and post-curing.

- Cleaning: It should be applied to the part that is covered with uncured polymer resin. Isopropanol is commonly used for cleaning to remove the excess resin from the sample. In this step, cleaning the surface is performed to be sure that any additional leftover resins are removed from the part.
- Post-curing: Depending on the resin, it should be cured by UV light or heating in a high-temperature oven.

Post-processing ensures that no reactive resin residue is left on the samples, and optimal mechanical properties are achieved. Unlike FDM, UV post-treatment, as one of the post-processing techniques applied to SLA or DLP, is usually required to complete the photopolymerization process to enhance the strength of the material. Although some of the post-processing methods used in FDM technology can also be applied to 3D prints of SLA, post-curing is generally needed on SLA-printed parts to achieve the highest possible stability and strength. Afterwards, the parts can be painted, machined, and assembled for specific finishes.

SLA or DLP uses a variety of acrylate resins that are classified according to their characteristics and applications (e.g. medical, dental, or jewelry). The prototyping resins are low in cost and can produce a high resolution and high stiffness in prints with a smooth injection molding–like finish, making it ideal for prototyping applications. Some elastomeric resins of low hardness are popular but may not be easily finished by sanding or machining.

The most common post-processing techniques applied in SLA and DLP prints include (Ryan et al. 2021):

1. *Washing:* After 3D printing, parts are usually slimy due to the uncured resin still being attached to the surface of printed parts. Isopropyl alcohol (IPA) and water are commonly used for washing.
2. *Basic support removal and sanding of support nibs:* After washing, supports have to be removed for parts made in SLA and DLP printers before any other post-processing can be done.
3. *Wet sanding/mechanical polishing:* Wet sanding of the support nibs can be done in flat surfaces. Because the surface can only be sanded at the support nibs, the overall geometry of the parts will not be affected. Wet sanding provides a smooth surface, making it ideal for complex geometries. The use of water while sanding may result in some light spots on the 3D prints.
4. *Chemical mechanical polishing (CMP):* This is a hybrid process that flattens the surface of the part through a mechanical material abrasion and chemical surface reactions. This method uses abrasive particles located

on the contact area between the part to be polished and a polishing pad. The main purpose of this process is to reduce the surface roughness and increase the glossiness of SLA printed resins.

5. *Mineral oil finish:* This process is ideal for removing white marks and uneven spots on the printed parts.

6. *Spray paint (clear UV protective acrylic):* This is done by spray painting the model to help conceal layer lines. This reduces the need to sand the unsupported side of the model. The varnish may also protect the model from yellowing by limiting UV exposure. However, it increases the overall dimensions and may result in an "orange peel" effect on the surface.

Aside from common post-processing techniques applied to usual 3D prints for ideal functions and aesthetics, there are post-processing techniques that are applicable and required for some SLA-printed models to increase their mechanical properties and maintain stability. These include the following:

- *UV-post-curing:* This post-processing improves the strength of 3D prints due to the complete curing of the resin. Moreover, post-curing at higher temperatures may lead to a shorter curing time, which results in higher mechanical properties. However, it is time-consuming and requires more facilities. Some companies develop post-curing devices specially designed for their printers. UV post-curing significantly increases the elastic modulus and the ultimate tensile strength; however, it decreases the ultimate strain (Garcia et al. 2020).

- *Thermal Post-Curing:* Uzcategui et al. (2018) reported that UV post-curing and thermal post-curing are the most common post-processing methods being employed to improve conversion as well as the mechanical properties after SLA printing. However, UV post-curing does not improve the interior sections of the prints, which is the reason for their inferior properties compared with fully converted materials. For that, it is recommended to perform both types of curing to ensure high conversion of the entire material.

8.4.3 Post-Processing in Selective Laser Sintering (SLS)

As previously mentioned, SLS does not need supports because the unsintered powder surrounding the parts provides the support structures during the whole process, and thereby, complex geometries (e.g. complex designs and components with overhangs) can be produced. However, the use of post-processing techniques is still advisable to improve the surface finish, aesthetics, and mechanical properties of the 3D prints. The first step is the complete removal of all powder particles. Secondly, depending on the materials' properties (commonly polyamides (PA)), the parts can also be annealed or heat-treated to provide an equilibration of the polymer. Finally, the surface can be treated similarly to FDM parts in terms of coating methods.

The materials usually used in SLS are thermoplastic polymer granules/powders. Nylon is the most common material for SLS, as it is ideal due to its flexible, strong, and lightweight properties. Nylon or PA is stable against heat, chemicals, impact, water, dirt, and UV light, which makes it ideal for rapid and successful prototyping. Unlike other printing technologies, SLS post-processing requires minimal labor and time, and provides consistent results. Some of the recommended post-processing applied in SLS includes:

1. ***Standard finish:*** An air jet is used to remove excess powder. For additional cleaning, plastic bead blasting may be used to remove the remaining unsintered powder from the surface.
2. ***Dyeing:*** This is a fast and cost-effective method to color SLS 3D printed parts. The dyeing process starts by immersing the 3D printed part in the dyeing tank, which contains the dye and an auxiliary agent. Dyeing is ideal for SLS 3D printed parts due to their porosity.
3. ***Spray painting or lacquering:*** Spray painting or lacquer coating (clear coat or varnish) is commonly used for SLS 3D printed parts. Various finishes can be achieved via lacquering, e.g. metallic sheen or high gloss. Lacquer coatings may increase water tightness, improve surface hardness, increase wear resistance, and limit smudges and marks on the surface of the SLS 3D printed parts.
4. ***Waterproof treatment:*** Coatings may be applied to enhance the inherent water-resistant properties of the SLS 3D printed parts. The surface of the part is covered by dip-dyeing, spraying, or even coating with a layer of epoxy resin. Vinyl-acrylates and silicones have been reported to provide excellent results.
5. ***Roller polishing:*** This is a polishing process that may provide a smooth finish on the surface of nylon 3D printed parts. Roller polishing starts by putting small stones or steel balls in a drum; the drum then vibrates at a high frequency to provide the grinding effect. This process results in the rounding of sharp edges and may also slightly affect the part dimensions. It is therefore not recommended for parts with complicated shapes or fine details.

Other post-processing techniques after SLS 3D printing of polymers may also include the use of a tumbler, bead blasting, brushing, and manual water jets. These post-processing methods may have the following drawbacks:

- labor-intensive
- increased risk of wearing down fine feature details
- damaging fragile geometries
- producing inconsistent final parts

Note that the use of each post-processing method discussed here is viable for all 3D printing technologies of polymers.

8.4.4 OTHER FINISHING METHODS FOR POLYMERS

Other, more sophisticated methods are also used for polymer 3D printed parts. These are:

- *Physical vapor deposition (PVD):* This is a process wherein the surface of the polymer 3D printed part is coated with a metal or ceramic material by ionizing the atoms of the coating material. This process does not require any medium to transfer the coating but is usually done under a vacuum. These coatings change the surface characteristics, producing a part that is more robust against wear, friction, heat, chemicals, and so on.

- *Chemical vapor deposition (CVD):* This is a process wherein the surface properties of common 3D printing polymers can be functionalized, thereby allowing the bulk properties of 3D printed objects (e.g. strength) can be manipulated/designed separately from surface properties. Figure 8.3(1) shows a schematic of the initiated chemical vapor deposition (iCVD) process used to coat 3D printed PLA and ABS parts/substrates. Parts/substrates such as a bolt, nut, and comb were printed with ABS material. It can be seen in Figure 8.3(2) that the uncoated ABS part surfaces were hydrophobic. These parts were then coated with the hydrophilic P(HEMA-co-EGDA) and were readily wetted (Figure 8.3(3)). After this, the parts were coated with poly (1H,1H,2H,2H-perfluorodecyl) acrylate (PPFDA), and it can be observed that the surfaces regained hydrophobicity (Figure 8.3(4)). This shows that different surfaces of 3D printed parts can be functionally tuned using the CVD process.

- *Electroplating:* This is the process of depositing a thin layer/coating on the surface of another part for functional or decorative purposes through the electrodeposition process. 3D printed parts manufactured using SLA, FDM, PolyJet, and others may be electroplated with different metallic alloys. In such a case, the surface will have enhanced properties, such as improved electrical properties, increased strength, improved heat reflection, better chemical resistance, aesthetic value, and a smooth finish.

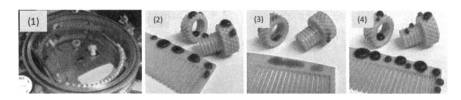

FIGURE 8.3 Sequential deposition of P(HEMAco-EGDA) and PPFDA. (1) Nut, bolt, and comb in the iCVD reactor. Water droplets (colored with blue food coloring) on ABS substrates (2) without coating, (3) coated with P(HEMA-co-EGDA), and (4) coated with PPFDA. (Extracted from Cheng, C. and Gupta, M., (2017) *Beilstein Journal of Nanotechnology*, 8, 1629–1636)

- *Electroless deposition or plating, also known as autocatalytic deposition of metals:* This is the uniform coating of thick metallic layers on the surface of non-conductive parts (substrates) through the reduction of metallic ions from a liquid electrolyte. 3D printed parts via SLA used this method to coat nickel platinum, copper, and palladium metallic materials. The resulting parts with functionalized surfaces may be used in the fields of electronics such as MEMS, micro-robots, metamaterials, as well as other chemical and mechanical engineering systems.
- *Post-processing to protect from electrostatic discharge (ESD):* Many electric/electronic devices require component materials to be ESD-safe in order to prevent damage from the build-up of electric potential (static discharge). Usually, several post-processing methods are applied, namely:
  1. Painting/coating
  2. Covering with conductive tape
  3. Wrapping with aluminum-coated or carbon-filled films

Some companies have developed ESD-safe 3D printing materials, which eliminates the need for post-processing.

8.5 POST-PROCESSING TECHNIQUES IN AM OF METALS AND ALLOYS

8.5.1 CLASSIFICATION SCHEMES OF POST-PROCESSING TECHNIQUES FOR METALS AND ALLOYS

Post-processing for AM of metals and alloys can be categorized into two main categories, as illustrated in Figure 8-4.

1. Mechanical finishing methods, including:
   - Powder removal (depowdering)
   - Support removal
   - Machining
   - Mechanical super-finishing
2. Thermal methods to modify the surface texture of parts; these include:
   - Heat treatment
   - Stress relieving
   - Thermal super-finishing

8.5.2 POST-PROCESSING STEPS IN AM OF METALS AND ALLOYS

On the other hand, post-processing techniques for AM of metals and alloys can be categorized stepwise in the following post-processing steps:

1. Powder removal (depowdering)
2. Stress relief

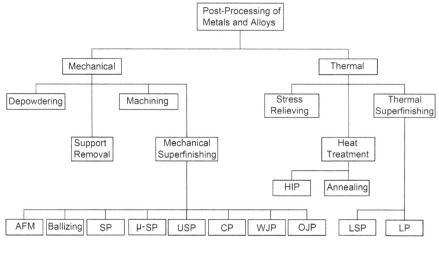

FIGURE 8.4 Classification of post-processing operations for additively manufactured metals and alloys, as based on mechanical and thermal aspects.

3. Support removal
4. Machining
5. Thermal post-processing
6. Surface finishing

1. Powder removal (depowdering): First, the powder that remains on the build plate after the printing process needs to be removed. This can be achieved with standard cleaning procedures. Excess powder is often saved or sent back to the supplier for recycling purposes. Depowdering is performed by water or air blasting. Depowdering is related to the mechanical post-processing category (Figure 8.4).

Excess powder can be removed manually, but automated solutions have arrived on the market that vibrate or rotate to remove excess powder (Figure 8.5). This 3D printing post-processing technique works like a sort of centrifuge, rotating the part in all three dimensions. Models printed using powder bed fusion are fabricated using plastic or metal powders. Residues of powder, especially in the case of holes or more complex internal channels inside the model, can cling to or remain in the model.

2. Stress relief step: Heat treatment is developed to relieve stresses accumulated in the metal, and hence, it is related to the thermal post-processing category. This step typically involves the cooling and heating of the piece inside a furnace and is

FIGURE 8.5 SFM-AT800S depowdering machine. (Courtesy of Solukon Maschinenbau GmbH in collaboration with Siemens)

performed when the piece is still attached to a support structure, or the laser heats the material and subsequently cools it during the build process; without it, there is a risk of a part warping or even cracking once it is removed from the build plate.

3. Support removal: This is a mechanical post-processing category. Many pieces need a support structure to be printed, so before any heat treatment or finishing can be effected on the piece, support structures need not to be removed to help reduce distortion of features on the part; without supports, there is a risk of having a build failure if the part warps too much and causes an issue. Although a band saw or files might be enough to remove parts and support structure, in the case of harder materials that resist intense strain (such as Inconel superalloy), it might be necessary to use CNC machining, or even wire electrodischarge machining (EDM).

4. Machining: Machining and abrasive finishing are conventional manufacturing techniques to improve the form accuracy and surface finish of functional parts in various industries. They are used as common post-processing methods for AM parts because of their high maturity and good accessibility. It's not uncommon to finish metal 3D prints with some CNC machining, milling, turning, engraving, or tapping. Bai et al. (2019) employed CNC milling to post-process ASTM A131 steel parts processed by directed energy deposition (DED) to reduce the surface roughness of the workpiece from 23 μm to 0.6 μm, and the high cutting speed contributed to a more favorable surface finish. Besides, it was found that the milling procedure hardly changed the microhardness of the DED samples. However, the noticed anisotropic machinability of the workpiece was due to the anisotropic microstructure during the printing process, as suggested by the authors (Bai et al. 2019).

5. Thermal Post-Processing: The thermal post-processing methods for additively manufactured parts can significantly alleviate residual stresses, reduce cracking, and homogenize the microstructure that has been initiated during the printing process due to rapid heating and cooling of the material. They also serve to improve mechanical properties, including hardness, fatigue life, and ductility, help to homogenize the microstructure, and reduce material heterogeneity in the part; without this, there is a risk of part failure due to build orientation effects or other microstructural weaknesses that can arise during the build.

Direct metal laser sintering (DMLS) printed Ti–6Al–4V parts were found to be impacted by residual stresses due to the high cooling rates and steep temperature gradient, which can affect the fatigue life. Tensile residual stress (TRS) imparted to the material can initiate fatigue crack growth. Stress relief annealing, at suitable temperatures, helps to reduce the residual stresses due to printing. Imparting compressive residual stress (CRS) will help to increase the fatigue life of the material.

The most common heat treatment process operations include hot isostatic pressing (HIP) and annealing:

1. **HIP** is a frequently used thermo-mechanical treatment method, which combines high-temperature and high-pressure technologies. Its heating temperature usually reaches 1000–2000°C for 2–4 h to help close and fuse internal pores, voids, and defects. High-pressure inert gas is employed as the pressure medium in a closed container, where the working pressure can reach 200 MPa. The manufactured parts are pressed evenly in all directions with high temperature and pressure. Therefore, the manufactured parts have high density, good uniformity, and excellent performance. HIP has the characteristics of a short production cycle, low energy consumption, maximizing material utilization by improving material properties, and allowing smaller, lighter-weight, high-strength parts. HIP can heal or eliminate the inherent defects and pores in the parts (superalloy Inconel 718) produced by powder bed fusion (PBF)–electron beam melting (EBM) (Figure 8.6).

   HIP can significantly strengthen the fatigue strength of Ti–6Al–4V prepared by EBM. This improvement is due to the reduction of crack initiation points in the material. The mechanical properties of the EBM products have been optimized by HIP processes. Excellent mechanical properties can be achieved with HIP treatment because of the reduction in porosity and unmelted material, as well as the coarsening of microstructure during the high operating temperature of the HIP process. HIP is especially suited for demanding industries such as aerospace and biomedical, and even energy and automotive. The equipment is large and costly and may require a specialty service provider. HIP pressure chambers typically are limited in size, ranging from ϕ75 mm to2 meters.

2. **Annealing** is a heat treatment process performed by controlled heating and cooling for increased stress relief and improved mechanical properties. Such treatments may require 2–4 h in an inert or vacuum furnace at

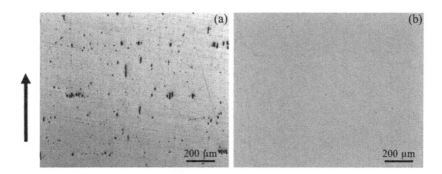

FIGURE 8.6 Optical micrographs presenting defects. (a) Electron beam melting (EBM) part with obvious defects. (b) Optical microstructure after the hot isostatic pressing (HIP) treatment. The arrow indicates the build direction. (Reprinted from Goel, S., Sittiho, A., Charit, I., Klement, U. and Joshi, S., Effect of Post-treatments under Hot Isostatic Pressure on Microstructural Characteristics of EBM-built Alloy. *Additive Manufacturing*, 28, 718, 727–737, Copyright (2019), with permission from Elsevier.)

temperatures ranging from 650 to 1150°C. Heat treatments may be required to improve the desired strength, hardness, and ductility, fatigue, or bulk properties. Annealing, homogenization, or recrystallization may be needed to achieve uniform bulk properties and to realize the desired microstructures. The layer-by-layer deposition can result in directionally dependent properties and can vary with respect to the orientation of the part within the build chamber. Heat treatment furnaces used for treating metal parts may need to operate at high temperatures and use inert atmospheres, such as argon or vacuum, when processing certain materials. Relief annealing of residual stresses present in an AM part may be required to ensure dimensional stability. Research has indicated that different heat treatments of the same material were required to optimize either hardness or wear resistance.

6. Surface Finishing: This is the "last touch" to further smooth and polish the piece, optimize aesthetics, and if the application requires it, reduce roughness to a minimum. In particular, surface treatments take care of problems induced by powder accumulation (partially melted powder) and staircasing or layering defects.

8.5.3 TECHNOLOGIES TO IMPROVE SURFACE QUALITY OF METALS AND ALLOYS PROCESSED BY AM

These technologies comprise operations based on both mechanical and thermal post-processing (Figure 8.7).

8.5.3.1 Surface Finishing Operations Based on Mechanical Post-Processing

The mechanically based technologies comprise abrasive flow machining (AFM), roll burnishing and ballizing, shot peening (SP), micro-shot peening (μSP), ultrasonic

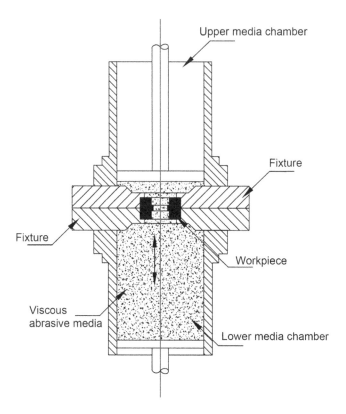

FIGURE 8.7 AFM schematic. (From Youssef, H. and El-Hofy, H., (2020) *Machining Technology and Operations, 2-Volume Set.* CRC Press, USA. With permission from CRC Press)

shot peening (USP), cavitation peening (CP), water jet peening (WJP), and oil jet peening (OJP) (Figure 8.7).

1. **Abrasive flow machining (AFM):** This is a purely mechanical process that finishes surfaces and edges by extruding a viscous abrasive medium flowing, under pressure, through or across a workpiece. This process provides a high level of surface finish and close tolerances with an economically acceptable rate of surface generation for a wide range of industrial components. A common setup is to position the work part between two opposing cylinders, one containing media (visco-elastic polymer) and the other empty. The polymer has the consistency of putty. The medium is forced under pressures ranging between 0.7 and 20 MPa to flow through the part from the first cylinder to the other, and then back again, as many times as necessary to achieve the desired material removal and finish (Figure 10.1). Many types of abrasives are used in AFM: mostly, B_4C and SiC; however, Al_2O_3 or diamond abrasives are sometimes used. The most abrasive action occurs during the

process if the hole changes size or direction. AFM is capable of removing EDM, EBM, and laser beam machining (LBM) damaged surfaces and significantly improving surface roughness (Youssef et al. 2020).

When finishing additively manufactured metals using abrasive flow machining, the surface roughness of additively manufactured parts was about 10–50 μm, which was significantly improved when processed by AFM to a value of 0.94 μm. Tan and Yeo (2017) investigated this post-processing technique and realized a high surface quality on Inconel 625 parts manufactured by DMLS.

2. **Roll Burnishing and Ballizing:** This is a surface hardening and finishing process, which involves rolling or rubbing smooth hard objects under considerable pressure over minute surface irregularities that were previously produced by machining, AM, or other operations. It is also used to improve the size and finish of internal and external cylindrical and conical surfaces. Since surfaces are cold worked and in residual compression, they possess improved wear and fatigue resistance. Also, the surfaces produced have high corrosion resistance due to their improved surface quality. Both soft and hard surfaces can be burnished (Youssef et al. 2023). The boss and tapered parts of holes are roller burnished, as shown in (Figure 8.8a), thus improving the surface quality by smoothing and leveling scratches, tool marks, and pittings. In ballizing, Figure 8.8b, a slightly over-dimensioned hard ball (D) is axially pushed (forced) into the hole (d) through its entire length. Burnishing and ballizing are applicable for smoothing hydraulic components, valves, sealing journals, shafts, plungers, and fillets produced conventionally or by AM.

3. **Shot peening (SP):** Shot peening is a cold working process, which is used to impart a compressive residual stress (CRS), which will suppress the TRS,

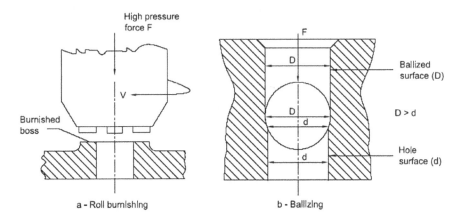

FIGURE 8.8 Roll burnishing (a) and ballizing (b). (From Youssef, H.A., El-Hofy, H.A. and Ahmed, M.H. (2023) *Manufacturing Technology: Materials, Processes, and Equipment.* CRC Press, Florida, USA. With permission from CRC Press)

decreasing the chances of crack propagation. In shot peening, solid objects such as metal shots, glass, or beads are used to impact on the solid part (Figure 8.9); as a result, a plastic deformation occurs on the surface of the material and the material just below. In consequence, a CRS is induced in the material.

Zhang and Liu (2015) studied the improvement of the surface quality and mechanical properties of additively manufactured stainless steel parts fabricated using DMLS. In general, shot peening effectively enhances the roughness, hardness, compressive yield strength, and wear resistance of the part. Accordingly, it was noted that due to shot peening, grain size refinement occurred in the material. In this study, the initial residual stress (with a value of −119 MPa) was increased up to a value of −700 MPa.

4. **Micro-shot peening (μ-SP):** Micro-shot peening is carried out using micro-spherical particles called micro-shots. The main difference from the SP is the sizes of shots, which range from 30 to 150 μm in diameter. Generally, the micro-shots are made of high-speed tool steel, cemented carbide, or ceramic, with high hardness. It has been found that μ-SP improves the wear resistance of 17CrNiMo steel, and a combination of SP and μ-SP gives even better results (Zhang et al. 2016).

5. **Ultrasonic shot peening (USP):** This method uses powerful ultrasonic vibrations of an acoustic horn with a high frequency, which sets spherical shots in motion. As a result, the intensity of pressing is very high. USP gives good results for soft materials and reduces the probability of crack initiation

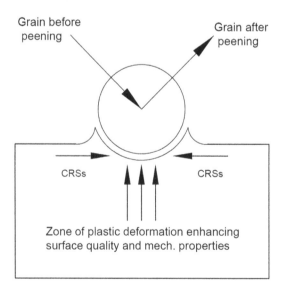

FIGURE 8.9 Schematic diagram of shot peening. (Adapted from Davim, J.P. and Gupta, K. (Eds.), *Handbooks in Advanced Manufacturing, Additive Manufacturing*, Edited by Pou Juan, Riveiro Antonio, and Paulo Davim J., Elsevier Inc, doi: 10.1016/C2018-0-00910-X2021).

and propagation. Positive aspects have also been found for the physical and mechanical properties of this treatment, such as improved corrosion resistance, increased contact, and fatigue strength (Wang et al. 2020).

6. **Cavitation peening (CP):** Cavitation peening (CP) does not use shots but produces cavitation, and the deformation is caused by pressure waves. A great advantage of cavitation peening is the absence of a shot peening medium that could contaminate the surface, as in other methods, e.g. SP or USP, where ceramic or metallic particles are used, or as in the case of laser shock peening (LSP), where surface contaminating products may appear. The process efficiency is three times higher for CP than for water jet peening. This method is effective for peening hard materials such as titanium and its alloys. Chromium molybdenum steel is also a topic for research due to its common use in industrial applications (Figure 8.10). Tan and Yeo (2020) show the mechanism of irregularity removal from the surface produced by means of powder bed fusion (PBF) under cavitation conditions.

7. **Water jet peening (WJP):** In this method, a high-velocity water jet hits the surface. The droplets generate a high peak pressure, which plastically deforms the surface. WJP is different from cavitation peening. Results on spring steel, titanium, and nickel suggest that this may be a valuable method for applications where increased roughness and compressive residual stress (CRS) are required (Zhang et al. 2016).

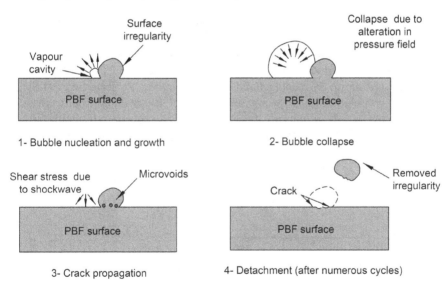

FIGURE 8.10 Cavitation peening (CP) mechanism of irregularity removal from the surface produced by means of powder bed fusion (PBF). (1) Bubble nucleation and growth; (2) bubble collapse; (3) crack propagation; (4) detachment (after numerous cycles) (Reprinted from Tan, K.L. and Yeo, S.H., Surface Finishing on IN625 Additively Manufactured Surfaces by Combined Ultrasonic Cavitation and Abrasion. *Additive Manufacturing,* 31, 100938, Copyright (2020), with permission from Elsevier)

8. **Oil jet peening (OJP):** The principle of the OJP method is the same as for water jet peening, the difference being the medium, which is hydraulic oil, used to create the jet. With this method, it is possible to introduce CRS without significant changes in the surface topography. By using OJP, it is possible to improve CRS in materials such as low carbon steel or aluminum alloys.

8.5.3.2 Surface Finishing Operations Based on Thermal Post-Processing

The thermally based technologies comprise laser shock peening and laser polishing.

1. **Laser shock peening (LSP):** In LSP, a laser beam is used to create a plasma shock wave, which is reflected on the surface of the component, inducing a CRS that deeply penetrates into the material, as shown in Figure 8.11. A transparent layer is required, although a non-transparent (opaque) layer is also often used. The transparent layer (usually water) is used to reduce the plasma produced on the surface and increase the shock wave intensity. The temperature at the heating point reaches up to 10,000°C, and the pulse length is about 0.15 to 0.30 ns. After the laser beam passes through the transparent layer, it hits an opaque layer, usually made of aluminum, zinc, or copper, which is also called a sacrificial coating. The opaque layer is used to prevent direct laser ablation on the surface. An impulse of laser reaches the absorption layer, and then ionization occurs in the heated zone, and a shock wave is generated, which plastically deforms the material. LSP is an appropriate method for improving the surface quality and mechanical properties of additively manufactured components.

   Sun (2018) studied the microstructural and mechanical properties of wire arc additively manufactured 2319 aluminum alloy subjected to LSP. LSP was able to impart a maximum CRS of 100 MPa up to a depth of 0.75 mm, which significantly improved the tensile properties. It significantly improved the microstructure and microhardness. LSP is widely utilized to

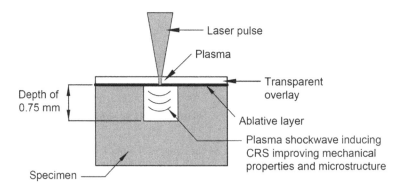

FIGURE 8.11 Schematic diagram of laser shock peening (LSP). (Adapted from Ding, K. and Ye, L., in *Laser Shock Peening*, Woodhead Publishing, Sawston, 2006)

improve the fatigue life of compressor blades, jet engine fans, and aircraft structures. It has also been applied to improve the surface properties of processed maraging steels.

2. **Laser polishing:** Laser polishing (LP) is considered a potential method for improving the surface roughness of AM parts. It is a noncontact surface finishing process that uses laser irradiation to achieve subsequent surface smoothening, as shown in Figure 8.12 (Marimuthu et al. 2015). In this process, a laser beam is used to melt the surface of the workpiece to reduce the surface roughness. During laser polishing, morphology apexes can reach the melting temperature. The liquid material redistributes because of the effect of gravity and surface tension. Once the laser beam stops scanning the surface, the temperature of the heat-affected zone (HAZ) drops rapidly, resulting in the solidification of the molten pool, and the surface roughness reduces accordingly. Laser polishing is an automated process that changes surface morphology by re-melting without changing or affecting the bulk properties.

During the past 20 years, laser polishing technology has been extensively utilized to post-process different materials. Mai and Lim (2004) applied laser polishing to reduce the roughness of 304 stainless steel from 195 to 75 nm, consequently increasing surface reflectivity to 14% and reducing diffusion reflectivity to 70%. Lamikiz et al. (2007) claimed that the hardness of a laser polishing post-processed surface was slightly higher and more uniform than other surfaces, with almost no cracks or HAZ. Laser polishing can also improve the surface properties of SLM parts.

Laser polishing of Ti6Al4V parts helps to reduce the surface roughness by the ability of re-melted liquid material to be uniformly spread. The formation of martensite during re-melting of titanium alloy resulted in higher hardness. Reduction in surface roughness also helped to improve the wear resistance of the components. It was noted that due to LP, the

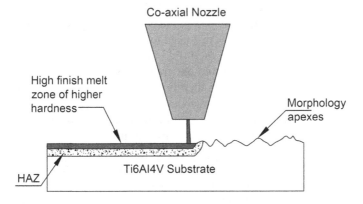

FIGURE 8.12 Schematic diagram of laser polishing. (Adapted from Marimuthu, S. et al., (2015) *International Journal of Machine Tools and Manufacture*, 95, 97–104)

surface roughness was reduced to 80% than the initial surface roughness (Dadbakhsh et al. 2010 and Ma et al. 2017).

8.5.3.3 Conclusions Regarding Finish Post-Processing Operations of Metal and Alloys

The performance of AM metallic materials, such as wear resistance, can be improved by applying the appropriate surface treatment. One of the traditional methods of increasing resistance to abrasive wear is shot peening (SP). However, the development of SP results in introducing new treatment methods, such as cavitation peening (CP), which is considered one of the most promising peening methods. The crucial application of CP is the removal of stresses after previous treatments. SP is generally applied to reduce the surface roughness, increase the hardness, and densify the surface layer microstructure, which leads to work hardening effects. In addition, the residual compressive stresses introduced into the material have a beneficial effect on the performance of the surface layer. Therefore, SP is considered a competitive process to heat treatment, enabling the required mechanical properties of steels and alloys produced by AM (Swietlicki et al. 2022).

Various peening methods have been used for finishing parts made of metallic materials produced by printing, and combinations of peening methods with heat treatment are also used by Soyama and Takeo (2020). Significantly, the use of cavitation peening seems to make the most sense, due to the possibility of processing locations for working fluid where solid shots cannot reach. Moreover, it is suggested that the combination of shot peening methods, namely, hybrid peening (SP + CP) in the case of additive manufactured components, and the use of light metals such as magnesium alloys and aluminum could pave the way for new, unprecedented solutions by increasing the fields of industrial applications (Bagherifard 2019).

8.6 POST-PROCESSING TECHNIQUES IN AM OF CERAMICS

As the additive manufacturing of ceramics (AMC) is now more widely adopted, the need to develop and optimize appropriate post-processing steps has emerged as one of the more critical challenges that need to be overcome. Post-processing steps of ceramics include pyrolysis (up to 700°C) and sintering (up to 2300°C), which are necessary to remove binders from the green body and consolidate the ceramic powder into its final geometry with final properties. One of the challenges is that green bodies formed by AM machines contain, at a maximum, 65–72 vol.% ceramic powder. This means that post-processing to full density will result in dimensional shrinkage exceeding 30%, which can lead to warping, cracking, and poor dimensional control.

Post-processing for AM of ceramics involves various techniques aimed at improving the mechanical properties, surface finish, and overall quality of the printed ceramic parts. Ceramics, being brittle materials, require specific post-processing methods. Here are some common post-processing techniques used in ceramic AM:

- **Debinding:** For ceramics produced via methods like binder jetting, vat polymerization, or material extrusion, debinding is a crucial step. It involves removing the binding material used during the printing process. This is typically done through thermal or chemical means to eliminate organic components from the printed part.
- **Sintering:** Sintering is a key step in ceramic AM, where the part is subjected to high temperatures to fuse the ceramic particles together, achieving the final density and strength of the part. This step is critical, as it transforms the printed part into a solid ceramic object.
- **Surface smoothing:** Techniques such as polishing, grinding, or chemical treatments can be employed to smoothen the surface of the printed ceramic parts, improving their appearance and functionality.
- **Polishing:** Similarly to metals, post-processing for ceramics may involve polishing operations to achieve precise dimensions or specific features that could not be achieved during printing.
- **Surface Coating or Glazing:** Applying coatings or glazes to ceramic parts to enhance properties such as surface hardness, wear resistance, and aesthetics. This can also help improve the part's resistance to environmental factors.
- **Quality control and inspection:** Inspection methods like computer tomography (CT) scanning, X-ray inspection, or other non-destructive testing methods are used to ensure the integrity and dimensional accuracy of the ceramic parts.
- **Assembly and joining:** Joining multiple ceramic parts through methods like adhesive bonding or mechanical fastening after printing if the design requires assembly.
- **Surface texturing:** Etching or other surface modification techniques to create specific textures or patterns on the ceramic surface for functional or aesthetic purposes.
- **Densification enhancement:** Some additional processes can be used to further enhance the densification of ceramic parts, such as HIP or pressureless sintering, especially if higher strength and density are required.

The post-processing steps for ceramic AM parts play a critical role in achieving the desired properties, dimensions, and quality required for specific applications. As previously mentioned, the selection of these post-processing techniques is often based on the type of ceramic material used, the final application of the part, and the required performance characteristics.

8.6.1 Post-Processing in Binder Jetting Technology (BJT)

The heat treatment performed on ceramics fabricated using BJT usually consists of a curing process, in which the green body is subjected to a temperature around 200°C to strengthen the polymer. Once the curing process is completed, the sample is transported to a high-temperature furnace. In this furnace, the printed sample is subjected to a binder burnout stage, during which the polymeric binder is removed from the sample. After the binder burnout stage, the sample is subject to a sintering stage, during which the ceramic is densified by subjecting it to high temperatures to allow the objects to achieve their highest mechanical properties. The density of the final parts is dependent on the sintering time and temperature. The BJ process is also not limited by any thermal effects, such as layer warping. The roughness and porosity of the final objects are relatively higher than those obtained through VP processes. To fill the pores, post-processing steps could involve infiltrating the porous objects with a solution in order to strengthen the parts. The surface roughness can be improved by manual operations like sanding or grinding.

8.6.2 Post-Processing in Vat Photopolymerization

After the fabrication of the green parts, they are subjected to debinding and sintering, consecutively. The debinding and sintering temperatures depend upon the binding material and the structure material used, respectively. For alumina and zirconia materials, the sintering temperature could range up to 1600°C. During debinding and sintering, it is necessary to sinter the parts under a controlled temperature to reduce the thermal stresses and avoid cracking. After sintering, the final objects obtained are highly dense ceramic objects with complex features. The final objects result in fine surfaces with a low Ra value and minimum porosity, which makes the parts ready for post-processing operations like grinding, sanding, or infiltrating the pores. In vat photopolymerization technologies used for ceramics, such as SLA or DLP, post-processing steps are essential to achieve the desired properties, accuracy, and surface finish of the printed ceramic parts.

8.6.3 Post-Processing in Material Extrusion (ME)

In ME, the green parts are formed by depositing the feedstock material as a semi-molten state through the liquefier's nozzle while the material is maintained between the temperatures of 200 and 265°C. Once the green parts are formed, they are subjected to debinding and sintering processes. The liquid binder components are first removed by capillary action when set inside the dry alumina powder at a lower temperature than 200°C. This is followed by heating the parts over 200°C, such that the remaining residual binder material is removed by evaporation and decomposition, leaving porous green parts, followed by sintering the porous green parts up to a temperature of 1600–1650°C for alumina-based feedstock, causing densification and linear shrinkage of the ceramic parts. Ceramics can be printed with 100% infill density using ME processes, but once the binder is burned out after sintering, the final workpieces tend to be porous.

Virtually all ceramic AM built parts require at least some post-processing, which is a step that is as critical as that of the respective AM process itself. Heat treatments to relieve stresses may also be important, because residual stresses readily result in cracking and/or warping of the product. Post-processing commonly involves clean-up, debinding of any support materials, and product densification. Clean-up of printed parts can be costly and time-consuming (depending on the AM process and the type of product). The debinding step, however, which is required for many, but not all, ceramic AM processes, can be a primary source of defects and associated failures. Sintering of a green-body AM build presents perhaps the most common challenge for delivering the desired shape and (full) density without introducing defects and part failures.

Most recently, a new densification technique called ultrafast high-temperature sintering (UHS) has been reported (Wang et al. 2020). In this method, the part to be sintered is sandwiched between two carbon tapes through which electric current is passed, providing rapid (in seconds) heating up to temperatures as high as 3000°C. The process is performed in an inert atmosphere. UHS has been demonstrated to achieve densities greater than 95% while yielding uniform compositions and favorable microstructures for a broad range of ceramics that can be difficult to densify with the same results using conventional approaches.

8.6.4 Post-Processing in Material Jetting (MJ)

The green bodies formed using MJ processes consist of wax or solvent, which needs to be removed prior to the sintering process. The dewaxing is done by various processes like drying in the dry chamber under a controlled temperature for a certain time or putting into carbon black powder that can remove wax through capillary action. After dewaxing, the objects are sintered at high temperature based on the filling material type. The sintering action is generally held for a certain time to burn the residual wax or any high-molecular-weight surfactant residues, followed by heating to the maximum temperature of the structure material in consecutive steps. Post-sintering, the objects obtained tend to achieve highly dense ceramic components. Post-processing steps for ceramic parts produced via MJ technologies play a significant role in achieving the desired properties, dimensional accuracy, and surface finish required for various applications.

8.6.5 Post-Processing in Powder Bed Fusion (PBF)

The post-processing steps for ceramic parts produced via PBF technologies play a crucial role in achieving the desired properties, dimensional accuracy, and surface finish required for various applications. The selection of these techniques is typically based on the type of ceramic material used, the intended application of the part, and the desired performance characteristics.

The fabricated parts are nearly densified in the SLS process, but the residual binder material can be removed by heat-treating the specimen. After sintering, the relative density of the parts is increased, and they are fully densified. Post-processing the green parts by methods such as an infiltration process, polishing, thermal treatment, and coating, can improve their mechanical strength, surface roughness, and structural integrity, and decreases porosity.

8.7   FUTURE OUTLOOK FOR POST-PROCESSING AND CONCLUSIONS

As a future outlook, there are three viewpoints to be highlighted. These are:

1. *Post-processing standardization:*

   Various researchers and scientists have carried out efforts to facilitate AM post-processing commercialization. Each part produced via an AM process contains multiple defects. These defects decrease the life expectancy of a produced part and limit its utilization. To prescribe a particular post-processing technique for specific defects, there is a need to establish standardization of the process. These efforts will guarantee a manufactured part's surface integrity and adequate mechanical characteristics, thus avoiding failure during application.

2. *Part complexity:*

   The post-processing, depending on the complexity of the printed features, has some challenges in both cleaning and post-curing steps, such as taking out the part from the build plate and removing the support structures. In addition, in the cleaning step for removing the excess resin between the features, mainly the high-aspect-ratio features are prone to be deformed. Thus, printing small-scale features presents more difficulties in cleaning and removing the leftover resin. Moreover, over-curing the parts may cause cracks or breakage of the printed part.

3. *Post-processing automation:*

   Post-processing covers a variety of stages that 3D printed parts have to undergo before being used for the final purpose, such as powder removal, stress relief annealing, wire cutting, other finishing, hot isostatic pressing (HIP), and so on. Some of these procedures still require manual operation, wherein skilled operators are necessary for key tasks. It may be cost-effective to complete the prototype or even dozens of parts manually, but if hundreds or even thousands of parts are produced, the demand for post-processing automation in AM becomes extremely urgent. Automated solutions for post-processing can improve the overall production efficiency of AM.

In conclusion, AM is no longer just a method for prototyping but has firmly established itself as a technology for volume production. However, components made by AM require extensive post-processing, a challenge that is frequently underestimated and not taken seriously. As early as during the design phase for a new product, AM users must take into account and resolve issues regarding suitable, cost-efficient post-processing, including surface refinement. This ensures that the expected workpiece characteristics can be achieved. Today, the post-processing of 3D printed components is still done with costly and highly erratic manual operations, which make a cost-effective, industrial production practically impossible. The right know-how and suitable equipment are key factors for an economically viable and successful use of AM.

During AM manufacturing, the post-processing techniques consume almost 43% of the total time (www.3dnatives.com/). Reducing the cycle time is one of the significant challenges. Therefore, there is an urgent need for process automation of post-processing techniques. These automated solutions can promote production effectiveness. This can be done via machine or deep learning techniques, useful in process automation, process control, and optimal solutions.

In order to automate these technologies, some companies have also begun to implement robotic solutions that can install printing substrates, clean powder, and unload and post-processing parts. Some companies, such as Dye Mansion, are focused on post-processing machines only. Others, such as Carbon and FormLabs, are 3D printer manufacturers that are adding post-processing systems to work seamlessly with their printing setups. In addition, some companies, like Stratasys, have introduced new technology like water-soluble supports, WaterWorks™, which requires users to simply immerse the prototype into a water-based solution and wash away the supports, thus eliminating the tediousness of spending several hours removing supports from newly created prototypes and cleaning them afterwards. In conclusion, the number of advanced automatic post-processing solutions will certainly increase in the future so as to adapt to the growing development of the AMT.

8.8   REVIEW QUESTIONS

1. What is post-processing of 3D printed components, and why is it an essential part of the additive manufacturing process?
2. What is post-processing in additive manufacturing?
3. Is there a technology available that can handle all post-processing tasks?
4. What is the importance of shot blasting and mass finishing for post-processing?
5. Which post-processing tasks can be handled by shot blasting?
6. Which post-processing tasks can be handled by mass finishing?
7. Enumerate some of the advantages of post-processing techniques.
8. Distinguish cleaning, post-curing, and finishing, which are the various tasks of post-processing. Name two AM processes that do not require post-curing and one that does not require cleaning.
9. What are the common post-processing techniques in AM?
10. Why is post-processing necessary in additive manufacturing?.
11. What considerations should be taken for selecting post-processing techniques?
12. How does post-processing affect the mechanical properties of AM parts?
13. Is post-processing always necessary for 3D printed parts?
14. Are there any automated post-processing techniques available?
15. How does post-processing impact the cost and lead time of additive manufacturing?
16. Can post-processing techniques be used to repair defects in 3D printed parts?

BIBLIOGRAPHY

Albright, B. (2021) Post-Processing 3D Printed Prototypes Getting to the Finished Part. Available online: https://www. digitalengineering247.com/article/post-processing-3d -printed-prototypes/ (accessed on 5 August 2021).

AM Post-Processing Cost Identified. Available online: https://www.materialstoday.com/additive-manufacturing/news/ampostprocessing-cost-identified/ (accessed on 10 November 2020.

Bagherifard, S. (2019) Enhancing the Structural Performance of Lightweight Metals by Shot Peening. *Advanced Engineering Materials*, 21(7), 1801140. https://doi.org/10.1002/adem.201801140

Bai, Y., Chaudhari, A., and Wang, H. (2019) Investigation on the Microstructure and Machinability of ASTM A131 Steel Manufactured By Directed Energy Deposition. *Journal of Materials Processing*, 276, 116410. https://doi.org/10.1016/j.jmatprotec.2019.116410

Bhandari, S., Lopez-Anido, R. A., and Gardner, D. J. (2019) Enhancing the Interlayer Tensile Strength of 3D Printed Short Carbon Fiber Reinforced PETG and PLA Composites via Annealing. *Additive Manufacturing*, 30, 100922. https://doi.org/10.1016/j.addma.2019.100922

Carlota, V. (2020) Post-Processing Trends Report Reveals Current Methods and Challenges—3D Natives. Available Online: https://www.3dnatives.com/en/postprocess-trends-2020-020920204/ (accessed on 12 December 2021).

Cheng, C., and Gupta, M. (2017) Surface Functionalization of 3D-Printed Plastics via Initiated Chemical Vapor Deposition. *Beilstein Journal of Nanotechnology*, 8, 1629–1636. https://doi.org/10.3762/Bjnano.8.162

Dadbakhsh, S., Hao, L., and Kong, C. Y. (2010) Surface Finish Improvement of LMD Samples using Laser Polishing. *Virtual Physical Prototyping*, 5(4), 215e221. https://doi.org/10.1080/17452759.2010.528180

Davim, J. Paulo and Gupta, Kapil. (Eds.). (2021) *Handbooks in Advanced Manufacturing, Additive Manufacturing*. Edited by Pou Juan, Riveiro Antonio, and Paulo Davim J., Elsevier Inc,. doi: 10.1016/C2018-0-00910-X

Ding, K. and Ye, L. (2006) General Introduction. In *Laser Shock Peening*. Sawston: Woodhead Publishing.

Garcia, E. A., Ayranci, C., and Qureshi, A. J. (2020) Material Property-manufacturing Process Optimization for Form 2 VAT-photo Polymerization 3D Printers. *Journal of Manufacturing and Materials Processing*, 4(1), 12. https://doi.org/10.3390/jmmp4010012

Goel, S., Sittiho, A., Charit, I., Klement, U., and Joshi, S. (2019) Effect of Post-treatments under Hot Isostatic Pressure on Microstructural Characteristics of EBM-built Alloy 718. *Additive Manufacturing*, 28, 727–737. https://doi.org/10.1016/j.addma.2019.06.002

Jorgenson, L. (2021) How Annealing Makes Your 3D Prints Better. Available Online: https://www.fargo3dprinting.com/annealingmakes-3d-prints-better/ (accessed on 6 August 2021).

Lamikiz, A., Sánchez, J., de Lacalle, L. L., and Arana, J. (2007) Laser Polishing of Parts Built up by Selective Laser Sintering. *International Journal of Machine Tools and Manufacture*, 47, 2040–2050. https://doi.org/10.1016/j.ijmachtools.2007.01.013

Ma, C. P., Guan, Y. C., and Zhou, W. (2017) Laser Polishing of Additive Manufactured Ti Alloys. *Optics and Lasers in Engineering*, 93, 171–177. https://doi.org/10.1016/j.optlaseng.2017.02.005

Mai, T. A. and Lim, G. C. (2004) Micromelting and its Effects on Surface Topography and Properties in Laser Polishing of Stainless Steel. *Journal of Laser Applications*, 16, 221–228. https://doi.org/10.2351/1.1809637

Marimuthu, S., Triantaphyllou, A., Antar, M., Wimpenny, D., Morton, H., and Beard, M. (2015) Laser Polishing of Selective Laser Melted Components. *International Journal of Machine Tools and Manufacture*, 95, 97–104. https://doi.org/10.1016/j.ijmachtools .2015.05.002

Pandey, P. M., Reddy, N. V., and Dhande, S. G. (2003) Slicing Procedures in Layered Manufacturing: A Review. *Rapid Prototyping Journal*, 9(5), 274–288. https://doi.org /10.1108/13552540310502185

Ryan, C. Dizon John, Gache, Catherine L. Ciara, Cascolan, Honelly Mae S., Cancino, Lina T., and Advincula, Rigoberto C. (2021) Review: Post-Processing of 3D-Printed Polymers. *Technologies*, 9(3), 61. https://doi.org/10.3390/technologies9030061

Soyama, H. and Takeo, F. (2020) Effect of Various Peening Methods on the Fatigue Properties of Titanium Alloy Ti6Al4V Manufactured by Direct Metal Laser Sintering and Electron Beam Melting. *Materials*, 13(10), 2216. https://doi.org/10.3390/ma13102216

Sun, R. (2018) Microstructure, Residual Stress and Tensile Properties Control of Wire-arc Additive Manufactured 2319 Aluminum Alloy with Laser Shock Peening. *Journal of Alloys Compound*, 747, 255–265. https://doi.org/10.1016/j.jallcom.2018.02.353

Swietlicki, Aleksander, Szala, Mirosław, and Walczak, Mariusz. (2022) Effects of Shot Peening and Cavitation Peening on Properties of Surface Layer of Metallic Materials—A Short Review. *Materials (Basel).*, 15(7), 2476. https://doi.org/10.3390/ma15072476.

Tan, K. L. and Yeo, S. H. (2017) Surface Modification of Additive Manufactured Components by Ultrasonic Cavitation Abrasive Finishing. *Wear*, 378(379), 90e95. https://doi.org/10 .1016/j.wear.2017.02.030

Tan, K. L. and Yeo, S. H. (2020) Surface Finishing on IN625 Additively Manufactured Surfaces by Combined Ultrasonic Cavitation and Abrasion. *Additive Manufacturing*, 31, 100938. https://doi.org/10.1016/j.addma.2019.100938

Tijing, L. D., Dizon, J. R. C., Ibrahim, I., Nisay, A. R. N., Shon, H. K., and Advincula, R. C. (2020) 3D Printing for Membrane Separation, Desalination and Water Treatment. *Applied Materials Today*, 18, 100486.

Uzcategui, A. C., Muralidharan, A., Ferguson, V. L., Bryant, S. J., and McLeod, R. R. (2018) Understanding and Improving Mechanical Properties in 3D-printed Parts Using a Dual-Cure Acrylate-Based Resin for Stereolithography. *Advanced Engineering Materials*, 20(12), 1800876. https://doi.org/10.1002/adem.201800876

Wang, C., Weiwei, P., Bai, Q., Cui, H., Hensleigh, R., Wang, R., Brozena, A. H., Xu, Z., Dai, J., Pei, Y., Zheng, C., Pastel, G., Gao, J., Wang, X., Wang, H., Zhao, J.-C., Yang, B., Zheng, X. (R)., Luo, J., Mo, Y., Dunn, B., and Hu, L. (2020) A General Method to Synthesize and Sinter Bulk Ceramics In Seconds. *Science*, 368(6490), 521–526. https:// doi.org/10.1126/science.aaz7681

Wohlers Associates. (2021) Wohlers Specialty Report on Post-Processing. https://wohlersas-sociates.com/.

Youssef, Helmi and El-Hofy, Hassan. (2020) *Machining Technology and Operations, 2-Volume Set*. USA: CRC Press.

Youssef, Helmi A., El-Hofy, Hassan A., and Ahmed, Mahmoud H. (2023) *Manufacturing Technology: Materials, Processes, and Equipment*. Florida, USA: CRC Press.

Zhang, J., Li, W., Wang, H., Song, Q., Lu, L., Wang, W., and Liu, Z. (2016) A Comparison of the Effects of Traditional Shot Peening and Micro-Shot Peening on the Scuffing Resistance of Carburized and Quenched Gear Steel. *Wear*, 368, 253–257. https://doi.org /10.1016/j.wear.2016.09.029

Zhang, P. and Liu, Z. (2015) Effect of Sequential Turning and Burnishing on the Surface Integrity of Cr-Ni-based Stainless Steel Formed by Laser Cladding Process. *Surface and Coatings Technology*, 276, 327–335. doi.org/10.1016/j.surfcoat.2015.07.026

9 Design for Additive Manufacturing

9.1 INTRODUCTION

Unlike traditional manufacturing processes, additive manufacturing (AM) is a technique that builds parts layer by layer. AM presents several advantages over traditional manufacturing. These are mass customizations, aesthetics, part consolidation, weight reduction, functional customizations, and so on (Quarshie et al. 2012). This chapter discusses design for AM and then presents technological and computational challenges in developing tools that can aid in design for AM. Design for AM is a notion that implies considering AM constraints early in the design; that is, preliminary design. Design for AM, if implemented correctly, will reduce the iterations a design needs to undergo before manufacturing, thus saving time and money.

Due to different capabilities of the AM process as compared with traditional manufacturing, the design methods need to be adapted for AM. Specifically, new design rules and tools are needed to:

- utilize topology optimization directly for part design
- use lattice and cellular structures in part design
- use multi-material and different material distributions in part design

Current methods of specifying and verifying product quality are not sufficient to specify and verify the shapes and parts that can be produced using AM. Therefore, product quality specification and verification methods should be adapted to include the capabilities of AM techniques.

The commonly used standard format for digital mock-ups used in AM is called stereolithography (SLA) file format (Three D Systems 1988). This format basically consists of triangulated geometry for the part to be created. Each triangle has unit normal and three vertices ordered to follow the right hand rule to find the direction of the normal. The coordinates of the vertices are represented in 3D Cartesian coordinates.

The precision of the triangulation and the number of triangles in the Standard Triangle Language (STL)define the precision with which the part will be produced. Usually, large numbers of triangles are needed to approximately represent a freeform surface. Figure 9.1 shows an example of coarse and fine triangulations approximating a section of an adapter tube in an STL file. The coarse (Figure 9.1a) triangulation has 1658 triangles, while the fine (Figure 9.1b) triangulation has 19,320 triangles. Variations of STL format include color information in the file for each triangle.

DOI: 10.1201/9781003451440-9

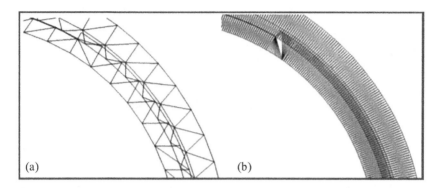

FIGURE 9.1   (a) Coarse (1658 triangles) and (b) fine triangulations (19,320 triangles) approximating a section of an adapter tube in an SLA file. Usually, triangulations work well with flat surfaces. For freeform surfaces, large numbers of triangles are needed to approximate the surface. (From Bandyopadhyay, A. and Bose, S., (2016) *Additive Manufacturing.* CRC Press, Taylor and Francis, Boca Raton. With permission from CRC Press)

To overcome the limitation of the STL format, the ASTM (American Society of Testing and Materials) standards committee has developed a new standard called AM file (AMF) format. AMF is an extended Markup Language (XML)-based format and is part of ASTM 2915 standard 2013. XML has the advantage that it can be easily interpreted by computers and can be expanded without affecting the compatibility of the files.

The basic advantage of AMF over SLA is the capability to include the following additional information regarding the object:

- Color specification
- Texture maps
- Material specification
- Constellations
- Additional metadata
- Formulae
- Curved triangles

Traditionally, AM parts are created by sending digital mock-ups to the AM machine, which then creates the parts. The digital mock-up is then processed by the AM machine in order to create the product layer by layer based on the specific technology used by the AM machine.

The overall steps required to create a given part after receiving the SLA file are shown in Figure 9.2. The digital mock-up in the form of an SLA file is imported in the native application for the AM technology. Usually, the user can now visualize in the native application to input the desired orientation and location of the part with respect to the build volume in the particular AM process. The SLA file and the user input are taken together in order to build slices of the digital mock-up. These slices create layers to be used in the AM process for producing the part. Although there are

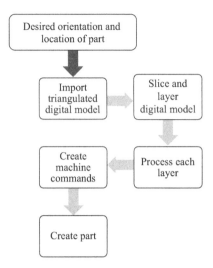

FIGURE 9.2 Steps for processing a digital model into a physical part using AM technology ((From Bandyopadhyay, A. and Bose, S., (2016) *Additive Manufacturing.* CRC Press, Taylor and Francis, Boca Raton. With permission from CRC Press)

techniques to create the slices directly from a computer-aided design (CAD) model without creating an STL file, these techniques are not commercially popular. Further processing of each layer may be required based on the particular AM technology. Some AM technology requires building support structures with the same materials as the part, while others can use different materials.

The processed layers and related information are then used to create machine commands that will be passed to the AM technology in order to produce the part. In this process, there are usually issues with *slicing* and *support structure building* from the SLA file.

Slicing the SLA representation of the digital model is done to create layers. The basic method of slicing is by using two parallel planes based on the orientation of the digital model. The SLA representation is then intersected with these two parallel planes. The distance between the two parallel planes is usually the desired layer thickness for the AM part.

After generating the layers using slices as discussed, the next task involves generating support structures based on the particular AM technology (e.g., fused deposition modeling [FDM], SLA). For generating a support structure, information regarding the material properties (usually, strength and weight) and geometry of subsequent layers at the time of layer construction has to be computed beforehand. If there are overhang features in consecutive layers, the weight and strength are used to identify whether a support structure will be needed or not (Figure 9.3).

As previously mentioned, AM is unique in the fact that it can produce surfaces and features that are not feasible using traditional manufacturing techniques. Examples of such features include lattice structures, internal cavities of different shapes and forms, assembly parts manufactured at once, and porous parts (Roundtable Forum

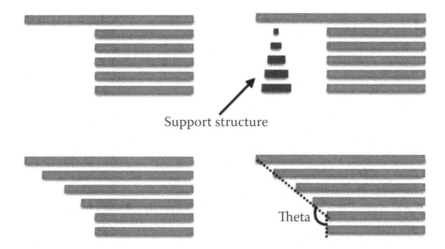

FIGURE 9.3 Examples showing how decisions are made regarding support structure creation. Either needed or may be avoided. (From Bandyopadhyay, A. and Bose, S., (2016) *Additive Manufacturing*. CRC Press, Taylor and Francis, Boca Raton. With permission from CRC Press)

2013). Due to these additional capabilities, designers need tools to be able to utilize the capabilities in optimizing their design. One such technique used by designers is topology optimization (Bendsoe and Sigmund 2011, in addition to Brackett et al. 2011). Topology optimization utilizes different mathematical techniques to identify the location of material and holes from a preliminary identification of the location of loads and constraints in a part or assembly. Various examples of application of topology optimization for AM are present in the literature (Keulen et al. 2014, Meisel et al. 2013, and Vayre et al. 2012). The surfaces of this topology-optimized part would be very time-consuming and highly cost-inefficient using traditional manufacturing techniques.

9.2 BENEFITS OF AM

Design for Additive Manufacturing (DfAM) is the art, science, and skill to design for manufacturability using the AM methods. This design process helps engineers to create more intricate shapes while reducing weight and material consumption. On the other hand, DfAM allows manufacturers to achieve highly complex geometries suitable for assembly and creates parts that would otherwise be difficult or costly to produce with traditional manufacturing methods. Among the benefits of AM are (https://www.desktopmetal.com/):

1. **Complex geometry:** Among the chief benefits DfAM brings to engineers and designers is the ability to produce parts with greater complexity than can be achieved using traditional manufacturing methods.

2. **Generative design:** Generative design uses software tools to create highly optimized parts backed by complex computational simulations. These parts are built to withstand specific loads and designed according to user-defined constraints, putting material where it's needed, resulting in parts that are as much as 50% lighter than conventional designs but with equal strength.

3. **Tooling-free manufacturing:** DfAM eliminates the need for hard tooling such as jigs, fixtures, or molds. Designers can produce multiple versions of a part, then quickly iterate on the design based on functional testing results. The speed and accessibility of AM allows designers to go through many design iterations in a short time.

4. **Assembly consolidation:** Creating complex new designs, combined with AM capabilities, opens the door to use DfAM principles to explore assembly consolidation. In this case, combining several parts into fewer, multifunctional assemblies significantly increases production efficiency, saves money by reducing waste, and lightens the weight of parts by eliminating the need for screws and other fasteners, thus reducing the production cost and simplifying the assembly procedures.

5. **Internal features and channels:** By building parts layer by layer, AM creates parts with internal features such as conformal cooling channels that would be impossible to produce by conventional machining or forming methods. Incorporating such complex internal features results in improved performance, increased production, and reduced time and cost.

6. **Fine features:** AM allows manufacturing engineers and designers to create highly complex parts with fine features that would be too time-consuming and expensive by conventional methods such as machining and forming. Using AM, these parts can be produced without the need for fixtures or other special tooling.

7. **Light weighting:** AM has the ability to create complex geometries which opens the door to light weighting of parts through the use of lattice-like designs and closed-cell infill and light weighting features that normally couldn't be produced using the traditional manufacturing processes. Hence, through light weighting, AM reduces waste, thus leading to shorter processing times and lower production costs.

AM design optimization can be done in several directions (European Powder Metallurgy Association, 2017):

- Reduce the total number of parts.
- Design for functionality.
- Design parts to be multifunctional.
- Lightweight.
- Topological optimization.
- Design for ease of fabrication.

DfAM covers both the design aspects, including part features, process parameters, and design considerations of any resulting outcome, such as accuracy, anisotropy, surface roughness, build time, and cross section (Figure 9.4; Bikas et al. 2019).

9.3 DESIGN ASPECTS FOR AM

The design aspects are defined as any particular features that can be quantified at the design phase. These include the geometric features of the part's shape (overhangs, bores, channels, bridging, minimum wall thickness, minimum size, and supports) and process parameters (layer thickness and build orientation) (Figure 9.4).

9.3.1 GEOMETRIC FEATURES

There are some restrictions regarding the geometries that can be built using AM. The layer-by-layer principle followed by AM machines has its limitations, since each layer must be built directly above the previous one. Therefore, not every geometry/ shape is possible to manufacture, as each geometrical feature must obey a certain geometrical continuity. Once this geometric continuity is not considered in the part design, the resulting part will suffer in its integrity (e.g. deformation, porous mass, or reduced density). The design aspects that determine the quality of the parts are as follows.

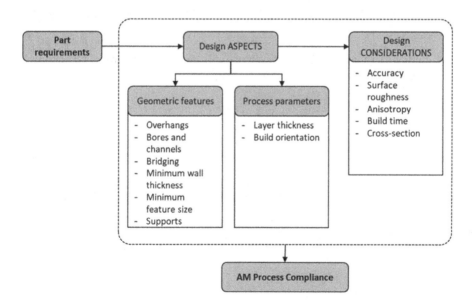

FIGURE 9.4 AM design aspects and considerations. (With kind permission from Springer Science+Business Media: Bikas, H., Lianos, A.K. and Stavropoulos, P. (2019) A Design Framework for Additive Manufacturing. *The International Journal of Advanced Manufacturing Technology*, 103, 3769–3783.)

9.3.1.1 Overhanging Geometries

An overhanging geometry is any geometry whose orientation is not parallel to the build vector. Manufacturing overhanging geometries is a challenging feature for the majority of the AM processes. In this respect, the ability of the AM process to manufacture a layer of material displaced to the previous defines its ability to create overhanging geometries. The magnitude of this layer's parallel shift sets the limit for the maximum overhang length (d) and the maximum slope angle (θ) (Figure 9.5). However, overhanging geometries can be successfully additively manufactured with the addition of support structures of poor layer adhesion or low post-processing demands. In such a case, a support structure guarantees a successful build of the part. It decreases the efficiency of the AM process through extra build time, material, post-processing operations and equipment, and final process cost. Therefore, in order to increase AM manufacturability, it is desirable that the part only has self-supported geometrical features. The most important overhanging geometries are presented in Figure 9.5. Support methods are based on the following rules (Jiang et al. 2018):

1. Avoid large-size holes parallel to the printing surface.
2. Avoid surfaces with a large overhang angle.
3. Avoid trapped surfaces where support materials are difficult or impossible to remove.
4. Optimize support structure to reduce the support material used.

Moreover, the design of support structures should be based on the following principles:

- Supports should be able to prevent parts from collapse/warping, especially the outer contour area, which needs support.

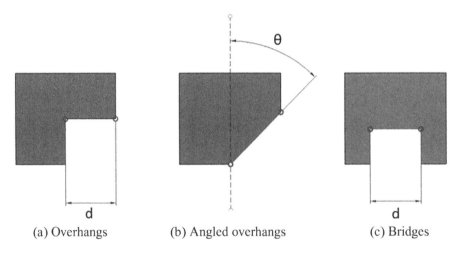

(a) Overhangs   (b) Angled overhangs   (c) Bridges

FIGURE 9.5 Overhanging categories. (From Bikas et al. 2019)

- The connection between the support and final parts should be of minimal strength to perform the supporting function, with the aim of easily removing them through post-finishing processes.
- The contact area between the support and the AM parts should be as small as possible to reduce surface deterioration after support removal.
- The material consumption and build time should be considered as a significant factor, as well as the trade-off between them and the final printed quality.

a. **Overhangs:** Overhangs are one-sided abrupt geometry changes Figure 9.5a. The horizontal distance an AM machine can build without supports is limited, and if it is exceeded, the whole build could fail. The limit of an overhang length is affected by the type of the AM process, the material used, and the actual AM machine. When part specifications call for a greater overhang, the decision is whether to alter the part's geometry or maintain it and add supporting structures.

b. **Angled overhangs:** The simplest way to resolve this is to replace horizontal overhangs with angled ones Figure 9.5b. In correlation with the length of the overhangs, the overhanging angle is a geometrical limiting factor to most AM technologies. Some AM technologies can produce angled overhangs of a certain gradient, where others cannot. When the angled overhang adaption is not feasible, due to the specifications and the geometry of the part, a support structure needs to be introduced to support the overhanging feature.

- For extrusion AM technologies, extreme angled overhangs cannot be created, as material cannot be deposited in mid-air.
- For powder bed fusion AM processes, the powder surrounding the part acts as a support, and thus, steeper angled overhangs can be realized. In such a case, there is a drawback regarding surface roughness, as the surrounding powder is sintered unevenly on the downward-facing areas of the part.

c. **Bridging:** A bridge is a horizontal geometry between two or more non-horizontal features Figure 9.5c. It is defined as any surface in the part geometry that is facing down between two or more features. The designer must take into consideration the maximum length that the AM machine can bridge; otherwise, the part will not be successfully manufactured.

9.3.1.2 Bores and Channels

Manufacturing internal bores that are impossible with traditional methods is feasible with AM without additional cost (Figure 9.6) (Bikas et al. 2019). This enables the creation of parts with internal concave channels, thus achieving great heat convection capabilities or optimum fluid flow or structural reinforcement lattice structures. Cooling channels that follow complex paths through the volume of the part increases the heat transfer capacity, which provides efficient cooling to mold cavities and reduces cooling time. Table 9.1 shows bore and channel dimension limits for AM technologies.

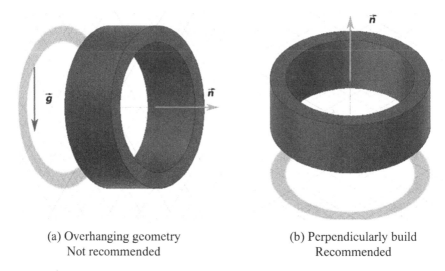

(a) Overhanging geometry
Not recommended

(b) Perpendicularly build
Recommended

FIGURE 9.6 Bores or channels: (a) overhanging geometry, (b) perpendicularly manufactured. (From Bikas et al. 2019)

9.3.1.3 Wall Thickness

There is a minimum feasible wall thickness for each AM process. This is due to the building threshold determined by the AM machine diameter of laser beam and flow focal point. Indicative minimum wall thicknesses are shown in Table 9.2. Below the lower limit of the allowed thickness, the wall feature either cannot be produced or when built, will suffer from deformation. An integer multiple of the fundamental tool path width must be used for the particular design. When the width of the geometry that must be manufactured is not an integer multiple of the fundamental tool path width, the slicer software will have to compensate for that issue. Another accountable parameter for thin walls is the height-to-thickness ratio.

9.3.1.4 Smallest Features

There are small features that challenge the ability of the AM machine when it comes down to manufacturability. A 2D thin feature an AM machine can manufacture is usually described by the diameter of the smallest possible pin. It also refers to the side of a rectangular or complex-curved geometry. This aspect should be considered at the design phase, as it defines the detail that can be introduced into the part. The smallest features of AM technologies can be seen in Table 9.3.

9.3.2 Process Parameters

The process parameters are selected at the slicing phase3 of the AM process. These are highly interconnected with the AM technology and the individual machine. The proper AM design considers both the layer thickness and the build orientation.

TABLE 9.1

Bores and Channels for AM Technologies

Bores and channels	Minimum diameter in (mm)	Vat polymerization			Extrusion	Material jetting			Binder jetting	Powder bed fusion				Direct energy deposition		Sheet
		SLA	DLP	CDLP	FDM	MJ	NPJ	DOD		MJF	SLS	SLM	EBM	LENS	EBAM	LOM
				N/A				N/A		N/A			N/A	N/A	N/A	N/A
	0.4															
	0.5															
	0.75															
	1.25															
	1.5															
	2.0															
	3.0															
	4.0															
	5.0															

Source: From Bikas et al. 2019

TABLE 9.2
Minimum Wall Thickness

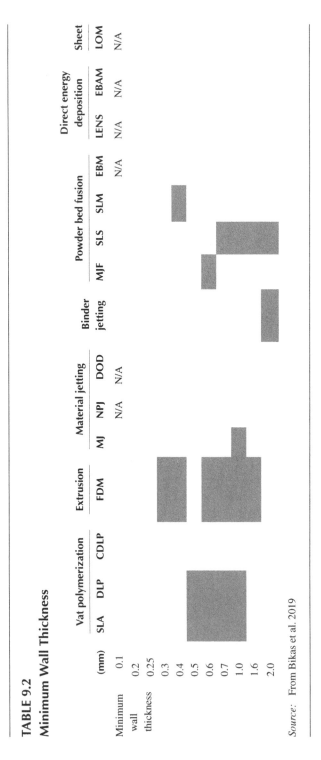

	Vat polymerization			Extrusion	Material jetting			Binder jetting	Powder bed fusion				Direct energy deposition		Sheet
(mm)	SLA	DLP	CDLP	FDM	MJ	NPJ	DOD		MJF	SLS	SLM	EBM	LENS	EBAM	LOM
						N/A	N/A					N/A	N/A	N/A	N/A

Minimum wall thickness (mm): 0.1, 0.2, 0.25, 0.3, 0.4, 0.5, 0.6, 0.7, 1.0, 1.6, 2.0

Source: From Bikas et al. 2019

TABLE 9.3

Smallest Geometric Feature for AM Technologies

	Vat polymerization			Extrusion	Material jetting			Binder jetting	Powder bed fusion				Direct energy deposition		Sheet
(mm)	SLA	DLP	CDLP	FDM	MJ	NPJ	DOD		MJF	SLS	SLM	EBM	LENS	EBAM	LOM
0.1							N/A				N/A	N/A	N/A	N/A	
0.2															
0.25															
0.3															
0.5															
0.6															
0.7															
1.0															
1.6															
2.0															

Source: From Bikas et al. 2019.

9.3.2.1 Layer Thickness

Layer thickness is a factor that affects both the quality of the print and the build time needed to complete the part. With smaller layer thickness, more detailed parts are produced, and the staircase effect is minimized. Additionally, with smaller layer thickness, potential voids and gaps are eliminated, as the CAD file is sliced with more precision, and the geometrical accuracy is maintained. On the other hand, with thicker layers, the printing time is reduced. The slope angle is the major factor causing the staircase effect. As the slope angle increases, the stair size is increased. The layer thicknesses for AM technologies are presented in Table 9.4. A proposed solution to the staircase problem is adaptive slicing. In this case, the areas where details are needed are sliced with thin layer height, whereas areas whose quality will not be affected are sliced with thicker layer height to contribute to an effective build-up time and energy consumption.

9.3.2.2 Build Orientation

The build orientation is one of the most crucial AM process parameters. The orientation of the part relative to the build vector of the fundamental build unit determines which geometrical features are overhanging geometries. Subsequently, the build orientation determines the volume of support structures needed to successfully manufacture the part (Figure 9.7). Moreover, it sets the axis on which the mechanical properties show anisotropic behavior.

9.4 DESIGN CONSIDERATIONS FOR AM

A design consideration is anything that will result in an effect on the finished product, including the mechanical properties of the part. The most important design considerations include the following.

9.4.1 Anisotropic Mechanical Properties

AM technologies produce parts with anisotropic mechanical properties. The anisotropic behavior is due to the nature of the AM process, which can be related to the lamellar nature, cylindrical extrusion shape, short fibers within the raw material, and scaffold and lattice structures within the volume of the part. Reducing the anisotropy with heat treatment improves, to some extent, the mechanical properties for components that fit into a furnace. There are two approaches to design a part taking into account the anisotropic mechanical properties:

- Orient the designed part in such a way that the loads are received in the direction in which the AM technology has the greatest mechanical strength.
- Optimize the part shape with the mechanical strength anisotropy in mind.

9.4.2 Accuracy (xy Plane versus z Axis)

Another important design consideration is to distinguish between the machine's accuracy on the xy plane and the z axis. The accuracy of the AM machine used to

TABLE 9.4
Layer Thickness for AM Technologies

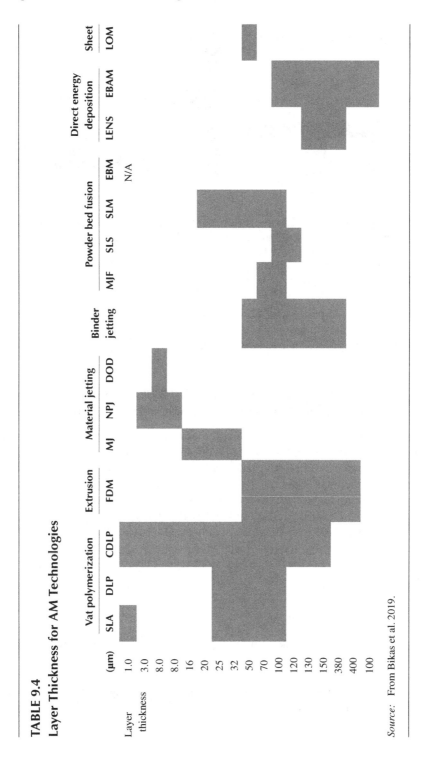

	Vat polymerization			Extrusion	Material jetting			Binder jetting	Powder bed fusion				Direct energy deposition		Sheet
(μm)	SLA	DLP	CDLP	FDM	MJ	NPJ	DOD		MJF	SLS	SLM	EBM	LENS	EBAM	LOM
1.0												N/A			
3.0															
8.0															
8.0															
16															
20															
25															
32															
50															
70															
100															
120															
130															
150															
380															
400															
100															

Layer thickness

Source: From Bikas et al. 2019.

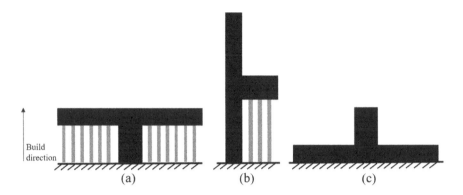

FIGURE 9.7   (a) T part needs the most support in this direction; (b) T part needs less support than (a); (c) T part does not need support in this direction. (With kind permission from Springer Science+Business Media: Devine, Declan M. (2019) Polymer-Based Additive Manufacturing (Biomedical Applications). 10.1007/978-3-030-24532-0. doi:10.1007/978-3-030-24532-0)

produce the desired part is crucial for the designer at the design phase. For assemblies in general, the dimensional accuracy with which the machine can manufacture has to be considered for the build to be successful.

9.4.3 TOLERANCE QUALITY OF SELECTED AM TECHNIQUES IN COMPARISON TO TRADITIONAL MANUFACTURING TECHNIQUES

AM process quality, consistency, and capabilities are continuing to improve. Existing standards will be applied more to AM. AM specific standards will become more relevant and complete, and new standards will be developed. Lieneke et al. (2015) recently classified the achievable tolerances of several AM processes according to ISO 286-1, taking into account part orientation (Figure 9.8). Similar work has been done by Griesbach (2016) for SLA, material jetting, material extrusion, and SLS, and by Mintetol et al. (2016) for FDM. Such efforts will enable standards organizations to bring researchers and industry together to establish standards that can be built upon to support process-specific DfAM, more general process selection, and process chain development.

9.4.4 SURFACE ROUGHNESS

The surface roughness of the completed part determines the post-processing steps needed to achieve the desired surface quality. The surface roughness after AM is not uniform throughout the entire printed surface. This due to the geometry's slope angle and the unintentional sintering under angled overhangs. Another reason for surface non-uniformity is the gaps resulting from insufficient planning for filling of the path.

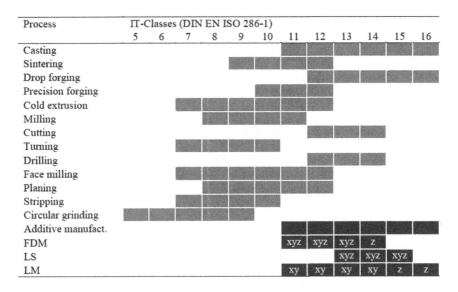

FIGURE 9.8 Achievable tolerances of traditional and AM processes. (From Lieneke, T. et al., in *26th Annual International Solid Freeform Fabrication Symposium*, 371–384, Austin, Texas, 2015)

9.4.5 BUILD TIME

This is the total time required for an AM machine to manufacture a part. The build time and part orientation are highly related, since the material deposition speeds on the platform xy plane and the normal z axis are not the same. The build unit (e.g. nozzle and laser) moves, thus building the part, with larger speed on the xy axis than the speed at which the layers are adding up in the z axis. Changing the build orientation affects the time needed for the AM machine to complete the part. In this respect, horizontally oriented parts will, in general, be printed faster than vertically oriented ones.

9.4.6 PART'S CROSS-SECTIONAL AREA

The part's cross-sectional area (normal to build vector) affects the AM process in two ways depending on the AM technology. The first is related to the machine's build base, while the second is related to the stresses developed at the rest of the part's volume, which are related to its mechanical properties.

9.5 CHALLENGES AND OPPORTUNITIES OF AM

To represent research challenges and opportunities in AM, two aspects are considered in this section: first, design rules and tools for AM and second, specification and verification for AM (Bandyopadhyay and Bose 2016).

9.5.1 DESIGN RULES AND TOOLS FOR AM

These rules are focused on the stage of the design where the designers need tools that can model and optimize part topology, model material distributions, and model complex shapes and structures. These tools need to include AM process-specific constraints while designing parts.

9.5.2 POROUS PARTS AND LATTICE STRUCTURES

Porous parts and scaffolds have been demonstrated for biomaterials applications using AM techniques for more than a decade. Lattice-based structures have recently been proposed, with the main consideration in the automotive and aerospace industry.

9.5.3 MULTI-MATERIAL PARTS

Multi-material parts include different materials at different locations of a single part. Chen et al. (2013) produced parts of multi-materials to follow a given deformation profile with a given surface texture. They demonstrated a simple but powerful design tool to create these multi-material basic parts. Design tools that can provide the flexibility of using different materials and different lattice/cellular structures at different locations of complex parts are not yet available. There is a need for such tools in order to facilitate further innovations in the application of AM technology.

9.5.4 QUALITY SPECIFICATION AND VERIFICATION FOR AM

AM can produce shapes, materials, and structures that are not feasible in traditional manufacturing. New methods of specifying quality (by a designer) and verifying the quality (by a part/product inspector) are needed for AM parts. Furthermore, in-process AM quality techniques need to be developed so that manufacturers can keep their parts within the specified quality required

9.6 DESIGN GUIDELINES FOR LASER BEAM MELTING (LBM)

LBM is currently used in many industrial applications, since it offers high potential for weight saving in lightweight applications of aerospace industry. Most design engineers have limited experience in designing products for AM. The absence of comprehensive design guidelines is limiting the further applications and spread of laser additive manufacturing (LAM). The following design guidelines are relevant only for laser beam melting (i.e. selective laser melting) (www.epma.com/am).

9.6.1 HOLES AND INTERNAL CHANNELS

The recommended minimum standard hole size is 0.4 mm. Holes and channels with a diameter below 10 mm usually do not require support structures, while for diameters above 10 mm, support structures are needed, which can be difficult to remove in the case of nonlinear channels. In such a case, in order to avoid support structures, a possible option is to modify the channel profile from a circular one to an elliptical profile, which minimizes the overhang area, as shown in Figure 9.9.

9.6.2 MINIMUM WALL THICKNESSES

The recommended minimum wall thickness is usually 0.2 mm. However, it depends on the machine and powder material used. Figure 9.10 shows cubes with thin wall thicknesses. It is important to avoid too thin or unsupported walls where there is a chance of buckling in the surface, as shown in Figure 9.11.

9.6.3 MAXIMUM (HEIGHT-TO-WIDTH) ASPECT RATIO (AR)

The classical definition of AR is the ratio of height to width for a rectangle shape. The recommended AR ratio should not exceed 8:1. If the part has a reasonable section or supporting geometry, then it is possible to build at a higher aspect ratio. Typical high-AR features are shown in Figure 9.12 (Demir 2018).

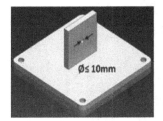

FIGURE 9.9 Holes and internal channels. (Courtesy of Renishaw)

FIGURE 9.10 Cubes with thin wall thicknesses. (Courtesy of Fraunhofer IFAM)

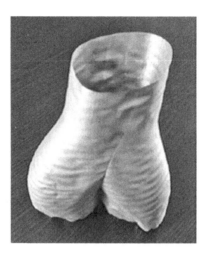

FIGURE 9.11   Ni718 thin walls manifold with buckling effect. (Courtesy of Renishaw)

FIGURE 9.12   Examples of manufactured thin walls showing high-quality conditions. (Adapted from Demir, A.G., (2018) *Optics and Lasers in Engineering*, 100, 9–17)

9.6.4   Minimum Strut Diameter and Lattice Structures

The minimum strut diameter is usually 0.15 mm, and complex lattice structures can be produced, which are impossible to produce by any other conventional technologies. Lattice structures offer the major advantage of reducing part weight while keeping the part strength, which is very important in the aerospace and transportation industries (Figure 9.13).

9.6.5   Part Orientation

The orientation of parts in the powder bed affects both product quality and cost. It influences the build time, the quantity of supports, the surface roughness, and residual stresses. The best suitable part orientation should provide:

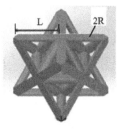
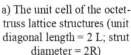
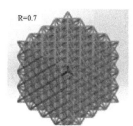
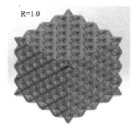

a) The unit cell of the octet-truss lattice structures (unit diagonal length = 2 L; strut diameter = 2R)

(b) Two structures with different relative densities (radii of struts R = 0.7, and 1 mm respectively)

FIGURE 9.13   Minimum strut diameter and lattice structures. (Adapted from Chen, L. et al., (2019) *Materials and Design*, 162, 106–118)

- short build time by minimizing the number of layers and part height
- minimum number of supports
- easy access to supports for ease of removal
- best surface roughness
- minimum staircase effect
- minimum residual stresses that lead to part distortion

9.6.6   OVERHANGS

When building parts layer by layer, it is necessary to avoid having too small an overhang angle (α), since each new layer must be supported at least partly by the previous layer. When the angle (α) between the part and the build platform is below 45° (Figure 9.14), support structures are needed to avoid poor surface roughness, distortion, and warping, which lead to build failure. During the SLM process, the poor surface roughness is the result of building directly onto the loose powder instead of using the support structure as a building scaffold. In such a case, the area melted at the focal point cools very quickly, and the stress generated curls the material upward. Supports would act as an anchor to the build plate, tying parts down to the plate in order to avoid upward curl. Additionally, the very poor surface consists of melted and partially melted/sintered powder because the laser penetrates the powder bed and starts to agglomerate loose powder particles surrounding the focal point instead of dissipating the excessive heat through the support structure (www.epma.com/am).

9.6.7   SUPPORTS

Most of the AM processes are unable to build overhang structures without proper support. The position and orientation of the part on the build platform have a significant impact on the need and type of support structures. This, in turn, affects the quality of the part and the post-processing operations needed to finish the part.

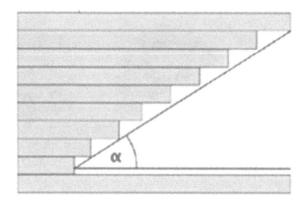

FIGURE 9.14   Overhang angle between build platform and part.

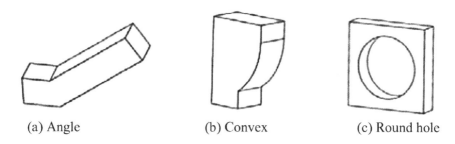

    (a) Angle   (b) Convex   (c) Round hole

FIGURE 9.15   Overhang structures. (From Ameen, W. et al., (2019) *International Journal of Mechanical and Industrial Engineering*, 13, 4, 265–269)

SLS is a powder bed fusion AM process in which the build parts are immersed on loose powder in the powder bed, which provides support to the overhang surfaces. The overhang structures include angle overhang with varying angles, convex overhang with varying radius, and hole overhang with varying diameters, as shown in Figure 9.15. The main function of support structures is to support the part in case of overhangs, strengthen and fix the part to the building platform, conduct excess heat away, and prevent warping or failure. Additionally, they should be easy to remove mechanically and have a minimal weight.

9.6.8   Surface Roughness

With SLM, the achievable maximum surface roughness (Rz) is usually between 25 and 40 µm in the as-built state, which is higher than the average values obtained by conventional milling and turning processes. Polishing helps reach much lower values (Figure 9.16). However, the part's design complexity may affect its ability to be polished efficiently. In addition to the surface roughness, other surface defects that must be avoided include:

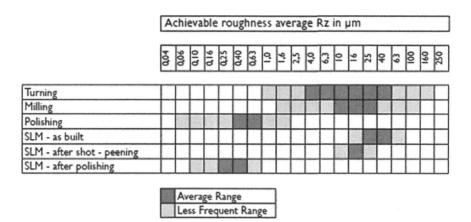

Achievable roughness average Rz in µm																			
	0,04	0,06	0,10	0,16	0,25	0,40	0,63	1,0	1,6	2,5	4,0	6,3	10	16	25	40	63	100	160	250
Turning								▓	▓	▓	■	■	■	■	▓					
Milling								▓	▓	▓	■	■	■	■	▓					
Polishing		▓	▓	▓	▓	■	■	▓												
SLM - as built											▓	■	■	▓						
SLM - after shot - peening											▓	■	▓							
SLM - after polishing			▓	▓	▓	■	▓													

■	Average Range
▓	Less Frequent Range

FIGURE 9.16 Standard roughness of parts made by SLM versus machining. (Courtesy of Hoiches Zeichennen and Electro Optical Systems (EOS), Germany)

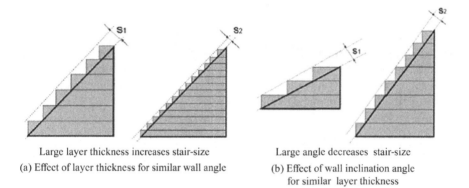

Large layer thickness increases stair-size	Large angle decreases stair-size
(a) Effect of layer thickness for similar wall angle	(b) Effect of wall inclination angle for similar layer thickness

FIGURE 9.17 Influence of layer thickness and inclination angle on the stair effect. (From Yasa, E. and Kruth, J., (2011) *Advances in Production Engineering and Management (APEM) Journal*, 4, 259–270)

- The staircase effect observed on a curved surface, which is more pronounced when the surface angle increases versus the vertical axis.
- Poor down-skin surface roughness and decreased dimensional accuracy, which are linked primarily to the fact that the heat generated by the laser beam does not evacuate quickly on down-facing surfaces.

The surface roughness of the SLM parts is highly dependent on many factors and process parameters. These include the grain size of powder particles, layer thickness, wall angle, and melt pool size. The wall angle parameter is defined as the angle between a specific surface and the horizontal plane. Figure 9.17 shows the influence of the wall angle and layer thickness on the surface roughness. The combination of wall angle and layer thickness produces the well-known staircase effect. The stair

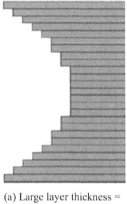
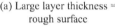
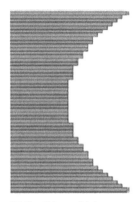

(a) Large layer thickness =
rough surface

(b) Small layer thickness =
smooth surface

FIGURE 9.18   Schematic illustrating the effect of layer thickness on surface finish. (Adapted with kind permission from Springer Science+Business Media: Devine, Declan M. (2019). Polymer-Based Additive Manufacturing (Biomedical Applications). 10.1007/978-3-0 30-24532-0. doi:10.1007/978-3-030-24532-0)

size decreases proportionally with the cosine of the wall angle. Consequently, the staircase effect can be reduced by decreasing the layer thickness or by increasing the wall angle (Yasa and Kruth 2011). The effect of layer thickness and inclination angle on the stair effect is illustrated in Figure 9.18.

9.6.9   THERMAL STRESSES AND WARPING

Warping is due to thermal stresses caused by the rapid solidification process. This, in turn, leads to part distortion and bad junctions between supports.

Ameen et al. (2019) showed that in EBM:

- Minimum round through hole diameter: 0.5 mm
- Minimum wall thickness: 0.6 mm
- Minimum round bar: 0.65 mm
- Minimum round slot: 0.1 mm
- Minimum cubed slot: 0.4 mm
- Lowest self-supporting angle overhang: 50°
- Smallest radius of self-supporting convex curve overhang: 7 mm
- Smallest self-supporting hole: 18 mm diameter

9.7   DESIGN GUIDELINES FOR FUSED DEPOSITION MODELING (FDM)

The following information explains the design considerations for AM high-quality FDM parts: https://grabcad.comand (https://3dprintergeeks.com/).

9.7.1 SHRINKAGE

In FDM, the material shrinks to a degree during the printing process because the selected material (ABS, ASA, polylactic acid (PLA), nylon, etc.) is melted down to a liquid, extruded to build the part, and finally cooled to be a solid. Some printers have features that allow users to compensate for the natural shrinkage of material to ensure the accuracy of every part. The shrinkage rate of PLA is typically between 0.2 and 0.25%, which is considerably lower than that of ABS (0.8%). The shrinkage rate of nylon stands at 1.5%, which is considerably higher than that of PLA and ABS.

9.7.2 WARPING

Warping occurs when a part is made with too thin walls or too high without adding supports to the part, such as ribs. It occurs because the part contracts/shrinks as it cools, which is dependent on the type of 3D printer (Figure 9.19). However, to avoid potential warping (deformation of vertical walls) when building thin-walled sections of a model, designers might add ribs to the walls, as shown in Figure 9.20.

9.7.3 HOLES

Holes are subject to many factors that affect their general usability when printed (Figure 9.21):

- Shrinkage can cause holes to be undersized.
- Printing a hole in a vertical orientation causes a misshapen hole because of the way FDM printing has a stair stepping effect on the part.
- Holes can also be prone to being abnormally weak because of weak neighboring walls.
- When tight tolerances are required, holes can be drilled or reamed to ensure accurate diameters.

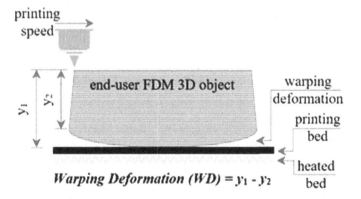

FIGURE 9.19 Warping deformation and measurements. (From Alsoufi, M.S. et al., (2019) *American Journal of Mechanical Engineering*, 7, 2, 45–60)

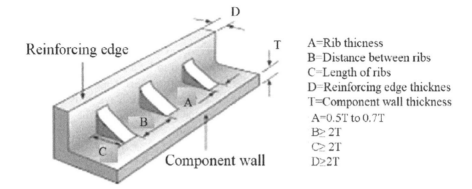

FIGURE 9.20   Adding ribs to avoid warping.

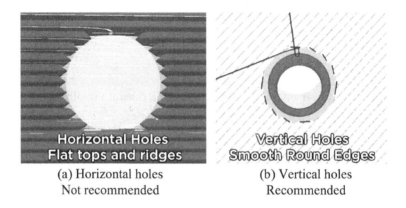

(a) Horizontal holes
Not recommended

(b) Vertical holes
Recommended

FIGURE 9.21   FDM of (a) horizontal holes and (b) vertical holes. (From https://coreelectronics.com.au/)

9.7.4   WALL THICKNESS

This is the distance between one surface of the part and the opposite surface. In general, printing a part with thinner walls makes a part more susceptible to shrinkage and warping. The recommended minimum thickness of an unsupported wall is 1.2 mm, while that for a supported wall is 0.8 mm. Figure 9.22 shows the recommended wall thicknesses.

9.7.5   THREADS

When designing and printing threads, it's important to keep in mind the contour width, slice thickness, material choice, and part orientation. When designing built-in threads, avoid sharp edges and include a radius on the root, because sharp edges cause stress concentrators in plastic parts. Creating an Acme thread design with rounded roots and crests has been found to work well when using FDM. Also,

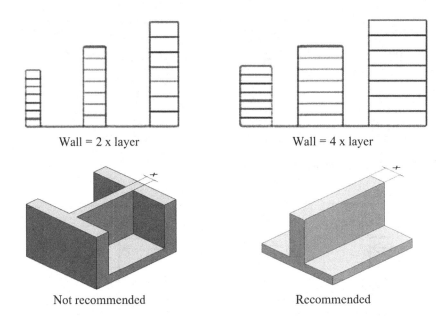

Wall = 2 x layer   Wall = 4 x layer

Not recommended   Recommended

FIGURE 9.22 Recommended wall thickness. (From https://forerunner3d.com/)

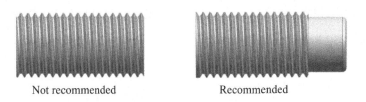

Not recommended   Recommended

FIGURE 9.23 Dog point thread. (From https://forerunner3d.com)

using a dog point head of at least 0.8 mm makes starting the thread much easier (Figure 9.23). Small threads produced from the FDM process are not recommended and are not possible for holes or posts smaller than 1.6 mm diameter. In this case, an easy alternative is to use a tap or die to thread holes or posts.

9.7.6 UNDERCUTS/OVERHANGS

As can be seen in Figure 9.24 and Figure 9.25, overhangs that are greater than 45 degrees need support material so that the part doesn't flop over while printing. If surface finish is a concern, printing orientation plays a large role. Stair stepping can negatively alter how well the part fits if that is its purpose.

9.7.7 FILLETS

Although fillets are not necessary in FDM parts, they are used to reduce stress concentrations and increase the overall strength of the part. As can be seen in Figure 9.26,

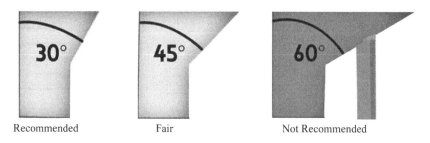

Recommended   Fair   Not Recommended

FIGURE 9.24 Overhanging angles. (From https://all3dp.com/)

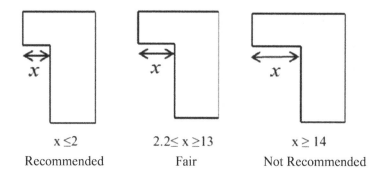

$x \leq 2$ $2.2 \leq x \geq 13$ $x \geq 14$
Recommended   Fair   Not Recommended

FIGURE 9.25 Recommended length of overhang structure. (From Mazlan, N.H. and Sudin M. N. (2018) *Journal of Mechanical Engineering*, SI 5, 3, 98–122)

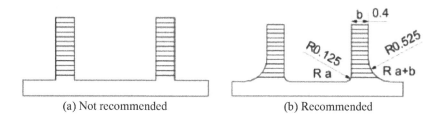

(a) Not recommended   (b) Recommended

FIGURE 9.26 FDM part fillet design. (From https://forerunner3d.com)

it is recommended to design fillets with an outer radius equal to the inner radius plus the wall thickness to maintain consistent thickness.

9.7.8 SIZE

Making very large or exactly-to-scale parts may use up large amounts of material, which increases part cost. When making prototypes, it is beneficial to print out a test part at a smaller scale to save cost.

9.7.9 ORIENTATION

Designers should note that extruded plastic has its strongest strength in the tensile mode along the xy plane. Since the layers are held together by "hot flow" across the strands (one strand is cooling while the other is laid upon it), the lowest strength is in the z direction for both tensile and shear modes. Additionally, changing the part orientation has an effect on the amount of material being used and the time spent to print your part.

9.7.10 ASSEMBLIES

In the case of assembly, proper clearance should be given between mating parts to prevent them from fusing together. The standard guideline for creating clearances on assemblies being produced and assembled is the minimum z clearance of the slice thickness. The xy clearance is, at least, the default extrusion width based on a minimum wall thickness. The minimum clearance needed for mating parts, when not producing the components fully assembled, is equal to the tolerance of the FDM machine itself. Under time constraints, a minimum clearance of 0.4–0.5 mm is a safe estimate to start with (https://grabcad.com/).

9.8 DESIGN GUIDELINES FOR STEREOLITHOGRAPHY (SLA)

9.8.1 WALL THICKNESS

In 3D printing, the wall thickness refers to the distance between one surface of the part and the opposite surface. A part made using stereolithography has a minimum wall thickness that is dependent on its overall size. Small objects, where the sum of x, y, and z dimensions is below 200 mm, need a minimum wall thickness of 1 mm. For medium-sized parts, where the sum of the x, y, and z dimensions is between 200 and 400 mm, the minimum wall thickness is 2 mm. For larger parts, a wall thickness of 3 mm is recommended. For parts with a high aspect ratio (height/thickness), it's recommended to increase the wall thickness to increase the part strength. The larger the wall thickness, the higher the stress will be on the 3D printed part, which eventually, will cause major surface and internal cracks. Therefore, hollowed objects with quite thin walls are preferred. Thinner parts weigh less and use less resin per 3D printed part. On average, the wall thickness should range from 1 to 2 mm (Figure 9.27).

9.8.2 UNIFORM WALLS

It is strongly recommended to keep walls consistent and uniform (Figure 9.28). The thin sections that consist of less material will shrink less than thicker ones, which leads to warping of the part that eventually causes cracks.

9.8.3 HOLES

Holes should have a minimum diameter of 0.5 mm. Holes smaller than 0.05 mm may be closed by the printing process.

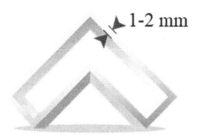

FIGURE 9.27 Recommended wall thickness. (From www.materialise.com/)

Recommended   Fair   Not Recommended

FIGURE 9.28 Ensuring uniform wall thickness. (From https://medium.com/)

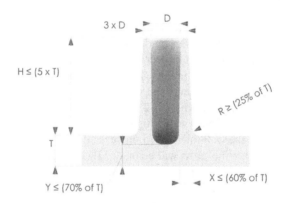

FIGURE 9.29 Guidelines for boss design. (From https://medium.com/)

9.8.4 Bosses

Bosses are used for attaching fasteners or accepting threaded inserts. The boss diameter should be between 2.0 and 3.0 times the diameter of the insert to provide sufficient strength and to minimize hoop shrink. The height of the insert should not exceed the height of the boss, as hoop shrinkage may occur below the level of the boss. As with injection-molded parts, ribs and gussets can be added to the boss to increase its strength. It is not necessary to add draft to the boss. Figure 9.29 presents the general guidelines for boss design.

9.8.5 SURFACE FINISH/TEXTURE

Generally, stereolithography produces parts with a smooth surface finish. Depending on the layer thickness, slow sloping surfaces may have small stair stepping appearance. SLA part surfaces are easily hand sanded, blasted, or tumbled for a smoother finish. Because the part is printed layer by layer, the orientation on the build platform influences both the surface quality and the strength. Figure 9.30 shows two examples of the same part built in two different orientations. The horizontally printed part clearly shows evidence of the "staircase" effect of the printing process. If the part is printed vertically, the surface quality will be better.

9.8.6 PART SIZE

The maximum build envelope for the SL machine is $25'' \times 29'' \times 21''$; however, parts can be built in segments and accurately bonded in post-processing. For high-definition SLA, segmenting and bonding is not recommended due to the small size of part features and the chance of corrupting those features.

9.8.7 RIBS

Ribs are generally used to increase the bending stiffness of the designed part without adding any additional thickness. They increase the moment of inertia, which increases the bending stiffness. Their thickness should be lower than the wall thickness to minimize the risk of potential cracks and additional stresses after 3D printing. Figure 9.31 shows the general guidelines for rib design. There is one more important facet to consider if the part has intersecting ribs, which will have greater thickness and thus, more material. This can be easily solved by simply hollowing the intersection of the ribs to preserve uniform wall thickness and material volume, as shown in Figure 9.32.

(a) Horizontally printed
Not recommended

(b) Vertically printed
Recommended

FIGURE 9.30 Effect of part orientation on surface finish. (From www.materialise.com/)

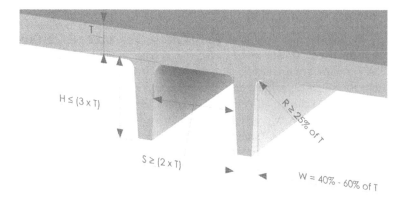

FIGURE 9.31   Guidelines for rib design. (From https://medium.com/)

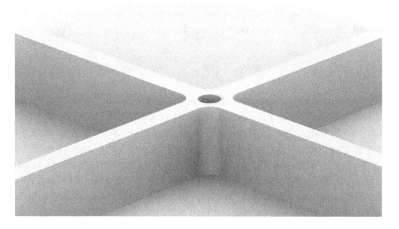

FIGURE 9.32   Hollowing the intersection of ribs to preserve uniform wall thickness. (From https://medium.com/)

9.8.8   GUSSETS

Gusset are considered as subsets of ribs that are used to support structures in order to minimize warping of the part by increasing the stiffness of 3D printed structures. If a gusset is attached to the boss, its height can be as much as 95% of that boss. However, its height should be less than four times the nominal wall thickness, and the preferred height is typically two times that of the nominal wall. Figure 9.33 shows a supporting gusset.

9.8.9   SHARP CORNERS

The presence of sharp corners increases stress concentration for 3D printed parts, which leads to cracks. The radius of sharp corners has to be closely considered

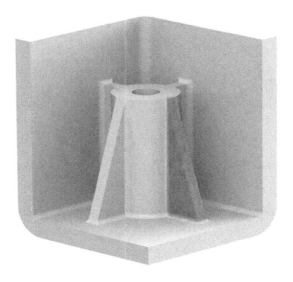

FIGURE 9.33   Supporting gusset (https://medium.com/)

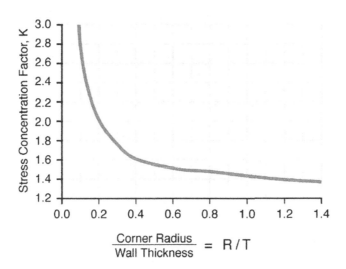

FIGURE 9.34   Stress concentration against radius to wall thickness ratio. (From https://medium.com/)

because the stress concentration varies with that radius for a given part thickness, as indicated in Figure 9.34. Accordingly, the stress concentration decreases as the ratio of the corner radius to the part thickness increases. At corners, the recommended inside radius is 0.5 times the material thickness, while the outside radius is 1.5 times the material thickness, as shown in Figure 9.35.

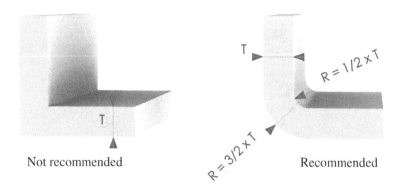

Not recommended   Recommended

FIGURE 9.35 Recommended corner radius. (From https://medium.com/)

FIGURE 9.36 Safe supporting angle (30°–150°). (From www.materialise.com/)

9.8.10 Support

SLA takes place in a liquid resin tank. Hence, parts need to be attached to the supporting platform to prevent them from floating away. This attachment is referred to as "support" and is required for all parts built using SLA. In addition to keeping the part in place, supports also make it possible to construct overhanging elements. Figure 9.36 shows a part that needs support. The self-supporting zone that does not require any support to print the part using SLA ranges from 150° to 30°, as shown in Figure 9.37 Based on these assumptions, Figure 9.38 shows external and internal supports. Supports can be avoided by applying a fillet, which can solve the issue in most cases. However, the horizontal surfaces will still need support if they hang out more than 2 mm, Figure 9.39. Figure 9.40 and Figure 9.41 show other examples of the effect of part orientation on adding supports.

9.8.11 Hollowing

Making hollow parts avoids extra charges and shrinkage issues in the thicker sections. In this case, the integration of one or more drainage or escape holes is needed.

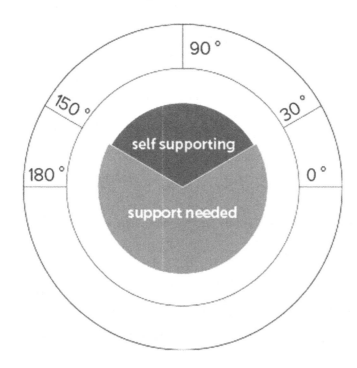

FIGURE 9.37 The self-supporting zone. (From https://i.materialise.com/)

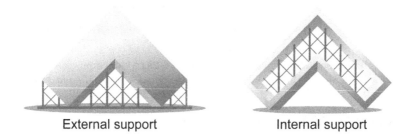

FIGURE 9.38 External and internal supports. (From www.materialise.com/)

Usually, escape holes are placed at the lowest point(s) of the part once it has been oriented and positioned on the build platform. These holes ensure that the pressure of the liquid resin inside and outside the part remains at the same level, which prevents the deformation of the design. Moreover, these holes are used to remove the excess resin inside the part once the printing process has been finished and the part has been removed from the 3D printing machine. The part is then emptied, cleaned, and cured in a UV oven to achieve optimal strength (Figure 9.42). Some hollow parts require support material on the inside to reinforce the structure. This support structure might not be removed if it cannot be accessed.

Sharp corner	Adding fillet with extension > 2 mm

FIGURE 9.39   Avoiding supports by incorporating a fillet with overhang less than 2 mm. (From www.materialise.com/)

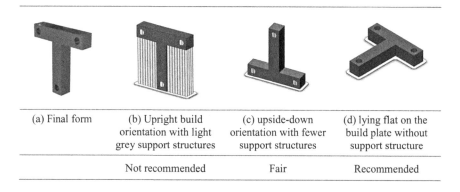

(a) Final form	(b) Upright build orientation with light grey support structures	(c) upside-down orientation with fewer support structures	(d) lying flat on the build plate without support structure
	Not recommended	Fair	Recommended

FIGURE 9.40   Schematic of simple "T" shaped-bracket. (With kind permission from Springer Science+Business Media: Hinchy, E.P. 2019. *Polymer-Based Additive Manufacturing*)

9.9   REVIEW QUESTIONS

1. Explain what is meant by DfAM.
2. What are the main AM design aspects and considerations?
3. List the main design guidelines for FDM.
4. Show the rules for FDM part fillet design.
5. Explain why ribs are added to structures made by FDM.
6. List the main design guidelines for SLA.
7. Explain why a corner radius is added to SLA parts.
8. Show the self-supporting zone for SLA processes.
9. Explain how to support by incorporating a fillet with overhang less than 2 mm.
10. Give an example showing the effect of part orientation on adding supports.

Horizontal
Not recommended

Vertical
Recommended

FIGURE 9.41 Effect of part orientation on adding supports. (From www.3dbeginners.com/)

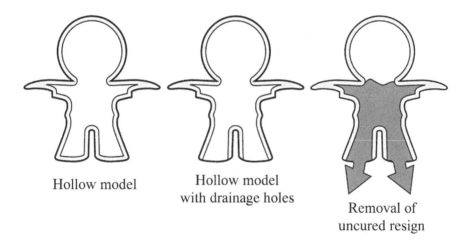

Hollow model

Hollow model
with drainage holes

Removal of
uncured resign

FIGURE 9.42 Position of drainage holes. (From https://i.materialise.com/)

11. Explain why drainage holes are necessary for powder-based AM. Give an example.
12. Explain the different machining processes used for post-processing of AM parts.
13. Explain how to improve the surface quality of AM parts.

BIBLIOGRAPHY

Alsoufi, M. S, Alhazmi, M. W., Suker, D. K, Turki, A. Alghamdi, T. A, Sabbagh, R. A, Felemban, M. A. and Bazuhair, F. K. (2019) Experimental Characterization of the Influence of Nozzle Temperature in FDM 3D Printed Pure PLA and Advanced PLA+. *American Journal of Mechanical Engineering*, 7(2), 45–60. https://doi.org/10.12691/ajme-7-2-1

Ameen, W, Al-Ahmari, A, and Abdulhameed, O. (2019) Design for Metal Additive Manufacturing: An Investigation of Key Design Application on Electron Beam Melting. *International Journal of Mechanical and Industrial Engineering*, 13(4), 265–269.

ASTM ISO/ASTM 2915–13. (2013) *Standard Specification for Additive Manufacturing File Format (AMF) Version 1.1*. West Conshohocken, PA: ASTM International. http://www.astm.org/

Bandyopadhyay, Amit and Bose, Susmita. (2016) *Additive Manufacturing*. Boca Raton: CRC Press, Taylor and Francis Group.

Bendsoe, M. P. and Sigmund, O. (2011) *Topology Optimization: Theory, Methods, and Applications*, Berlin: Springer.

Bikas, H., Lianos, A. K., and Stavropoulos, P. (2019) A Design Framework for Additive Manufacturing. *The International Journal of Advanced Manufacturing Technology*, 103, 3769–3783. https://doi.org/10.1007/s00170-019-03627-z

Brackett, D., Ashcroft, I., and Hague, R., (2011) Topology Optimization for Additive Manufacturing. In *Proceedings of the Solid Freeform Fabrication Symposium*, 348–362. Austin, TX: University of Texas.

Chen, D., Levin, D. I. W., Didyk, P., Sitthi-Amorn, P., and Matusik, W. (2013) Spec2Fab: A Reducer-tuner Model for Translating Specifications to 3D Prints. *ACM Transactions on Graphics*, 32(4), 1–135.

Chen, Ling, Cernicchi, Alessandro, Gilchrist, Michael D. and Cardiff, Philip. (2019) Mechanical Behaviour of Additively-manufactured Polymeric Octet-Truss Lattice Structures under Quasi-Static and Dynamic Compressive Loading. *Materials and Design*, 162, 106–118. https://doi.org/10.1016/j.matdes.2018.11.035

Demir, A. G. (2018). Micro Laser Metal Wire Deposition for Additive Manufacturing of Thin-Walled Structures. *Optics and Lasers in Engineering*, 100, 9–17. https://doi.org/10.1016/j.optlaseng.2017.07.003

Devine, Declan M. (2019). *Polymer-Based Additive Manufacturing: Biomedical Applications*. https://doi:10.1007/978-3-030-24532-0. Springer Cham.

Griesbach, V. (2016) *Rapid Technologien Toleranzmanagment*, Berlin: DIN Beuth Verlag Gmbh. ISBN 978-3-410.25776-9.

Hinchy, Eoin P. (2019) Design for Additive Manufacturing. In *Polymer-Based Additive Manufacturing*, edited by D. M. Devine. New York: Springer. https://doi.org/10.1007/978-3-030-24532-0

https:// www.3dbeginners.com/

https://3dl.tech/en/design-guidelines-fdm-technology/

https://3dprintergeeks.com/

https://coreelectronics.com.au/media/

https://www.desktopmetal.com/

https:// www.epma.com/am.

https://forerunner3d.com/ fdm-part-design-guide/

https://grabcad.com/

https://i.materialise.com/

https://www.materialise.com/

https://medium.com/

Jiang, J., Xu, X., and Stringer, J. (2018). Support Structures for Additive Manufacturing: A Review. *Journal of Manufacturing and Materials Processing*, 2(4), 64. https://doi.org /10.3390/jmmp2040064

Keulen, F. V., Langelaar, M., and Baars, G. E. (2014) Topology Optimization and Additive Manufacturing, Natural Counterparts for Precision Systems—State-of-the-art and Challenges. In *29th American Society of Precision Engineering.* http://www.aspe.net/ publications/

Lieneke, T., Adam, G., Leuders, S., Knoop, F., Josupeit, S., Delfs, P., Funke, N., and Zimmer, D (2015) Systematical Determination of Tolerances for Additive Manufacturing by Measuring Linear Dimensions. In *26th Annual International Solid Freeform Fabrication Symposium*, 371–384. Austin, Texas.

Mazlan, Nur Humaira and Sudin, M N (2018) Manufacturability of Mechanical Structure Fabricated using Entry Level 3D Printer. *Journal of Mechanical Engineering*, SI 5(3), 98–122.

Meisel, N. A., Gaynor, A., Williams, C. B., and Guest, J. K (2013) Multiple-Material Topology Optimization of Compliant Mechanisms Created Via Polyjet 3d Printing. In *24th International Solid Freeform Fabrication Symposium.* http://utwired.engr.utexas.edu/

Mintetol, P., Iuliano, L., and Marchiandi, G. (2016) Benchmarking of FDM Machines through Part Quality Using IT Grades. *Procedia CIRP*, 41, 1027–1032.

Quarshie, R., Machachlan, S., Reeves, P., Whittaker, D., and Blake, R. (2012) *Shaping Our National Competency in Additive Manufacturing: The Additive Manufacturing Special Interest Group for the Technology Strategy Board.* https://connect.innovateuk.org/doc-uments/2998699/ 3675986/UK+Review+of+Additive+Manufacturing+-+AM+SIG+Re port+-+September+2012 .pdf/a1e2e6cc-37b9-403c-bc2f-bf68d8a8e9bf.

Roundtable Forum. (2013) Additive Manufacturing: Opportunities and Constraints: Hosted by the Royal Academy of Engineering. http://www.raeng.org.uk/publications/reports/ additive-manufacturing.

Three D Systems. (1988) *Stereolithography Interface Specification.* Valencia, CA: 3D Systems.

Vayre, B., Vignat, F., and Villeneuve, F. (2012) Designing for Additive Manufacturing. *Procedia CIRP*, 3, 632–637.

Yasa, E. and Kruth, J. (2011) Application of Laser Re-melting on Selective Laser Melting Parts. *Advances in Production Engineering and Management (APEM) Journal*, 4, 259–270.

10 Impact of Additive Manufacturing on Conventional Manufacturing Processes

10.1 INTRODUCTION

Conventional manufacturing processes can normally be classified into five basic groups according to their nature, the equipment used, and the parts produced. These groups are:

1. *Solidification processes*, where the material is melted or heated to the semi-fluid state, poured into a mold, and left to solidify, taking the mold shape. These processes can be applied for most metals, ceramic glasses, and plastics. Most processes that operate in this way are called casting or molding. Casting is the name used for metals, and molding is the common term used for plastics.

2. *Particulate processing*, where the starting materials are powders of metals or ceramics or combinations of both. These powders are commonly pressed into a die cavity to form a green shape (compaction). Then, the shape is heated in the solid state to consolidate particles (sintering). These processes include powder metallurgy (PM), hot isostatic pressing (HIP), and metal injection molding (MIM).

3. *Deformation processing*, where ductile materials are deformed permanently to the required shape by applying mechanical forces that exceed their yield stress. Heating below the melting point may be used to improve the deformation ability of the material and reduce forces. These processes include forging, rolling, extrusion, drawing, and sheet metalworking, which involves bending, forming, and shearing operations performed on starting blanks and strips of sheet metal.

4. *Material removal processes*, where a solid part (which is manufactured by any of the previous processes) is reduced to a required geometry by removal of extra parts. The most common processes in this category are machining operations such as turning, drilling, and milling and grinding. The removed parts usually take the form of chips, which reduces material utilization efficiency; thus, the name "subtractive processes" is nominated for this category.

5. *Joining processes*, where more than one part are joined to form a more intricate part. The most common permanent joint is made by welding,

DOI: 10.1201/9781003451440-10

which includes a variety of processes such as arc welding, gas welding, laser welding, brazing, soldering, spot welding, etc. (Groover 2010).

The emerging additive manufacturing (AM) technology, which started about 30 years ago, is considered a revolutionary technology that incorporates sustainability into manufacturing processes and provides alternative paths for some of the above-mentioned conventional manufacturing processes. It can produce unique bodies integrating several parts into a single one and eliminating the number of components and assembly activities. AM can fabricate complex and efficient parts that could be impossible to manufacture with conventional manufacturing, reducing material waste. AM enables on-demand and on-site production, which contribute to cutting lead times significantly. It has proved its efficiency in resource and material allocation. The most important advantage of AM for producing complex products is that design complexity is not dependent on the cost, as it can be altered and optimized numerically without the need for prototypes or trial experiments before manufacturing. All these advantages recommend AM for adoption in industry at an escalating rate, which is expected to replace or devalue the role of other manufacturing processes. However, investing in AM technology is considered costly due to the high acquisition costs of 3D printers and the high costs associated with AM raw materials. Post-processing activities, including heat treatment and secondary machining, might be required for finishing, so they increase the energy and cost as well. Further, AM is only suitable for low-volume, highly complex, and customized batches (Mecheter et al. 2023). That is why manufacturing organizations still hesitate to invest in this technology and replace the traditional manufacturing methods with AM. Alternatively, AM is taking an effective role in developing the applicability, efficiency, and economic aspects of other conventional manufacturing processes, and its impact on them is progressive. In the following sections, the impact of AM on each category of conventional manufacturing is presented, along with the specific role it plays in its applications.

10.2 THE IMPACT OF AM ON CASTING

Casting is the oldest metal manufacturing process, which dates back to more than 3000 years BC. The earliest metal castings were made with gold and copper, and bronze soon replaced them due to its rigidity and strength, announcing the Bronze Age. The process started using stone molds and moved on to sand casting for expendable molds and die casting for permanent molds. Using the lost wax process, a few decades later, casters created plaster molds from wax models, which enabled them to create more intricate shapes, such as statues and jewelry.

At present, metal casting ranks second only to steel rolling in the metal producing industry. It is estimated that castings are used in at least 90% of all products and in all machinery used in manufacturing. The process is unique among metal forming processes for a variety of reasons. The most obvious is the wide variety of the process techniques, which provide the possibility of producing complicated shapes ranging in weight from a few grams to several hundred tons. The process is applicable and economically viable for a single product or for a small number of products, while some casting methods are quite well suited to mass production. Virtually any metal that can be melted can be produced by casting. These advantages are contrasted with some disadvantages. The most serious is the safety hazards and environmental problems associated with the processing of hot molten metal. Other disadvantages of some casting methods include porosity, limitations on mechanical properties, and poor dimensional tolerances and surface finish.

A casting factory is equipped for making molds, melting metals in furnaces, transferring molten metal to the molds, performing the casting process, and the cleaning and finishing of castings. The heart of the process is the mold, which contains the cavity where the molten metal is poured. Molds are made of a variety of materials, including sand, plaster, or ceramic. Casting processes are numerous and vary in equipment, materials, and manufacturing procedures, but the most often applied are sand casting for expendable molds, investment casting (a development of the lost wax process) for ceramic shell molds, and die casting for permanent molds.

The impact and role of AM in casting processes is significant; it has the potential to supplement or marginally replace them. Now, some castings can be directly printed using metal powders, such as titanium alloys, nickel alloys, and steel parts. On the positive side, however, AM has been applied extensively in almost all procedures of the processes, from the prototype of a newly required product, to making patterns, cores, and molds, to the finishing of products. Further, AM is used in revolutionizing the design of products, assemblies, and parts.

10.2.1 SAND CASTING

The sand casting process requires a very long lead time to achieve a product. A lot of time is involved in making the pattern manually or by machining to shape the cavity of the mold and core, forming the core in the core box, assembly of the pattern and core, and making the sand mold. Different AM processes have been introduced to reduce this lead time and to improve the quality of patterns, molds, and cast products, as shown in Figure 10.1. The application of AM technologies for producing cast parts is frequently termed rapid casting.

Prototypes: The first impact of AM on this part of the process was the application of rapid prototyping to assist in the manufacturing of new products for demonstration, and to test geometry and dimensions. The prototype is directly used for making the mold to save time. There are several AM methods to make a prototype, such as fused deposition modeling (FDM), lamination object manufacturing (LOM), stereolithography (SLA), selective laser sintering (SLS), three-dimensional printing (3DP), etc. An example polyamide-based model (prototype) of a heat exchanger made by SLS is shown in Figure 10.2.

AM of sand core and mold: SLS can be used to directly create sand molds or cores using resin-coated sand. Usually, silica sand has been used in SLS to make molds or cores. Al_2O_3, zircon, and man-made ceramsite sand (an artificial foundry sand originating in China, used as a substitute for chromite sand and zircon sand) were also tested to improve the properties of the mold or core while reducing cost.

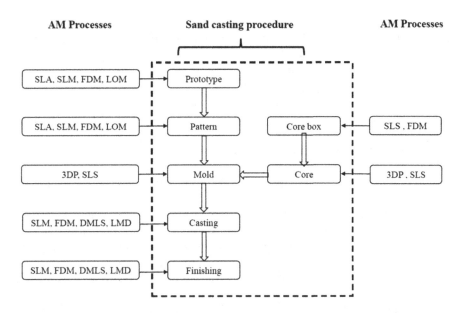

FIGURE 10.1   Application of AM technology in different stages of sand casting. (Adapted from Kang, J.-w. and Ma, Q.-x., (2017) *Special Report*, 14, 3)

FIGURE 10.2   SLS made polyamide-based model (prototype) of heat exchanger for a Pratt & Whitney PW6000 engine. (From Kang, J.-w. and Ma, Q.-x., (2017) *Special Report*, 14, 3)

The use of resin is usually slightly higher than that of a traditional molding method because of the slight burn-off during laser sintering and insufficient flow of the binder resulting from the high-speed moving laser spot. The strength of the formed molds or cores after SLS is not great enough for use, so they have to be baked in an oven. Agents can be used to facilitate the setting process, so the post-curing process may be omitted. The agent is sprayed by another nozzle, or it is mixed with sand in advance of sweeping on the sand bed.

ExOne and Voxeljet, ProMetal RCT technology, and 3DP-ZCast method are commercial 3D printer brands for sand core or mold printing. Currently, the forming size is as big as 4000 × 2000 × 1000 mm, which can increase the production rate of cores and molds for small castings by many units per batch and satisfy the requirement of large castings. To guarantee the operation of printers and the mechanical and performance properties during the casting process of sand core or mold, some 3D printer companies provide their own specific binders and sands for 3D printing. For example, ExOne provides PHENOLIC Binder (FB101) and SILICATE Binder (FB901). Phenolic binder is for castings with lower melting points, such as aluminum alloys, and silicate binder is for alloys with a higher melting point, such as cast steel. ExOne printers can take furan resin, phenolic resin, and sodium silicate binders for quartz sand, and the first also for aluminum oxide, while the latter two can be used for synthetic sand. For phenolic resin and sodium silicate, post-curing is necessary; however, the sand block made from furan resin is free from post-curing, and 3D Systems (formerly ZCorp) is currently utilizing a specific form of plaster together with olivine sand as a possible material system for rapid production of patternless molds for nonferrous casting. It needs post-curing to improve strength (Kang and Ma 2017).

The sand molds and cores produced by SLS and 3DP methods do not need to have parting lines or parting surfaces similar to those in conventional casting because of the unit piece forming in AM. They have now been successfully used to produce complicated castings such as cast iron and aluminum cylinder blocks, cylinder heads, gear boxes, etc. during the development of new castings and small batch production. Further, both these methods have been proved to greatly shorten the lead time and production time and reduce cost. Cores can be integrated in one design, as shown in Figure 10.3. The pattern and core can be designed and built by 3D printing in one step, with no parting lines (which required an extra material removal process from the cast product in conventional casting). Figure 10.4 shows a 3D printed sand core and complete sand mold developed by OK Foundry designs.

Direct AM of metal parts: laser engineered net shaping (LENS), direct metal deposition (DMD), direct metal laser sintering (DLMS), and selective laser melting

FIGURE 10.3   Integration of cores into a single piece for an airbrake casting. (From https://callisto.ggsrv.com/imgsrv/FastFetch/)

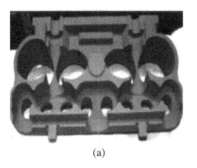
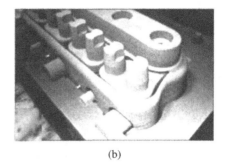

(a) (b)

FIGURE 10.4 3D printed cores and molds by OK Foundry designs for limited to volume production of gray and ductile iron castings, United States: (a) 3D printed core; (b) 3D printed mold. (Panel (a) from https://okfoundry.com/wp-content/; panel (b) from http://jameso22.sg -host.com/wp-content/uploads/)

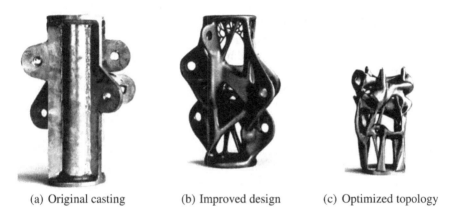

(a) Original casting (b) Improved design (c) Optimized topology

FIGURE 10.5 Optimization of a casting product for additive manufacturing. (From: Kang, J.-w. and Ma, Q.-x., (2017) *Special Report*, 14, 3)

(SLM) methods can be used to directly print metal parts from metal powder. The 3D direct making of metal parts is also called fast manufacturing, without the making of patterns, dies, or molds. Laser melting can be used for superalloys, titanium alloys, steels, especially stainless steels, and aluminum alloys. Mixed alloy powders can be used to obtain mechanical properties that could not be achieved by conventional casting. A borescope boss of nickel-based alloy has been produced by SLM instead of casting and was fitted in an Airbus turbine engine. GE Aviation produced the 19 fuel nozzles in its LEAP jet engine via powder bed fusion (PBF) in large quantities.

The solid structure of castings and molds has been redesigned into truss or spatially open and skeleton structures for higher mechanical properties and lower material consumption. A typical example is shown in Figure 10.5. This kind of innovation is just beginning, but it will have an unimaginable impact on manufacturing, including casting production.

If limited only to the shape forming process, casting is much faster than AM, where metal is printed point by point, profile by profile, and layer by layer, so it is very time-consuming, while tons, or hundreds of tons, of liquid metal can be poured into a mold cavity in less than an hour. Therefore, AM will never replace casting, but its role in developing casting technology will keep growing.

10.2.2 INVESTMENT CASTING

The role of AM in investment casting has revolutionized the capabilities, the effectiveness, and accordingly, the fields of application of the process. Now, new terms are being used in this context, such as rapid investment casting (RIC) or "AM assisted investment casting". The conventional method of investment casting is less effective in terms of cost and time for developing new wax patterns and hard tooling for low-volume production and prototypes. The implementation of AM results in a reduction of casting time while providing the same quality to the final product as well expanding the range of applications in many industries. It is cost-effective even for single or small-scale production (Sigirisetty 2022).

Investment casting or lost wax casting is a process supported by molding wax patterns. Beeswax was primarily used to form patterns necessary for the casting process, but modern blends for investment casting wax are compounds of hydrocarbon wax, natural ester wax, synthetic wax, natural and synthetic resins, or organic filler materials and water. Wax patterns can be easily made in large quantities by injection molding in metallic dies and assembled with sprues. The delicate wax patterns must have the strength to withstand the forces encountered during the mold making. Molds are made of refractory ceramic materials (water glass quartz sand or silica sol zircon sand) that form a sufficiently thick layer that completely covers the pattern. The mold is allowed to dry in air for several hours to allow hardening of the binder, and is then held in an inverted position in an oven at a temperature of 90–175°C to allow the wax pattern to run out of the cavity, and the heat is increased to 700 to 1000°C to ensure high strength of the mold and eliminate the danger of gas formation during casting. The molten metal is poured and left to solidify and cool down, the mold is broken up, and castings are separated from the sprue. Investment castings are usually small in size (less than 35 kg), but with neat intricate geometries. All types of metals, including steels, stainless steels, and other high-temperature alloys, are suitable for investment casting. Typical products include: turbine blades and other components of turbine engines, complex machinery parts, jewelry, gears, cams, valves, and dental fixtures. However, it is difficult to cast objects requiring cores. Holes produced by this method cannot be smaller than 1.6 mm and should be no deeper than about 1.5 times the diameter. This technology requires a longer production cycle as compared with other methods.

RIC is changing the field of investment casting. There are three approaches for this change:

1. Rapid prototyping of investment casting patterns based on an exact copy of the final product created in computer-aided design (CAD) software.
2. Direct fabrication of ceramic molds for case fabrication.

3. Rapid prototyping (RP)-manufactured injection molds, further classified into wax and non-wax, and RP-manufactured wax injection molds for tooling (Ripetskiy et al. 2023).

Pattern materials: The printed pattern in RIC is usually made of polymeric materials, which have higher strength compared with waxes. This enables suitability for the manufacturing of higher-accuracy, intricately shaped, thin-walled structures. The most commonly used materials are:

1. *Polylactic Acid (PLA)* is the thermoplastic material most suited to the investment method due to its low melting point temperature, with a glass transition temperature of 60–65°C and a melting temperature of 130–180°C. It does not produce hazardous toxins such as those produced by acrylonitrile butadiene styrene (ABS). The ability of PLA to melt and be processed again is what has made it so prevalent in this process.
2. *Acrylonitrile butadiene styrene (ABS)*: is a standard thermoplastic polymer commonly used for injection applications. It has a boiling point of 145.2°C and a density of 1.060–1.080 g/cm^3, and it is insoluble in water.

Types of AM techniques for RIC:

1. *Fused deposition modeling (FDM):* This is a 3D printer in which heat is used as a source to melt down the polymer material that comes out of the nozzle in its plastic state. The temperature required is 180 to 200°C for PLA material and 230 to 240°C for ABS. This material is deposited on the printer bed with the in-sync movement of the xyz coordinates of the machine.
2. *Multi-jet printing or MJM:* This is a material jetting printing process that uses the piezo printhead technology to deposit materials layer by layer during printing. MJM has higher dimensional accuracy as compared with silicone rubber molding. This technology has been found to have better ability to make patterns of parts with freeform surfaces, such as gas turbine blades.
3. *Stereolithography or SLA 3D printing:* This type of 3D printing technology targets creating models, prototypes, and patterns for RIC. The principle is focusing a UV laser on photopolymer resin. The resin, which is sensitive to UV light, will be solidified, form a single layer, and ultimately, make up the body of a three-dimensional solid object. SLA produces patterns with high accuracy, isotropic and watertight objects with excellent qualities and smooth surface finish. It can print the pattern with both a cellular internal structure and a smooth and compact surface. When the pattern is heated for burn-out of the mold, the resin will collapse inward and avoid the failure of shell cracking.

Applications of RIC:
Investment casting is used in almost all industrial sectors, especially electronics, petroleum, chemical, energy, transportation, light industry, jewelry, textile,

pharmaceutical, medical equipment, pumps and valves, and other sectors. Some of these applications are discussed in this section.

1. Automotive industry: Precision RIC provides a wide variety of components for the automotive industry, especially those with intricate geometries, thin sections, and high dimensional accuracy. It is extensively used for manufacturing complex engine components such as turbocharger housings, intake manifolds, and valve bodies. Other applications include suspension parts, door handles, grill emblems, etc. The process provides wide material versatility, design flexibility, and cost efficiency. A typical application is the inlet manifold to attach triple 45-mm Weber carburetors onto a vintage slant-six Chrysler engine, shown in Figure 10.6. The figure represents the CAD model, the SLA printed pattern, and the final product (Lumley 2018). Figure 10.7 represents three high-precision automotive components made by the leading British Lestercast Ltd.

2. Aerospace industry: The major requirements of parts for this industry are light weight and higher mechanical properties at high temperatures (refractory materials), in addition to complex geometries with internal channels, cavities, and fine structures. All these prerequisites can be covered by RIC. Typical investment casting products include:

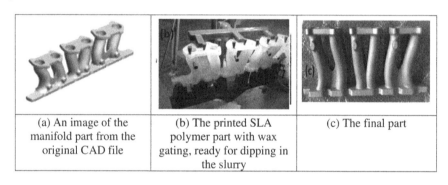

(a) An image of the manifold part from the original CAD file	(b) The printed SLA polymer part with wax gating, ready for dipping in the slurry	(c) The final part

FIGURE 10.6 RIC of a prototype casting without hard tooling. (a) An image of the manifold part from the original CAD file; (b) the printed SLA polymer part with wax gating, ready for dipping in the slurry; (c) the final part. (Courtesy of AWBell Pty Ltd)

FIGURE 10.7 Three high-precision automotive components. (From https://lestercast.co.uk /investment-casting-automotive/)

Engine turbine blades with complex blade shapes and internal cooling channels designed to improve cooling and the mechanical performance of the blade.

Aero engine casing: complex casing structures, such as turbine nacelles and combustor casings.

- Gas turbine generator components that operate in a high-temperature, high-pressure, and high-speed environment. Investment casting can provide superalloy materials and complex internal cooling channels to meet their special working requirements.
- Structural parts such as aircraft fuselage parts, wing ribs, connectors and brackets, and suspension parts.
- Structural supports including frames, brackets, and connectors, etc., which need to have high strength, stiffness, and durability to withstand the gravity and mechanical stress of the spacecraft.
- Rocket engine parts for spacecraft, propulsion system parts, attitude control devices, etc. (Wang 2023).

Figure 10.8 shows examples of RIC applications in aerospace, including (a) turbine blades made of superalloys based on nickel and cobalt for gas turbines by "PBS, Precision Castings" and (b) ultra-lightweight aircraft seat frame using lattice optimization, 3D printing, and investment casting, made by Michigan's Aristo Cast. The latter is made of magnesium because it weighs 35% less than conventional aluminum for seat frames and has a higher strength-to-weight ratio (current metal additive printers cannot print magnesium). Due to its material and latticed design, the resulting seat frame weighs 56% less than typical current models, which saves millions of dollars in fuel for a single year of 615-seat Airbus A380 flights. It would also

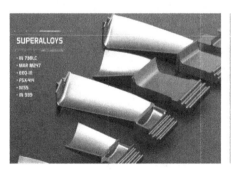
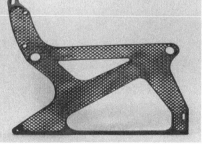

(a) Turbine blades for gas turbines   (b) Magnesium airplane seat frame

FIGURE 10.8 Applications of investment casting in aircraft: (a) turbine blades for gas turbines; (b) magnesium airplane seat frame. (Panel (a) from https://i.ytimg.com/vi/fP -wT8zJ2zs/; panel (b) from Harris, A., Modern Metal Casting plus Additive Manufacturing Equals Paradise Foundry, Design and Make with Autodesk, 2017. www.autodesk.com/design -make/articles/metal-casting-and-additive-manufacturing)

translate to a footprint-reducing 140,000-plus fewer tons of carbon in the atmosphere (Harris 2017).

3. Medical applications: There are numerous RIC applications in the medical field, including medical implants such as hip and knee joints, surgical instruments, keel punches, injury stabilizing devices, surgical room equipment, etc. Sizes in the range of 5–250 mm and wall thickness about 0.5 mm are common for implants made by this technology. There must be special physical, mechanical, and medical characteristics for the materials used in casting these parts. Commonly used alloys include nickel-, cobalt-, and copper-based alloys, low alloy steels, stainless steels, and the recently dominating titanium alloys. Figure 10.9 shows examples of hip and knee RIC manufactured prosthetic implant parts made by ICAST ALLOYS LLP.

10.2.3 Die Casting

Die casting is a process used for a large proportion of mass-produced components worldwide. The basis of die casting is that molten metal is poured under high pressure into a mold cavity, where the cavity must withstand the pressure and temperature, which sets high requirements for the manufacturing of the die as well as the material properties. There is significant cost related to both the casting equipment and the metal dies, which primarily makes the technology economical only for mass production with relatively high integrity. Die-casting dies produced with traditional manufacturing technologies have shape complexity limitations that necessitate the use of different inserts and components, and assembly complications.

A die consists of two halves: the cover die, which is fastened, and an ejector die, which moves in and out each casting. The ejector half has ejector pins to push the cast part out of the die, and it's important that the holes for the ejector pins are accurate to prevent molten metal from penetrating and destroying the die. The parting line between the two halves needs to be parallel with a fine surface to ensure

FIGURE 10.9 RIC manufactured medical implants. (Courtesy of ICASTALLOYSLLP)

sufficient sealing when the dies are closed to cast high-quality parts. Tools for die casting are made of high-quality tool steel with good strength and hardness at high temperatures. The most influential failure mode in die casting is thermal fatigue, and the die material needs to be chosen dependent on the thermal stress and the thermal conductivity.

In die casting, cooling is of great importance for the quality of the part, the productivity, and the lifetime of the die. The cooling time in a casting cycle is important to ensure sufficient cooling of the melt to prevent warp when ejecting the casted part. Uniform thermal control can considerably improve the die-casting process and the final quality of the part. With traditional die design, the cooling channels can only be designed with straight line drilling using a computer numerically controlled (CNC) machine, resulting in an uneven cooling process and causing internal stress that adversely affects the quality of the part as well as the productivity. By reducing the cooling cycle in a controlled manner, the scrap rate can potentially be reduced in parallel with increased productivity (Ringen et al. 2022).

Impact of AM: It allows new design solutions for dies that can improve quality and productivity in the die casting processes. It enables complex designs without adding extra time to the build process and is used as an alternative or complementary technology for the production of more advanced tools, which further enhances the control of the casting process. Some of the greatest advantages of implementing the technology are weight reduction and reduction in number of components and time to market. On the outside, the die would have only a small number of connectors, which would simplify the setup and maintenance of the casting machine.

Conformal cooling channels: The most important role of AM for die casting dies is that it enables the use of cooling channels that follow the shape of the mold cavity and give fast and uniform cooling and a more accurate control of the solidification, usually called conformal cooling channels. These channels are produced through the SLM building of the die. Their structure and form are optimally adapted to the geometry of the specific insert and its function in the die-casting mold. The technology is already used in the making of tools in plastic injection molding, blow molding, sheet and metal forming dies, and to some extent, bulk forming such as extrusion, rolling, and forging dies. Conformal cooling enhances thermal control and can potentially drastically lower cycle times. A larger cooling area can easily be achieved, and the channels could be designed to have relatively low flow resistance. In addition to advanced cooling channels, AM can be used to produce die inserts with venting slots to evacuate air and core gas (Hovig et al. 2016). In a case study by Oskar Frech, they developed a die-casting die with conformal cooling channels close to the surface (the contour) of the component for efficient tempering. They found that the cooling time was reduced from 12 to 5 seconds, a reduction of nearly 60% achieved with the help of the cooling channels. Consequently, the total process cycle was shortened by over 12%, while the manufactured parts exhibited no material defects. Figure 10.10 shows the CAD file of a die with conformal cooling channels inside.

AM techniques and die materials: The most commonly applied AM processes for die manufacturing are SLM and direct metal laser sintering (DMLS). In

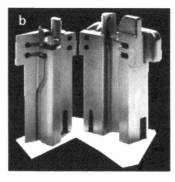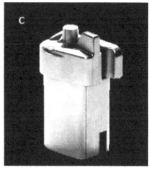

FIGURE 10.10   (a) The CAD file of a die with conformal cooling channels inside the insert, (b) the channels' profile inside the die' and (c) the assembled die. (From Oskar Frech GmbH)

aluminum die casting, the most frequent die material conventionally manufactured is H13 Hot Worked Tool steel, which has high strength at elevated temperatures and high hardenability. This tool material has a thermal stress of 2604 MPa and thermal conductivity k = 27 W/mK at 500°C. For AM of the die, hot-work steel powder, EN 1.2709, with almost the same properties as H13, can be used. This metal powder is used to produce tool inserts with conformal cooling for injection molding of die-casting as well as functional components. The surface structure of die-casting components is improved through AM in addition to reduction of mold separating agent and increased mold service life.

10.3   THE IMPACT OF AM ON POWDER METALLURGY

10.3.1   AM AS PART OF POWDER METALLURGY (PM) TECHNOLOGY

PM generally means the processing of powder metals to manufacture components, using pressure and temperature to consolidate them in the form of a net shape product. The process has been used since the 1920s, when pressure was applied by a press with special tooling to form a green compact body, followed by consolidation of particles in a furnace at a temperature below the melting point, known as sintering. With time, the process has been developed to incorporate new techniques for applying the pressure and temperature, which led in the 1970s to MIM. A further diversification in the pressure-heating cycle resulted in HIP. The emergence of AM is considered a recent development of PM technology. Major international associations such as the European Powder Metallurgy Association (EPMA) classify the PM industry into five classes, the latest being AM. Similarly, the American Metal Powder Industries Federation (MPIF) classifies PM processes as including the AM class. Therefore, the presently known PM technology covers the following processes:

1. Manufacturing of powders: Metal powders are a versatile and indispensable cornerstone of modern manufacturing, enabling the production of intricate and customized components that were once deemed unattainable. With ongoing innovations in material science, from high-strength alloys to environmentally friendly powders, the metal powder market continues to evolve, empowering industries to push the boundaries of what's possible in terms of efficiency, sustainability, and design freedom (Gorges 2023). The preparation of powders for all other PM processes has become a huge global industry that shares a growing portion of metal manufacturing. There are various techniques to produce metal powders, and numerous characterization techniques to assess their quality and adaptability for subsequent processing, all discussed in Chapter 7 (Sections 7.3.2 and 7.3.3).

2. Compaction-sintering PM: The process aims to manufacture a wide range of structural PM components with unique materials and properties, high quality, and superior performance. It is of critical importance to a number of sectors, including automotive, aerospace, biomedical, and construction. Processing involves mixing the powders with suitable lubricants, pressing the mixture in a die to produce a compact (green) part having sufficient cohesion to enable safe handling, and then heating, usually in a protective atmosphere, where the compressed particles weld together and confer sufficient strength on the material for the intended use (sintering). A recently developed high-velocity compaction technique reduces compaction time to less than 20 milliseconds by a high-energy impact. Further densification is possible with multiple impacts at intervals as short as 300 ms. This approach is potentially particularly useful for the production of large components. PM can produce parts from below 1 gram up to 30 tons in weight and in huge production volumes.

In special cases, compaction is done at an elevated temperature, such that sintering occurs during the process. This is termed hot pressing or pressure sintering. In many cases, the sintered part is subjected to additional processing such as forging, heat treatment, plating, coating, or machining to get a product with the required quality. The process is applied for the production of parts having a significant, carefully controlled porosity designed to serve a useful purpose, such as filters and self-lubricating bearings.

3. Hard metals and diamond tools: Hard metals is the term used to signify a group of sintered, hard, wear-resisting materials based on the carbides of one or more of the elements tungsten, tantalum, titanium, molybdenum, niobium, and vanadium, bonded with a metal of lower melting point, usually cobalt. Tungsten carbide is the most widely used. These materials are commonly referred to as cemented carbides or simply as carbides, as, for example, carbide tools. By varying the carbide particle size, the amount of binder metal, and the sintering conditions, properties such as wear resistance, impact strength, resistance to cratering, and hot hardness may be optimized for a given application. For example, in the case of a wire drawing die, wear resistance is the major requirement, but for a cutting tool, especially if subject to intermittent loading, high impact strength is required.

Diamond cutting tools are made by similar processes. In this case, it is important to provide a matrix, which gives maximum support to the diamonds in order to keep the tool sharp. Each cutting application therefore requires separate consideration, and matrix materials range from bronzes of different compositions to cemented carbides.

4. Metal injection molding (MIM): The process commenced in the early 1970s with ceramic powder and was then adapted to metal powders (as metal injection molding) at the end of that decade. In principle, metal powder is intimately mixed with a thermoplastic binder and worked in an extruder to produce a plasticized feedstock (usually in a granulated form), which is then injected at a slightly elevated temperature into a mold with the required shape and finally, sintered. In order to get a readily injectable feedstock and a uniform powder loading, the powder has to be very fine. The carbonyl iron and nickel powders used in low alloy steel blends are spherical and have particle sizes between 2 and 10 μm. High alloy powders, such as stainless steels with particle sizes less than 40 μm, represent a dominant material type in current MIM production. The process is significantly more expensive than the simple compact and sinter technique; however, it produces parts of very complex geometry. By reason of their complexity, such parts would be very expensive to produce by machining. The process is growing at an increasing rate in various fields of industry (EPMA 2008).

5. Hot isostatic pressing (HIP): The powder is contained in a mold made in the form of a metal container, referred to as a can or capsule, and is compacted under hydrostatic pressure at high temperatures. The pressurizing medium is inert gas (most commonly argon) to avoid chemical reaction with the material. No lubricant is needed for the powder particles; therefore, high and uniform density can be achieved. After densification, the capsule has to be stripped off. The main advantage of HIP is the ability to reach the full density of the product, thus accomplishing compaction and sintering in one step. However, the process is relatively expensive and is limited to relatively simple shapes. Therefore, it is mainly used for producing billets of superalloys, high-speed steels, ceramics, etc., where the integrity of the materials is a prime consideration. The process has a wide range of capabilities, including large and massive near net shape metal components such as parts for the oil and gas sectors weighing up to 30 tons, or net shape impellers up to 1 meter in diameter. Equally, it can be used to make small high-speed steel (HSS) cutting tools, or even very tiny parts such as dental brackets. As a result, HIP has developed over the years to become a high-performance, high-quality, and cost-effective process for the production of many metal components. HIP can be used as a post-sintering operation for other PM processes to eliminate flaws and microporosity in cemented carbides when high-quality parts are needed for special applications (EPMA 2008), and a post-processing operation in AM.

6. Additive manufacturing (AM): Being based on building physical models, prototypes, patterns, tooling components, and production parts by consolidating plastic, metal, ceramic, and/or composite powder materials, the technology is evaluated as a PM process. Generally, most of the AM techniques start with material powders and pass through compaction of layers and consolidation by laser, electron beam, or arc-type heat sources. AM processes allow lightweight structures, either by the use of lattice design or by designing parts where material is only where it needs to be, without other constraints. Further, they enable the production complex internal channels or several parts in one.

Talking about real manufacturing in terms of large-scale production of large numbers of products, only a few niche AM products have reached this state. AM is expected to continue growing at a strong double-digit growth rate over the next few years. Global companies such as Airbus and General Electric are using AM techniques to produce complex metal parts for next-generation aerospace and medical products. It opens up a whole new world in shape feasibility, but major improvements can and need to be made relating to precision, material properties, and production speed, among others (EPMA 2019).

Generally, there are three principal reasons for using a PM product: cost savings compared with alternative conventional processes, unique attributes attainable only by the PM routes, and environmental aspects. PM is a green technology; in particular, it allows high material yield utilization and low energy consumption. Around 80% by weight of the raw material used in the manufacture of PM parts is derived from recycled scrap, and over 95% is present in the final product, compared with levels often of only 50% for conventional processes, as shown in Figure 10.11.

AM works in conjunction with other PM technologies. Like HIP, AM is more suitable for the production of small or medium series of parts. While the HIP process is generally used for the manufacturing of massive near net shape parts of several hundred kilograms, the AM process is more suitable for smaller parts of a few kilograms, and it offers an improved capacity to produce complex metal parts thanks to greater design freedom. MIM and compaction-sintering technologies also offer the possibility to produce net shape parts, but they are recommended for large production volume of small parts. Figure 10.12, outlines the way in which the various PM processes, including AM, overlap so as to enable designers and engineers to cover their cost, performance, and material requirements.

10.3.2 THE ROLE OF AM IN POWDER METALLURGY

1. AM products have substituted a limited number of compaction-sintering PM products that need high quality characteristics for the aerospace

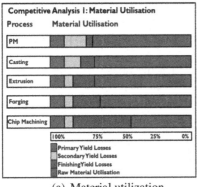
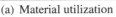
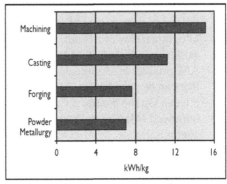

(a) Material utilization   (b) Energy consumption

FIGURE 10.11 Material utilization (a) and energy consumption (b) for different metal manufacturing processes. (From EPMA Vision-2025, Future Developments for the European PM Industry, 2015, European Powder Metallurgy Association, UK)

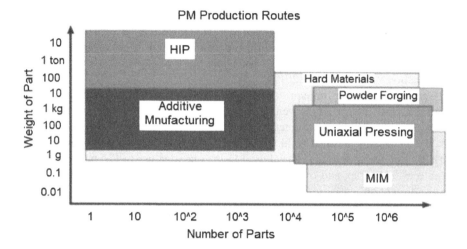

FIGURE 10.12 Relative position of the different segments of the PM industry per batch. (From EPMA Vision-2025, Future Developments for the European PM Industry, 2015, European Powder Metallurgy Association, UK)

industry and customized applications where cost is not a priority. However, for the automotive industry, and other industries based on large batches and economic consideration, the demand for PM products dominates.

2. AM technology has revolutionized the powder preparation and characterization techniques, and initiated unlimited novel material alloys and composites that were immiscible by conventional alloying techniques (mechanical alloying, functional materials, etc.).

3. AM processes are preferable to HIP products for smaller parts of a few kilograms, and they offer an improved capacity to produce complex metal parts thanks to greater design freedom, while the HIP process is generally used for the manufacturing of massive near net shape parts of several hundred kilograms. However, both categories are limited to small-batch production, as shown in Figure 10.11.

4. AM product sizes are relatively smaller than the range of product sizes that can be produced by MIM, which eliminates the possibility of negative effects. On the contrary, CAD virtual models and prototypes by AM have improved MIM by reducing the process cycle time and providing complex configurations.

5. Products of AM may need post-processing by PM techniques or their post-sintering processes, for example:
   • **HIP** is used as a post-sintering operation to eliminate flaws and microporosity in cemented carbides.
   • **Surface cold rolling** is a means of creating densification locally just in the areas where it is most often required, the surface layers of the component. The technology was originally applied to improve the rolling

contact fatigue resistance of PM bearing races but has subsequently been applied to the tooth profiles of both straight spur and helical gears to improve tooth bend fatigue strength and surface contact fatigue strength (pitting wear resistance).

- **Infiltration** is an alternative method of improving the strength of inherently porous sintered parts by filling the surface connected pores with a liquid metal having a lower melting point that propagates by capillary action without pressure. The process is used quite extensively with ferrous parts using copper as infiltrant, but to avoid erosion, an alloy of copper containing iron and manganese is often used.

- **Impregnation** is analogous to infiltration except that the pores are filled with an organic as opposed to a metallic material. An outstanding example is oil-impregnated bearing materials, but increasingly, impregnation is being done with thermo-setting or other plastic materials. The benefits include some increase in mechanical properties and sealing of the pores, which may provide pressure-tightness and will also prevent the entry of potentially corrosive electrolyte during a subsequent plating operation. Additionally, the machining of sintered parts is improved.

- **Heat treatment:** Although most of the bulk of sintered structural parts is used in the as-sintered condition, large quantities of steel parts are supplied in the hardened and tempered condition. Conventional hardening processes are used, but because of the porosity inherent in sintered parts, they should not be immersed in corrosive liquids—salt baths, water, or brine—since it is difficult to remove such materials from the pores. Heating should be in a gas atmosphere followed by oil-quenching. These restrictions may not apply to very high-density parts or to parts that have been infiltrated.

- **Steam treatment** involves exposing the steel part at a temperature around 500°C to high-pressure steam. This leads to the formation of a layer of magnetite (iron oxide) on all accessible surfaces, and a number of desirable property changes result. First, the corrosion resistance is increased by the filling of some of the porosity. Second, this reduction in porosity of the surface layer leads also to improved compressive strength. Third, the oxide layer significantly increases the surface hardness and more importantly, the wear resistance. Steam treatment is often followed by dipping in oil, which enhances the blue/black appearance and increases the corrosion resistance still further. The treatment is not generally applicable to hardened and tempered parts because the exposure to the high temperature would undo the hardening.

- **Blueing:** Heating in air at a lower temperature (200–250°C) can also be used to provide a thin magnetite layer that gives some increase in corrosion resistance, but it is much less effective than steam treatment.

- **Plating:** Sintered parts may be plated in much the same way as wrought or cast metals, and copper, nickel, cadmium, zinc, and chromium plating are all used. However, it is important to note that low-density parts should be sealed—e.g. by resin impregnation—before plating to prevent the electrolyte from entering the pores and causing corrosion. Parts that have been oil-quenched cannot be plated satisfactorily unless the oil is removed before resin impregnation. Work has shown that it may be possible successfully to plate unimpregnated porous parts with nickel by electroless plating.

- **Coatings:** A large percentage of hard metal cutting tool inserts are now coated using chemical vapor deposition (CVD) or physical vapor deposition (PVD). The lower-temperature PVD process also allows steels to be given a wear-resistant layer of TiC, TiN, Al_2O_3, or a combination of these materials, and sintered HSS tools are also now being coated.

10.4   THE IMPACT OF AM ON DEFORMATION PROCESSES

Metal forming processes are used to produce structural parts and components that have widespread applications in many industries, including automobile, aerospace, appliances, etc. The trace and influence of metal forming can easily be found in our daily life, as about 15% to 20% of gross domestic product (GDP) of industrialized nations comes from metal forming processes, ranging from 7% of the U.S. GDP to 28% of the German GDP, both exceeding a value of $1 trillion (Cao et al. 2019).

These forming processes include a wide range of primary operations that deform a bulk material into standard sections such as structural shapes, rods, tubes, wires, plates, and sheets. These sections are further deformed in secondary operations to reach a final product with the desired geometry and characteristics. Forming processes can generally be classified into bulk and sheet forming. Bulk forming includes forging, rolling, extrusion, and drawing processes. Sheet forming includes shearing, bending, and deep drawing. All these processes require expensive equipment investment to start running, in addition to complex designed costly tooling, which renders them suitable only for large-scale production.

10.4.1   THE ROLE OF AM IN FORGING

AM has been presented as a possible alternative to traditional manufacturing methods, including forging. It can also be regarded as a complementary technology to be used by forgers to reduce lead times or lower cost. The new technology has many potential applications in forging.

Forging dies wear relatively fast and require frequent welding repair. This is true in particular for forging dies that are made of tool steel. Stem dies used for forging crankshaft are susceptible to rapid deformation due to the prolonged contact with the high-temperature steel of the crankshaft. Welding repair of some stem dies is required after as little as 500–600 pushes. This fairly frequent repair work requires removing dies from operation and replacing them. It would benefit such forging operations if the life of forging tools could be extended by cladding selective areas with a layer of heat- and wear-resistant alloy. Flood welding is a type of welding that uses a unique method of depositing vast amounts of weld metal. This process, as well as metal inert gas (MIG) and tungsten inert gas (TIG) welding, is applied to large die sets and components such as bolster plates, forging dies, hammer rams, die holders, and ejector holders. These traditional methods have been joined lately by wire arc additive manufacturing (WAAM) and laser direct metal deposition (LDMD). In addition to repair, these methods can enable die life extension by enhancing the wear and heat resistance of the surface. Tooling materials are generally characterized by high hardness at high temperature and good resistance to wear and corrosion. Under severe service conditions, they can suffer excessive wear, which will affect performance and shorten tool life. Rather than replacing worn-out with new tools, a surface layer of wear-resistant alloy can be applied by direct metal deposition to restore the worn-out part to full functionality and extend die life. Typical deposited materials include Stellite-6 and Waspalloy on FX-1, M2, and CPM-V steel with the objective of enhancing die life. Cladding of forging dies with Waspalloy is one of the main alloy candidates to enhance the heat and wear resistance of forging dies. Waspalloy is a heat-resistant superalloy that can take much higher loads at high temperatures.

For direct metal deposition, Waspalloy, Cobalt 6, and Stellite alloys 21™ and 21A™ exhibit compressive strength characteristics that are significantly better than the 4140 and FXBM base metals and the currently used welding rod F45FC material. These alloys should provide significantly improved performance and should be considered in field trials to be benchmarked against the currently used materials. Laser scanning "before and after" is a powerful tool to capture dimensional changes in the dies during production. The Waspalloy cladding deformed less than CorMet F45 after 120 pushes.

In general, AM can produce parts with strength similar to the same parts made by forging. However, the ductility will usually be lower for the additively manufactured parts due to some residual porosity. There are methods by which this porosity can be "closed". The traditional method that has been used for many years in biomedical and aerospace casting is HIP. This method subjects the parts to high isostatic pressure at high temperature, thereby closing most of the pores and increasing ductility. A respective increase in fatigue properties is obtained.

Another application of AM in forging is prototyping and production of small runs by 3D printing. The obvious advantage is that parts can be made with a short lead time, without the need to fabricate expensive tooling. However, the time required to print each part is much longer. The choice of which process to use depends on how many parts are required. For a small number of parts, 3D printing may be more cost-effective. The assumption is that a 3D printed part can provide the properties and functionality obtained by forging. While this is not necessarily feasible if the same material is used, the designer can compensate for lower properties of 3D printing by using an alloy with higher mechanical properties.

Alcoa, the major aluminum producer in the United States, has recently announced a new process designated Ampliforge, in which preformed parts made by AM are forged to finished dimensions. This process is illustrated in Figure 10.13. The forging step is designated to close any porosity present in the parts and achieve "as-forged" properties (Schwam and Silwal 2017).

FIGURE 10.13   The Alcoa Ampliforge process.

10.4.2   THE ROLE OF AM IN METAL ROLLING

AM processes offer multiple benefits and operational improvements for heavy industry. They can help by reducing downtime, improving lead time, and extending the overall lifetime of machinery. The global industry that AM lends itself to is the steel manufacturing industry (mainly melting furnaces, rolling mills, galvanizing, tinning, etc.). Metal 3D printing produces parts with intricate, complex features and high mechanical properties that are difficult or even impossible to create using traditional manufacturing methods, making it an ideal manufacturing method for steel industry parts.

The following examples provide a fair assessment of the role of AM in this industry from case studies by a large Spanish 3D manufacturer (the steel printers).

- *Hot rolling mill:* A key example in this line is the production of a 3D printed heat exchanger used to cool the blast furnace shell. The optimized design of the 3D printed part enables greater heat transfer efficiency and the elimination of weld joints, resulting in improved component and system reliability. The manufacturing lead time was also reduced by 90%, since no complex machining or welding operations were required. This presented a fast solution that reduces the risk of a blast furnace shutdown and ultimately extends the operating life of the blast furnace shell.

- *Cold rolling mill:* At the production stage, process control and equipment reliability are critical to achieving high operational performance. Lasers and cameras are installed throughout the production line for measuring the position of moving strips and for detecting material defects. 3D printed laser and camera housings are often used to protect the sensitive instrumentation from the harsh operating environment. In doing so, the lasers can operate effectively, reducing the risk of strip misalignment and the development of defects that require reprocessing. As automation within cold rolling mills continues, the integration of sensors and electronics into the production line is growing, representing increasing opportunities for bespoke 3D printed housings.
- *Galvanizing production line:* A case example is the production of a metal 3D printed zinc pump turbine assembly used in the pickling stage. The original pump was made of cast steel, which would only last for two pickling cycles before having to be replaced due to the corrosive chemicals the components were subjected to. By optimizing the design of the components to remove weld joints and by 3D printing them in 316L stainless steel, the lifetime of the pump assembly was extended by over 300%. As a result, maintenance hours and spare parts in inventory could be reduced, representing significant operational savings.
- *3D Printing of spare parts:* The breakdown of a single piece of equipment can cause significant bottlenecks and lead to severe production delays, which in turn, increases operating costs. Further, 3D printed spare parts can be produced rapidly on demand, reducing the need for large storage areas. Examples of these spare parts are shown in Figure 10.14. In turn, metal 3D printing presents an opportunity to generate long-term savings through maintenance spend reduction, lifetime reliability improvements, and functional performance improvements.

10.4.3   THE ROLE OF AM IN EXTRUSION PROCESSES

Extrusion is the process of forcing a metallic material billet in a closed compartment to flow through a shaped die opening profile of the desired cross section. The process offers two main advantages over other forming processes, namely, its ability to produce very complex cross sections with high surface finish in large reductions, and deformation of materials that are relatively difficult to form, such as stainless steels and nickel-based alloys; however, aluminum is the most commonly extruded material. Both advantages are due to subjecting the material to compressive and shear stresses only (under hot or cold conditions). The process is commonly used for mass production of long, straight metal products such as bars, solid and hollow profiles, tubes, and strips. The process is carried out on presses with high tonnage capacity, but the product quality is mainly controlled by the die design and manufacture.

Closed impeller of centrifugal pump used in a tinplate steel mill, Size 230 x 230 x 110 mm, Material: 316L, the obsolete part was 3D scanned and reproduced in a matter of few days with minimal post process machining required.

Cooling Nozzle Body, Roll cooling spray nozzle used by steelworks temper mill. The design of the part was optimized to improve flow performance, and produced at reduced unit cost and lead time. Size 76 x 47 x 83 mm, Material: 316L, Lead-time reduction 50 %.

Casing of a periscope camera used to inspect the walking beam furnace of a rolling mill. The original part was redesigned to introduce internal cooling channels that resulted in improved reliability of the camera system. Size: 159 x 159 x 98 mm, Material: 316L, Life time improvement 400 %

Volute pump casing of a zinc pump located in a steelworks galvanizing line 3D printing in 316L enabled greater chemical resistance and reduced manufacturing lead time. Size: 276 x 144 x 320 mm, Material 316L, Lead time Reduction: 67 %.

FIGURE 10.14 Typical examples of 3D printed spare parts for steel manufacturing industry. (From the steel printers)

The extrusion die is simply a steel disk with a cut-through opening that includes all the geometric details and dimensional tolerances of the final product. Consequently, any geometrical or dimensional errors or surface flaws or faults are directly reflected in the profile. Extrusion dies can be designed and produced with a virtually limitless array of shapes and sizes. Die design and shape affect the extrusion pressure, speed, exit temperature, material flow, and friction. The pressure in the extrusion of fine-sectioned profiles is very high, and the die needs to be supported by a backing to prevent bending and cracking under high pressure. Therefore, dies are used together with components such as backer and bolster supports. These components are conventionally processed by machining, honing, shaving, etc. to reduce friction, thus enhancing material flow and profile surface quality. These cutting processes may cause faults such as cracks, stress accumulation, or improper processing (burning), which affects the product quality. Further, the two-dimensional perspective in die design limits the possibility of controlling the reduction flow lines of the material for complicated sections. Therefore, it was essential to introduce AM as an advanced technology to produce these dies with complex geometries, internal features, and passages that are not easy to produce with traditional production techniques.

SLM has been applied to manufacture extrusion dies from powder metals directly from the data in the CAD drawing of the parts, based on a novel three-dimensional design perspective. Also, DMLS provides the possibility to produce fine details with high precision and is a highly time- and cost-saving technology with the ability to replace conventional manufacturing processes such as die design, tool shaping, etc. and produce multiple parts simultaneously. Geometric forms of maraging steel dies used in aluminum extrusion were optimized by means of flow dynamics, as well as surface quality and mechanical properties of the final part. It is possible to decrease the pressure and temperature due to friction and geometry during extrusion in such dies by decreasing the set size. In addition, optimization of the flow dynamics of the extrusion dies improves the mechanical and surface properties of the products. Dies were not subjected to any conventional finishing post-processes. Extrusion dies with integrated local cooling channels are possible (Koc et al. 2017). Figure 10.15 shows

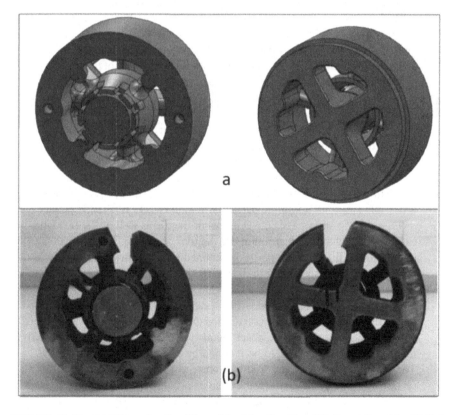

FIGURE 10.15 (a) Dimensional solid model of the die, (b) solid die produced by DMLS. (From Koc, E. et al., (2017) Selective Laser Sintering of Aluminum Extrusion Dies, *Technical Report*. https://doi.org/10.13140/RG.2.2.35134.89922)

(a) the three-dimensional design model and (b) the solid maraged steel die built by DMLS and age hardened.

10.4.4 THE ROLE OF AM IN SHEET FORMING TOOLS

Sheet metal forming is extensively adopted in most of the applications in aerospace, automotive, appliances, furniture, utensils, etc. As a mass production technology for thin metallic parts, there is no chance for AM to replace sheet forming products. The major role to be played by AM is improving the production of forming tools and dies, which represent the backbone of this industry. This involves redesign of the dies for higher performance, higher productivity, and lower cost. In this section, examples of different fields of application in sheet forming are discussed.

1. *AM integration of lubrication channels in sheet forming dies:* Laser beam melting (LBM) has the potential to change toolmaking. It allows tools of almost any complexity to be manufactured with very few geometrical limitations directly from 3D CAD data out of standard metal alloys, including hot-work steel. The focus is on tool optimization, which is addressed by incorporating novel features into tooling components, such as integrating a lubrication system in combination with tempering channels to improve the forming conditions. LBM technology facilitates a redesign of the tool's interior structure, thus leading to an approach that provides lubrication supply through the die instead of providing it externally. Structuring of the surface in combination with a corresponding distribution of the lubricant enables homogeneous wetting of the die's surface. Furthermore, this approach assures that lubricant is applied on the entire area where it is needed, i.e. the die's radius, where most of the metal forming takes place. This is a significant improvement compared with conventionally manufactured solutions, in which lubrication supply is merely achieved locally and in easy accessible areas but not in the particular regions where it is really needed.

   Figure 10.16 shows a 3D CAD model of the die as well as the two lubrication supply channels that follow the contour of the die. The open slots in the figure are filled with a porously scanned lamellar structure that facilitates lubricant transportation in the z direction from the supply channels to the outer die surface.

   An important design aspect is the orientation of the lamellae in relation to the direction of the sheet metal motion during the forming process. Since it has to be ensured that the sheet metal is directly lubricated in an areal manner, the lamellae have to be inclined by an angle between 0° and 90°. The manufacturing of the porous structure to fill the open slots requires a certain scan strategy, which ensures that adjacent scan tracks do not overlap each other but build a lamellar structure instead. These lamellae enable fluid transportation in their build direction (z direction, Figure 10.16), i.e. the lubricant is transported with high pressure from the supply channels through the lamellae and constantly wets the surface of the die during the

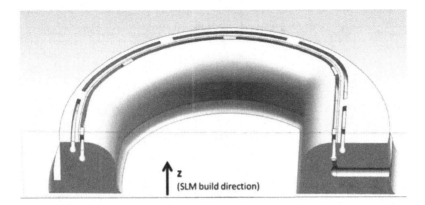

FIGURE 10.16 3D CAD model of forming die showing the lubrication channels. (From Gebauer, M. et al., (2016) in *iCAT 2016: Proceedings of the 6th International Conference on Additive Technologies*, ETH Zurich)

entire forming process. The lamellae consist of single weld lines on top of each other, which leads to low stiffness. Then, they are divided into small segments resulting in short lamellae and consequently, higher stiffness. The surrounding solid material further increases the lamellae's stiffness.

During the deep drawing process, a pump is connected to the die, ensuring the lubrication supply with pressures of up to 300 bar. Further, it is necessary to provide the tooling with a sealing ring, which is schematically visualized in Figure 10.17. This ring seals the tool at its outer diameter and therefore ensures that the lubricant is forced to enter the drawing clearance and hence, the zone where it is needed.

2. ***AM assisted construction of hot stamping dies:*** The die and mold industry plays a significant role in the manufacturing world due to the fact that nearly all mass-produced parts are manufactured employing processes that include dies and molds, directly affecting not only the efficiency of the process but also the quality of the product. Moreover, increasing demand in the automotive industry for high-strength and lightweight components has led to the promotion and development of hot stamping (also known as press hardening) processes. Through this technique, boron steel blank is heated until austenization at temperatures between 900 and 950°C inside a furnace and then transferred to an internally cooled die set, where it is simultaneously stamped and quenched. The transformation of austenite into martensite occurs thanks to a rapid cooling of the blank, at a temperature range of 420–280°C, through which the dies must be actively cooled at a minimum cooling rate of 27°C/s. The temperature of the hot stamping die must be kept below 200°C in order to ensure the cooling of the blank, achieve high strength, and prolong the lifespan of the tools. Therefore, hot stamping dies include cooling channels for controlling the thermal treatment of the formed sheet. The optimum cooling channels of dies and molds should

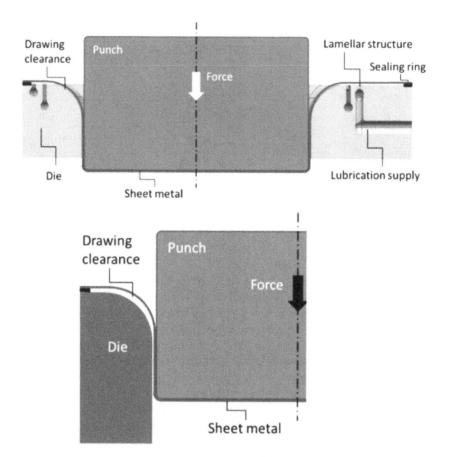

FIGURE 10.17 Scheme of tool design and deep drawing process (top); drawing clearance visualized in detail (bottom). (From Gebauer, M. et al., (2016) in *iCAT 2016: Proceedings of the 6th International Conference on Additive Technologies*, ETH Zurich)

adapt to the shape and surface of the dies so that a homogeneous temperature distribution and cooling are guaranteed; otherwise, the temperature of the tool can be increased during the productive process, the quenching may not be successfully achieved, and therefore, the final product will not meet requirements. Conventionally, cooling ducts are manufactured by deep drilling, attaining straight channels unable to follow the geometry of the tool. Recently, the integration of AM into die manufacture has played a revolutionary role in redesigning and manufacturing these dies via laser metal deposition (LMD), which is capable of fabricating nearly freeform integrated cooling channels and thereby shaping the so-called conformal cooling. These cooling ducts are additively built up on hot-work steel and then milled in order to obtain the final part. Conformal cooling in stamping dies targets shorter cycle times, improved mechanical properties of press hardened parts manufactured in the die, and a reduction of energy consumption for the cooling and idle times of forming presses. This cooling system is also applied in plastic injection molding and aluminum high pressure die casting. A typical example of a simple CAD model of a die set with conformal cooling where the cooling duct is perfectly adapted to the shape of the part is shown in Figure 10.18 (Cortina et al. 2018 (a), 8, 102).

In an experiment by Mueller et al. (2013), they manufactured hybrid hot stamping dies by machining and additively building up inserts with conformal cooling ducts by SLM. As a result, the additively manufactured channels cooled six times faster than the conventional drilled channels. They concluded that LBM proved to be a well-suited technology for manufacturing highly complex molds and tools that go beyond the limits of conventional production technologies. This innovative technology opens up ways to new design approaches of cooling systems in forming tools. Figure 10.19, shows the die structure for the hot stamping process (a), and the blank as well as a formed part (b), for a hot forming mold system composed of die, blank holder, punch, and cooling system. The die and punch are made by a number of insert blocks.

3. ***AM of tailored blank for sheet-bulk metal forming processes (SBMF):*** Functional integration and lightweight construction pose increasing demands on manufacturing processes and require innovative approaches. In this context, SBMF combines the advantages of conventional sheet and bulk forming processes, expanding the limitations in their flexibility, as they are based on costly and time-consuming tool manufacturing. However, the intended three-dimensional material flow presents major challenges regarding the material flow control. To enhance material flow control and part quality, the application of process-adapted semi-finished products is an expedient approach for these processes. So-called tailored blanks have a process-adapted sheet thickness profile or combine different mechanical properties within a single blank and were applied in the 1980s for the first time. Since then, tailored approaches have gained importance and found

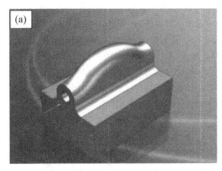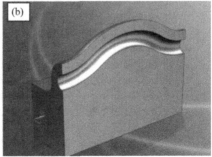

FIGURE 10.18 (a) Isometric view; (b) cross section of the 3D conformal cooling CAD model. (From Cortina et al. 2018b, 8, 102.)

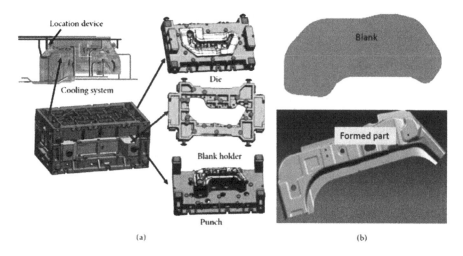

FIGURE 10.19 (a) Mold structure; (b) blank and formed part. (From Shang, X. and Pang, L., (2020) *Advances in Materials Science and Engineering*, 2020, 7621674)

broad applications in research and industry. At present, various technologies are used to manufacture tailored blanks. The tailored blanks applied in SBMF are often manufactured by combined sheet–bulk metal forming processes. The achievable gradient in thickness depends on various factors but is eventually constrained by the material volume of the initial blank. In this regard, AM offers new possibilities to overcome those limitations and to expand the limits of SBMF processes further, thus utilizing the advantages of both processes while reducing the major disadvantages. Additional material can be allocated with high geometric flexibility. Beside the allocation of additional material, this technology allows the manufacturing of discrete functional elements within the production of tailored blanks.

A representative application of the SBMF process involves deep drawing of a stainless steel blank within a single press stroke to a tailored blank and a subsequent upsetting operation to manufacture a functional component with an external gearing with different gear geometries (Figure 10.20). The drawing die and upsetting punch are part of the upper tool, whereas the drawing punch and upsetting punch are part of the lower tool. The cup wall is upset as soon as the cup comes into contact with the upsetting plate. The reduction of the cup height forces the material to flow radially into the gear cavity of the drawing die. A hydraulic press of about 4000 kN is needed for the dimensions presented in the figure. The functional component manufactured with this tool setup is an externally geared cup with 80 teeth.

The integration of AM with SBMF involves building a new layer with a high degree of geometric freedom on the outer circumference of the sheet by a laser beam PBF process to get a tailored blank, and subsequent SBMF (deep drawing + upsetting) to reach the final functional component

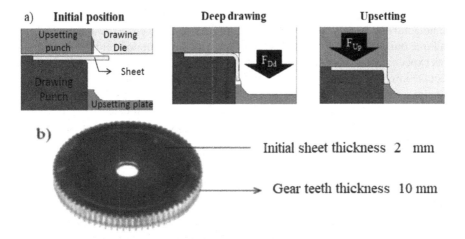

FIGURE 10.20   a) Setup of the conventional combined deep drawing and upsetting process; b) functional component manufactured within the process. (Adapted from Schulte, R. et al., (2020) *IOP Conf. Series: Materials Science and Engineering*, 967, 012034)

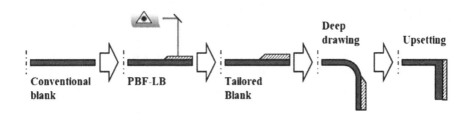

FIGURE 10.21   AM supported process chain to manufacture functional components by SBMF. (From Schulte, R. et al., (2020) *IOP Conf. Series: Materials Science and Engineering*, 967, 012034)

(Figure 10.21). The increased sheet thickness in the outer area provides an increased material volume for a higher die filling within the subsequent upsetting process that forms gear teeth with higher quality. The PBF-laser beam (LB) machine is equipped with a 600-W fiber laser. To reduce residual stresses and avoid distortion of the sheet metal, the heating of the build plate is set to 200°C (Schulte et al. 2020).

10.5   THE IMPACT OF AM ON MATERIAL REMOVAL PROCESSES

Material removal processes are intended to use cutting tools to eliminate parts from a manufactured body in the form of chips or minute particles to attain the required functioning and finishing of that body. Common processes are turning, milling, boring, drilling, grinding, polishing, lapping, etc., which mainly require cutting tools or abrasives. The removed part represents material and energy waste, which should be reduced to a minimum or avoided to improve the efficiency and productivity of manufacturing. In this context, AM, with the revolutionary free design capabilities and free building of parts in successive layers with any intended void configurations, was expected to mitigate the need for machining. However, both technologies have become interconnected and enhance the possibilities for one another. Similar mutual benefits are clear for drilling, boring, milling, and grinding tools and processes, and for the combined AM–subtractive hybrid machines that find extensive applications in modern manufacturing. In this section, examples of these mutual benefits are discussed.

10.5.1 THE ROLE OF AM IN DRILLING AND BORING

AM replaces drilling for internal channels: The demand for some machining processes, such as drilling and boring, in a wide range of products in different industries has been reduced as a direct effect of including AM products with intricate cavity configurations. A typical example is manufacturing manifolds. Rather than drilling holes from the part's exterior so that they intersect to form manifold passages, AM offers the chance to 3D print parts with precisely the internal passages, potentially with a curving path needed, and also with precisely the material needed to contain those passages (an optimized form rather than a block). This is shown in Figure 10.22, representing a manifold 3D printed using both LPBF and binder jetting made by Aidro (part of Desktop Metal Company; Zelinski 2022).

The reduced demand for drilling processes is, meanwhile, met by significant improvement of the design and manufacture of drilling tools (twist drills and boring drills) using AM equipment, mainly SLM, electron beam manufacturing (EBM),

FIGURE 10.22 Manifold component 3D printed by LPBF instead of drilling holes in a solid block. (From Modern manufacturing, India)

and directed energy deposition (DED). This, in turn, led to higher efficiency, productivity, and quality of conventionally machined parts.

Additive manufactured twist drills: With the constant development of AM technologies and easier access to equipment, it is now possible to design and fabricate solid tools suitable for various machining applications. Specially designed 3D printed twist drills seem more convenient for drilling operations than conventionally made tools. This can be shown when comparing the performance of a solid twist drill made from Maraging steel 1.2709 with that of a standard drill conventionally made of HSS. The 3D model of the drill was designed using a Siemens NX Concept Laser M2 using direct metal laser melting (DMLM) for the printing, and finished by grinding. The two drills are shown in Figure 10.23. The twist drill is an HK 11020480 HSS drill bit by Atorn. The hole drilling test was carried out on polyamide cylindrical specimens, which are commonly used in a variety of industrial applications including automotive and machine components, such as toothed wheels, bearings, bolts and nuts, pump components, couplings, cams, distributors, and drive shafts. The results proved that the dimensional and geometric accuracy of the holes made in the polyamide by the specially designed 3D printed drill bit were higher than those reported for the HSS drill bit, as were the mean cutting forces.

Considering drills with coolant ducts, conventional drills with inner coolant ducts may not be smaller than 13 mm in diameter. Using 3D laser printing, small diameter drills with spiral coolant ducts are fabricated directly from CAD models to obtain the designed geometry in a way that does not affect its stiffness and strength. These drills were proved experimentally to be of advantageous drill base bodies. Figure 10.24 shows a modular through-tool-coolant drill from Kennametal in sizes 10 mm diameter and smaller, made via laser PBF.

Additively manufactured boring tools: Large tools are required for precisely machining stator bores for the motors of electric vehicles. These tools are made by AM in order to realize a sufficiently lightweight tool. The length and cutting diameter needed to machine the large stator bore result in a large tool. If this tool had been produced conventionally from solid steel pieces, it would have been too massive for use in the tool changers of established machining centers used in automotive

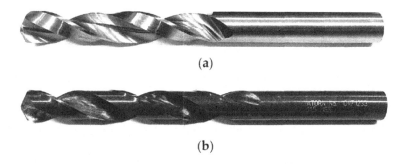

(a)

(b)

FIGURE 10.23   (a) 3D printed twist drill, (b) standard HSS twist drill. (From Nowakowski, L. et al., (2023) *Materials*, 16, 3035)

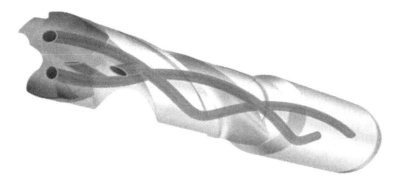

FIGURE 10.24   A twist drill 10 mm in diameter built by Kennametal using LPBF. (Courtesy of Kennametal)

FIGURE 10.25   3D printed tool for machining the stator bore of an electric car motor, built by Kennametal. (Courtesy of Kennametal)

production. But, using 3D printed geometric forms to reduce the mass, combined with polymer composite in place of metal for the shaft of the tool, together delivered a weight saving of 40 to 50% compared with what a conventional tool at this size would have required. Less mass means less energy used to accelerate the tool up to full rotational speed during machining and therefore, less cost. A typical 3D printed tool for machining the stator bore of an electric car motor produced by Kennametal is shown in Figure 10.25 (Zelinski 2022).

10.5.2 THE ROLE OF AM IN MILLING

Milling cutters consist of multiple blades to remove material from larger surface areas compared with other material removal processes. They are often made of hard, strong materials such as tool steels or cemented carbides that can withstand significant stresses without breaking or getting damaged. Some milling cutters are coated by polycrystalline diamond (PCD) to increase their lives tens of times longer than their uncoated counterparts. The limit for using PCD-coated milling cutters is that they can't be used in applications where the temperature exceeds 600°C.

AM is progressively being applied not only in manufacturing conventional milling cutters but in designing new ranges of special innovative cutting tools with more complex shapes and higher material properties to be generated. AM also introduced internal cooling channels in cutters that improve tool performance and productivity, as well as product quality. A good example is given in a case study by KOMET GROUP, a German global technology leader in the field of high-precision cutting tools. This group used Renishaw metal AM technology (UK) to produce new ranges of innovative milling cutters (Figure 10.26).

Renishaw's metal AM system uses laser powder bed fusion technology (LPBF) in an inert argon atmosphere. The system builds very thin layers (between 20 and 60 μm) of metal powder for better accuracy and surface quality of the finished part, and uses a high-performance ytterbium fiber laser for melting particles.

The use of this technology allows geometries to be produced that would be almost impossible by conventional means. It made it possible to place many more PCD blades on each tool and to change the arrangement of the blades to achieve a substantially greater axis angle. It has also greatly shortened the grooves. These changes mean that the tool is a lot more productive for the user. For example, with a 32-mm screw-in head, the number of grooves and blades has been increased from six to ten, achieving

FIGURE 10.26   Multiple cutter parts produced on a single build plate. (Courtesy of KOMET GROUP)

a feed rate that can be up to 50% higher. In addition, the ability to optimize the paths of the coolant channels ensures that each cutting edge is supplied precisely with coolant through a separate channel, while the external design of the bodies helps to ensure that chips are removed reliably from the face of the tool (KOMET GROUP 2017).

10.5.3 THE ROLE OF AM IN ABRASIVE MACHINING

Abrasive processes, such as grinding with the use of fixed-abrasive tools or lapping with the use of loose abrasive, are still the most popular and frequent methods applied for parts finishing. These processes are based on material removal from a workpiece through the action of hard, abrasive particles to improve the quality of the parts in terms of surface finishing and dimensional accuracy. Bonded abrasives act together as a grinding wheel made of a composite material designed for the particular application. The bonding material holds together the abrasive grains, and some porosity is necessary to provide clearance for the coolant and produced chips. The performance of grinding tools can be increased by increasing the tool porosity, especially for metal-bonded grinding wheels (GWs). The structure of electroplated tools is characterized by free spaces between abrasive grains determining free removal of chips during machining, as well as proper delivery of coolant to the working zone. Apart from that, electroplated tools are characterized by high versatility of their applications. These GWs are capable of covering tool bodies of any shape with an abrasive layer, although the internal porous structure cannot be fabricated by electroplating.

Intensive work on the development of the new constructions of the tool also involves lapping technology. Lapping is characterized by cutting speeds approximately ten times lower than in grinding, protecting the workpiece from burns and subsurface damage. Lower cutting speed results in a relatively small amount of the workpiece material being removed by loose abrasive suspended in oil or water slurry. It is used as the basic flattening process to obtain a high degree of flatness of a single surface (single-sided lapping), parallelism between two machined surfaces (double-sided lapping), or sphericity in the case of balls (Deja et al. 2021).

Currently, apart from the necessity of obtaining high dimensional and shape accuracy, the efficiency and economic aspects of the process are equally important. A longer effective grinding time and higher material removal rates can be obtained for wheels with controllable abrasive arrangement. Therefore, AM gives an opportunity to develop design and building sequences, as well as material combinations to fabricate GWs with precisely defined structures and also tools used in lapping and polishing. AM methods represented by PBF, such as DMLS and SLM, have been successfully used in the production of GWs. The following are different types of AM made abrasive tooling:

Metal-bonded abrasive tools: Metal-binding GWs have been widely applied to ceramic, optical glass, hard alloys, and other materials due to their excellent performance in holding force, binding, abrasion, molding, life span, and grinding press bearings, together with high precision and efficiency. Metal-based 3D printing methods, such as the SLM process, were successfully used to fabricate end face metal-bonded diamond tools with a regular distribution of grains, which improves the surface roughness of machined surfaces, tool life, and grinding efficiency. The application of incremental manufacturing enables the control of abrasive grain distribution in multiple layers and in three dimensions, as shown in Figure 10.27. Diamond abrasives (size 300–500 μm) are regularly distributed in metal binding powder (Ni75–Cr18–B2–Si5 alloy) on a substrate of AISI 1045 steel.

Resin-bonded abrasive tools: AM technology has been effectively used in manufacturing resin-bonded GWs. Metal-bonded GWs with a 3D controllable arrangement of abrasives using SLA equipment have proved more effective at material removal in comparison to wheels with random arrangement of abrasives. In addition, the uniformity of grinding trajectories can be optimized at the design stage of the tool by approving the spatial arrangement of abrasives (Figure 10.28).

10.5.4 Hybrid Machine Tools That Combine Additive and Subtractive Operations

Hybrid machine tools combining additive and subtractive processes have arisen as a solution to increasing manufacture requirements, boosting the potentials of both

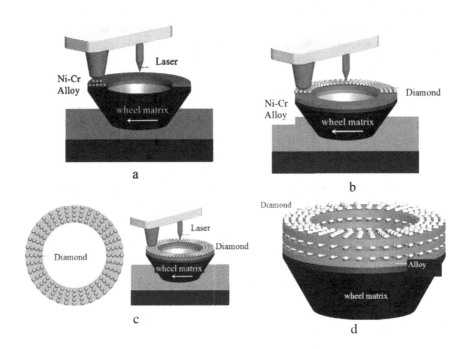

FIGURE 10.27 3D printing process of diamond grinding wheel with regularly distributed grains. (a) Ni based binder, 3D printer; (b) uniform spray of diamond grains; (c) regular distribution of grains; (d) diamond grinding wheel with regularly distributed grains. (Reprinted from Yang, Z., Zhang, M., Zhang, Z., Liu, A., Yang, R. and Liu, S. (2016) A Study on Diamond Grinding Wheels with Regular Grain Distribution using Additive Manufacturing (AM) Technology, *Materials & Design*, 104, 292–297, Copyright (2016), with permission from Elsevier)

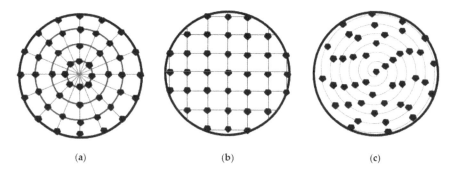

(a)   (b)   (c)

FIGURE 10.28 Schematic illustration of circular (a), rectangular (b), and spiral (c) patterns of abrasive grains arrangement in xy plane of grinding wheels made by SLA process. (With kind permission from Springer Science+Business Media: Qiu, Y. and Huang, H. (2019) Research on the Fabrication and Grinding Performance of 3-Dimensional Controllable Abrasive Arrangement Wheels. *The International Journal of Advanced Manufacturing Technology*, 104, 1839–1853.)

technologies while compensating and minimizing their limitations. In view of the market possibilities ahead, many machine tool builders have adopted this solution and started to develop different hybrid machines that combine additive and subtractive operations. Despite the numerous benefits of AM, the resulting parts usually require additional machining operations regardless of the additive approach. This way, hybrid machines have enabled overcoming the main drawbacks associated with AM, such as low accuracy and high surface roughness. The combination of both technologies in a single machine is therefore advantageous, as it enables one to build ready-to-use products with an all-in-one hybrid machine, which maximizes the strong points of each technology. Among the different metal AM technologies available, the industry has predominantly opted for PBF and DED processes.

Hence, hybrid machines give rise to new opportunities in the manufacturing of high-added-value parts, enabling the high-efficiency production of near net shape geometries, as well as the repair and coating of existing components. Besides, the capability to switch between laser and machining operations during the manufacturing process enables finishing by machining regions that are not reachable once the part is finished. In Figure 10.29, the main additive and subtractive process combinations are shown. They are divided into two groups according to the additive approach they are based on. It is worth mentioning that while PBF-based processes are mainly directed to producing complex whole parts, DED processes are more focused on the generation of coatings. That is why the latter can be combined with a wider range of subtractive processes.

The combination of additive and subtractive processes in a single hybrid machine is especially well suited to the manufacture of low-machinability materials, such as heat-resistant alloys and high-hardness materials, which are widely used in various industries, including aerospace and defense, automotive, medical, and oil and gas, among others. In fact, this hybrid manufacturing approach has already been used for remanufacturing existing high-added-value components, such as turbine blades,

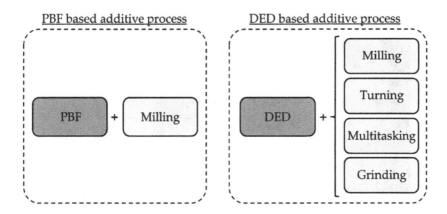

FIGURE 10.29 Different process combinations in existing hybrid additive–subtractive machines. (From Cortina et al. 2018a, 11, 2583)

integrally bladed rotors, gas turbine burner tips, or dies and molds. Nevertheless, the full integration of both processes is a complex task that must still overcome many difficulties.

A typical example is shown in Figure 10.30 for a hybrid machine by Phillips Corp., adding metal-deposition 3D printing heads to existing CNC machining centers. The same programmed motion that directs the cutting tool can also direct the nozzle and energy system for additive operations.

10.6   THE IMPACT OF AM ON WELDING

Welding is the process of using some form of heat energy to join metal parts. The heating source varies broadly from simple oxyacetylene fuel for limited manual parts, to electric arc for cost-effective accurate large-volume applications, to laser energy for small-size highly accurate but also more costly applications. Welding technology has recently reached a high level of automation through robots, which are an increasingly popular welding tool. Robotic welding brings new possibilities to the forefront, helping companies make products more efficiently and giving human welders the opportunity to remain focused on projects that would otherwise fail without their individual creativity, flexibility, and expertise. Just as the American Welding Society certifies people in manual welding, it also certifies people to be robotic welding arm operators. Using a controller device, a welding operator can program the robotic welder to move, heat the rod, and join two pieces of metal. Robotic welding is ideal for large manufacturing projects that require repeatability and safety by reducing the human factor. It also ensures higher accuracy, reduces waste, and improves production rates.

Generally, using welding to produce metal volume objects allows considerable lowering of their manufacturing cost at a simultaneous increase in productivity compared with the recently developed 3D printing technologies for metals, such as PBF and DED processes. However, the interactive role between AM and welding

FIGURE 10.30   Hybrid AM machine by Phillips Corp., made by adding metal-deposition 3D printing head to existing machining center. (From Modern machine)

technologies has introduced remarkable changes to both of them through enhanced performance, less lead times, and lower cost.

Specifically, the DED WAAM process is based mainly on welding technology. It combines gas metal arc welding (GMAW) techniques and AM. The process can be considered as an innovation of robotic welding or a hybrid of CNC milling and arc or plasma welding. WAAM has been revolutionary for the world of fabrication and design because it eliminates the need to manually cut and shape materials. Complex geometries and designs are now more possible than ever before, especially for complicated welding projects. Details of this process are discussed in Section 5.1.2.4 of this book. Figure 5.5 represents three arc welding technologies applied in WAAM processes, and Figure 5.6 shows a schematic of the WAAM system constituents.

The welding technology of AM of three-dimensional objects with the most prospects is plasma-arc technology (PAT) with the application of wires or powders. It allows part building at comparatively low heat input quality volumetric products with wall thickness from 3 to 50 mm from alloys based on Fe, Ni, Co, Cu, Ti, Al, as well as composite materials containing refractory components. Compared with SLM, the most promising AM technology for producing high-strength metallic volumetric products, PAM offers much higher productivity, more materials, higher material utilization, and much lower equipment cost, as shown in Table 10.1.

TABLE 10.1

Comparison of the Most Widely Accepted SLM Technologies of 3D Printing of Metallic Products with New Plasma-Arc Melting Technology Based on Plasma-Arc Welding

Characteristic	Technology of 3D printing of metallic products			
	SLM			PAM
	LENS company	POM company	AeroMet company	
Used equipment characteristic	Nd:YAG-laser, 1 kW power	CO_2-laser, 2 kW power	CO_2-laser, 14 kW power	Based on welding, 2–20 kW power
Productivity by incremental metal, cm³/h	8	8	160	>1000–15,000
Capability of processing along axes (degrees of freedom)	3 axes	3 axes	3 axes	4–5 axes
Type of material used for 3D-printing	Metal powder			Powders of metals, alloys, composite materials, powder mixtures. Solid and flux-cored wire
Material utilization factor, %	About 40	About 40	About 70	More than 90
Application fields	Manufacture and repair of small-sized expensive complex-shaped parts			Manufacture and repair of medium- and large-sized expensive complex-shaped parts for various applications
Tentative cost of main equipment units (per 1 kW of power)	Laser cost: 80,000–120,000 USD			Welding equipment cost: 1000–5000 USD

Source: Korzhik, V.N. et al., (2016) *The Paton Welding Journal*, 5–6, 117–123. With permission from the National Academy of Science of Ukraine.

Thus, consideration of welding technologies for 3D printing of complex-shaped metal products is urgent, as welding, at higher productivity, also allows realization of the principle of AM, namely, layer-by-layer formation of 3D structures. Moreover, welding technologies were developed long before the appearance of 3D printing and are much more mature and less costly. Therefore, it is highly relevant to use welding processes in the development of cost-effective methods of manufacturing maximum-density metallic parts and tools.

On the other hand, the role of 3D printing in welding is expanding. This exciting approach is not only flexible but also full of potential. Welders in the future may find even more new ways to leverage AM. For now, 3D printing is helping welders into more welding shops and trade jobsites in the following ways:

- Prototypes allow welders to test for fit and performance before fabricating a full-scale version.
- Customization is common in welding due to unique project requirements, and with 3D printing, tailor-made components can help welders achieve unmatched precision.
- Repairs are an easier option now thanks to 3D printed replacement parts, which give technicians the opportunity to avoid overhauling an entire component.
- Hybrid manufacturing combines 3D printing with traditional welding. This approach helps welding technicians improve efficiency and reduce wasted material.
- 3D printing gives welders the opportunity to try new designs that their traditional welding techniques can't accomplish alone, which may lead to increased creativity and time efficiency on the job (APEX 2023).

It's no surprise that 3D printing is already leaving a mark on the welding industry, and experts suggest that this trend will continue as both AM and modern fabrication techniques advance. Aspiring welders can embrace the benefits of this cutting-edge technology—from its creative possibilities to its ability to reduce waste and increase precision.

10.7   REVIEW QUESTIONS

1. Which of the five basic groups of conventional manufacturing is negatively affected by the development of AM? And why?
2. What are the limitations that restrict AM from overriding conventional manufacturing?
3. What are the main benefits and drawbacks of AM for casting processes?
4. Classify the specific AM techniques involved in the different stages of sand casting, giving reasons.
5. Discuss how PBF can build the molds for sand casting and eliminate the need for cores, and how the mold quality is affected.

6. How can topology optimization design techniques affect the different quality aspects of AM products? Give typical examples.
7. What are the techniques of rapid investment casting and their applications in different industries?
8. What is the role of cooling channels in metal injection molding dies, and how is it improved by AM?
9. Compare die casting and metal injection molding as manufacturing processes. How can AM assist in developing their products?
10. Which AM techniques are adopted in the Ampliforge process developed by Alcoa, the major aluminum producer in the United States? Is it considered a hybrid AM–forging process?
11. Suggest the sequence of forming processes in a rolling mill factory for the production of 1.5-mm thickness galvanized mild steel sheets starting from 10-mm thickness billets, and how AM techniques can be involved to develop the quality and productivity of this product.
12. What types of powder bed fusion processes are used in manufacturing extrusion dies for the production of tubular aluminum sections?
13. Using AM techniques, how can lubrication channels be integrated into deep drawing dies? And how can a controlled film thickness of the lubricant be ensured during the process?
14. Define tailored blanks. How can they be manufactured conventionally and as integrated with AM techniques? Give examples from the automotive industry.
15. How are 3D printed twist drills different from the conventional methods? And how do the differences affect the quality of drilling?
16. How can AM solve the problem of speed and inertia during precise boring of a large-diameter hole in an electric car motor housing's stator?
17. Compare SLM built milling cutters with those manufactured using traditional processes. How does this affect the quality of machining?
18. What are the differences between metal-bonded and resin-bonded abrasive tools built by AM processes? Suggest typical applications for each.
19. Compare the selective laser melting system and the recently developed plasma-arc melting system when building large-volume parts, considering productivity, range of materials, material utilization, and equipment cost.

10.8 BIBLIOGRAPHY

APEX Technical School. (2023) Exploring the Future of Welding: Additive Manufacturing and 3D Printing. https://apexschool.com/toolbox/

Cao, Jian et al. (2019) Manufacturing of Advanced Smart Tooling for Metal Forming, CIRP *Annals Manufacturing Technology*, 68(2), 605–628. http://dx.doi.org/10.1016/j.cirp.2019.05.001

Cortina, Magdalena, Arrizubieta Jon, Iñaki, Ruiz Jose, Exequiel, Ukar, Eneko and Lamikiz, Aitzol. (2018a) Latest Developments in Industrial Hybrid Machine Tools that Combine Additive and Subtractive Operations, *Materials*, 11, 2583. http://dx.doi.org/10.3390/ma11122583

Cortina, Magdalena et al. (2018b) Case Study to Illustrate the Potential of Conformal Cooling Channels for Hot Stamping Dies Manufactured Using Hybrid Process of Laser Metal Deposition (LMD) and Milling. *Metals*, 8, (2)102–117. http://dx.doi.org/10.3390/met8020102

Deja, Mariusz, Zieliński, Dawid, Abdul Kadir, Aini Zuhra, and Humaira, Siti Nur. (2021) Applications of Additively Manufactured Tools in Abrasive Machining: A Literature Review. *Material*, 14, 1318. http://dx.doi.org/10.3390/ma14051318

EPMA. (2008) Introduction to Powder Metallurgy — The Process and Its Products. UK: European Powder Metallurgy Association. https://www.epma.com/epma-free-publications/product/introduction-to-powder-metallurgy

EPMA. (2015) Vision-2025, Future Developments for the European PM Industry. UK: European Powder Metallurgy Association. https://www.epma.com/epma-free-publications/product/vision-2025-high-res

EPMA. (2019) Introduction to Additive Manufacturing Technology, A guide for Designers and Engineers. UK: European Powder Metallurgy Association. https://www.epma.com/

Gebauer, Mathias et al. (2016) High Performance Sheet Metal Forming Tooling by Additive Manufacturing. In *iCAT 2016: Proceedings of the 6th International Conference on Additive Technologies*. ETH Zurich. http://doi.org/10.3929/ethz-a-010802370

Gorges, Salvina. (2023) *Metal Powder Market Growth and Demand Analysis by 2030*. Pune, Maharashtra. Precedence Research (Canada-India) Company.

Groover, M. (2010) *Fundamentals of Modern Manufacturing, Materials, Processes, and Systems* 4th Ed. Hoboken, New Jersey: John Wiley and Sons, Inc.

Harris, Andy. (2017) Modern Metal Casting plus Additive Manufacturing Equals Paradise Foundry, Design and Make with Autodesk. https://www.autodesk.com/design-make/articles/metal-casting-and-additive-manufacturing

https://callisto.ggsrv.com/imgsrv/FastFetch/

https://i.ytimg.com/vi/fP-wT8zJ2zs/

http://jameso22.sg-host.com/wp-content/uploads/

https://lestercast.co.uk/investment-casting-automotive/

https://okfoundry.com/wp-content/

Kang, Jin-wu and Ma, Qiang-xian. (2017) The Role and Impact of 3D Printing Technologies in Casting, China Foundry. *Special Report*, 14(3). https://doi.org/10.1007/s41230-017-6109-z

Koc, Ebubekir, Akca, Yaşar, Oter, Z. Cagatay and Coskun, Mert. (2017) Selective Laser Sintering of Aluminum Extrusion Dies. *Technical Report*. https://doi.org/10.13140/RG.2.2.35134.89922

KOMET® GROUP. (2017) Case Study, Innovates Cutting Tools using Metal 3D Printing Technology. Renishaw, Case study, KOMET GROUP. www.renishaw.com/komet

Korzhik, V. N., Khaskin, V. Yu., Grinyuk, A. A., Tkachuk, V. I., Peleshenko, S. I., Korotenko, V. V., and Babich, A. A. (2016) 3D-Printing of Metallic Volumetric Parts of Complex Shape Based on Welding Plasma-Arc Technologies (Review). *The Paton Welding Journal*, 5–6, 117–123. https://doi.org/10.15407/tpwj2016.06.20

Lumley, Roger N. (2018) *Fundamentals of Aluminium Metallurgy*. Elsevier Ltd, https://doi.org/10.1016/B978-0-08-102063-0.00004-7

Mecheter, Asma, Tarlochan, Faris, and Kucukvar, Murat. (2023) A Review of Conventional Versus Additive Manufacturing for Metals: Life-Cycle Environmental and Economic Analysis. *Sustainability*, 15, 12299. https://doi.org/10.3390/su151612299

Metal Powder Industries Federation (MPIF). https://www.mpif.org/

Mueller, B. et al. (2013) Added Value in Tooling for Sheet Metal Forming through Additive Manufacturing, International Conference on Competitive Manufacturing. https://www.researchgate.net/publication/312164881

Nowakowski, Lukasz et al. (2023) Analyzing the Potential of Drill Bits 3D Printed Using the Direct Metal Laser Melting (DMLM) Technology to Drill Holes in Polyamide 6 (PA6). *Materials*, 16, 3035. https://doi.org/10.3390/ma16083035

Oskar, Frech (2015) GmbH, SLM solutions, Case study, Die-Cast Mold Insert with Conformal Cooling. https://www.slm-https://www.slm-solutions.com/fileadmin/Content/Case_Studies/

Qiu, Y. and Huang, H. (2019) Research on the Fabrication and Grinding Performance of 3-Dimensional Controllable Abrasive Arrangement Wheels. *The International Journal of Advanced Manufacturing Technology*, 104, 1839–1853. https://doi.org/10.1007/s00170-019-03900-1

Ringen, Geir, Welo, Torgeir, and Breivik Sissel, Marie (2022) Rapid Manufacturing of Die-casting Tools – A Case Study. *55th CIRP Conference on Manufacturing Systems, Procedia CIRP*, 107, 1565–1570. www.elsevier.com/locate/procedia

Ripetskiy, A. V., Khotina, G. K., and Arkhipova, O. V. (2023) The Role of Additive Manufacturing in the Investment Casting Process. *E3S Web of Conferences*, 413, 04015. https://doi.org/10.1051/e3sconf/202341304015

Schulte, R., Papke, T., Lechner, M., and Merklein, M. (2020) Additive Manufacturing of Tailored Blank for Sheet-Bulk Metal Forming Processes. *IOP Conf. Series: Materials Science and Engineering*, 967, 012034. https://doi.org/10.1088/1757-899X/967/1/012034

Schwam, D. and Silwal, B. (2017) Applications of Additive Manufacturing in Forging. *Final Report*. https://www.forging.org/uploaded/content/media/

Shang, Xin and Pang, Lijuan. (2020) Optimization Design of Insert Hot Stamping Die's Cooling System and Research on the Microstructural Uniformity Control of Martensitic Phase Transitions in Synchronous Quenching Process. *Advances in Materials Science and Engineering* 2020, 7621674. https://doi.org/10.1155/2020/7621674

Sigirisetty, Mohit. (2022) Role of Additive Manufacturing in Investment Casting Process. *International Journal for Research in Applied Science & Engineering Technology (IJRASET)*, 10(IV). www.ijrset.com

The steel printers, Spain, Understanding the Impact of Additive Manufacturing in the Steel Industry. https://www.thesteelprinters.com/news/understanding-the-impact-of-additive-manufacturing-in-the-steel-industry

Wang, Louie. (2023) Aerospace Investment Casting. https://www.linkedin.com/

Yang, Zhibo, Zhang, Mingjun, Zhang, Zhen, Liu, Aiju, Yang, RuiYun, and Liu, Shian. (2016) A Study on Diamond Grinding Wheels with Regular Grain Distribution using Additive Manufacturing (AM) Technology. *Materials & Design*, 104, 292–297. https://doi.org/10.1016/j.matdes.2016.04.104

Zelinski, Peter. (2022) 10 Ways Additive Manufacturing and Machining Go Together and Affect One Another. https://www.mmsonline.com/articles/10-ways-additive-manufacturing-and-machining-go-together-and-affect-one-another

11 Manufacturing Cost of Additive Manufacturing

11.1 BASIC CONCEPT OF MANUFACTURING COST

As a general concept, the manufacturing cost is the sum of costs of all resources consumed in the process of making a product. The manufacturing cost is classified into three categories: direct materials cost, direct labor cost, and manufacturing overhead.

- Direct materials cost: Direct materials are the raw materials that become a part of the finished product. Manufacturing adds value to raw materials by applying a chain of operations to maintain a deliverable product. There are many operations that can be applied to raw materials, such as welding, cutting, and painting. It is important to differentiate between direct materials and indirect materials.
- Direct labor cost: The direct labor cost is the cost of workers who can be identified with the unit of production. Types of labor who are considered to be part of the direct labor cost are the assembly workers on an assembly line.
- Manufacturing overhead: Manufacturing overhead is any manufacturing cost that is neither direct materials cost nor direct labor cost. Manufacturing overhead includes all charges that provide support to manufacturing.

Manufacturing overhead includes:

a. Indirect labor cost: The indirect labor cost is the cost associated with workers, such as supervisors and the material handling team, who are not directly involved in the production.
b. Indirect materials cost: The indirect materials cost is the cost associated with consumables, such as lubricants, grease, and water, that are not used as raw materials.
c. Other indirect manufacturing cost: includes machine depreciation, land rent, property insurance, electricity, freight and transportation, or any expenses that keep the factory operating.

11.2 COST ANALYSIS OF AM VERSUS SUBTRACTIVE PROCESSES (BREAK EVEN COST ANALYSIS)

In many instances, the cost of producing a product using additive manufacturing (AM) processes exceeds that of traditional methods. Figure 11.1 shows a typical

DOI: 10.1201/9781003451440-11

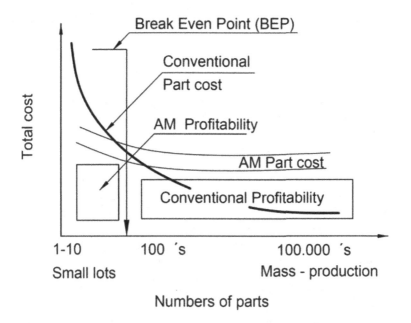

FIGURE 11.1 Breakeven analysis comparing conventional and additive manufacturing processes based on the Deloitte breakeven approach. (Adapted from Deloitte Review Issue, 2014)

analysis based on the Deloitte break even approach (Deloitte (2014), where the cost to produce a part by conventional methods is compared with the cost of producing an AM fabricated part. Conventional part production requiring significant upfront costs associated with tooling and process development along a long production stream must rely on production quantities and sales to recover the upfront costs and be profitable. AM part production can reduce upfront tooling costs and make sense for small lots of expensive parts. It is important to note that for producing parts by AM, a number of iterations may be required before a successful first part is produced, therefore increasing the cost of the first part produced. Another important consideration regarding the adoption of AM methods is the current uncertainty with total costs of AM parts, materials, design, and fabrication. It is important to note that the prices range widely with respect to powder quality, quantity, and purity. However, the price for AM powder feedstock will drop with greater AM adoption.

11.3 WELL-STRUCTURED AND ILL-STRUCTURED ADDITIVE MANUFACTURING COSTS

The costs of production in AM can be categorized in two ways (Young 1991). The first involves those costs that are "well structured", such as labor, material, and machine costs. The second involves "ill-structured costs" such as those associated with build failure, machine setup, and inventory. In the literature, there is more focus on well-structured costs of AM than ill-structured costs; however, some of the more significant benefits and cost savings in additive manufacturing may be hidden in the ill-structured costs.

11.3.1 Ill-Structured AM Costs

These are hidden in the supply chain, which is a system that moves products from supplier to customer. AM may, potentially, have significant impacts on the design and size of this system, reducing its associated costs. The elements to be considered are listed below:

- **Inventory:** The resources spent producing and storing AM products could have been used elsewhere if the need for inventory were reduced. Suppliers often suffer from high inventory and distribution costs. AM provides the ability to manufacture parts on demand. For example, in the spare parts industry, a specific type of part is infrequently ordered; however, when one is ordered, it is needed quite rapidly, as idle machinery and equipment waiting for parts is quite costly. Traditional production technologies make it too costly and require too much time to produce parts on demand. The result is a significant amount of inventory of infrequently ordered parts. This inventory is tied-up capital for products that are unused. They occupy physical space, buildings, and land while requiring rent, utility costs, insurance, and taxes. Meanwhile, the products are deteriorating and becoming obsolete. Being able to produce these parts on demand using AM reduces the need for maintaining large inventory and eliminates the associated costs.
- **Transportation:** AM allows the production of multiple parts simultaneously in the same build, making it possible to produce an entire product. This would reduce the need to maintain large inventories for each part of one product. It also reduces the transportation of parts produced at varying locations and reduces the need for just-in-time delivery. Traditional manufacturing often includes production of parts at multiple locations, where an inventory of each part might be stored. The parts are shipped to a facility where they are assembled into a product.
- **Consumer's proximity to production:** Three alternatives have been proposed. The first is where a significant proportion of consumers purchase AM systems or 3D printers and produce products themselves. The second is a copy shop scenario, where individuals submit their designs to a service provider that produces goods. The third scenario involves AM being adopted by the commercial manufacturing industry. One might consider a fourth scenario. Because AM can produce a final product in one build, there is limited exposure to hazardous conditions, and there is little hazardous waste, there is the potential to bring production closer to the consumer for some products (i.e. distributed manufacture). Further, localized production combined with simplified processes may begin to blur the line between manufacturers, wholesalers, and retailers, as each could potentially produce products in their facilities.

Case study of centralized distributed AM

Khajavi et al. (2014) compared the operating cost of centralized AM production and distributed production, where production is in close proximity to the consumer. This analysis examined the production of spare parts for the air-cooling ducts of the environmental control system for the F-18 Super Hornet fighter jet, where AM has already been implemented. The expected total cost per year was $1.0 million for centralized production and $1.8 million for distributed production. Inventory obsolescence cost, initial inventory production costs, inventory carrying costs, and spare parts transportation costs are all reduced for distributed production; however, significant increases in personnel costs and the initial investment in AM machines make it more expensive than centralized production. Increased automation and reduced machine costs are needed for this scenario to be cost-effective. It is also important to note that this analysis examined the manufacture of a relatively simple component with little assembly. One potential benefit of AM might be to produce an assembled product rather than individual components. Research by Holmström et al. (2010), which also examines spare parts in the aircraft industry, concurs that currently, on-demand centralized production of spare parts is the most likely approach to succeed; however, if AM develops into a widely adopted process, the distributed approach becomes more feasible.

- **Supply chain management:** The supply chain includes purchasing, operations, distribution, and integration. Purchasing involves sourcing product suppliers. Operations involve demand planning, forecasting, and inventory. Distribution involves the movement of products, and integration involves creating an efficient supply chain. Reducing the need for these activities can result in a reduction in costs. Some large businesses and retailers largely owe their success to the effective management of their supply chain. They have used technology to innovate the way they track inventory and restock shelves, resulting in reduced costs. Walmart, for example, cut links in the supply chain, making the link between their stores and the manufacturers more direct. This technology has the potential to bring manufacturers closer to consumers, reducing the links in the supply chain.
- **Vulnerability to Supply Disruption:** If additive manufacturing reduces the number of links in the supply chain and brings production closer to consumers, it will result in a reduction in the vulnerability to disasters and disruptions. Every factory and warehouse in the supply chain for a product is a potential point where a disaster or disruption can stop or hinder the production and delivery of a product. A smaller supply chain with fewer links means there are fewer points for potential disruption. Additionally, if production is brought closer to consumers, it will result in more decentralized production, where many facilities are producing a few products rather than a few facilities producing many products. Disruptions in the supply chain might result in localized impacts rather than regional or national impacts. Figure 11.2 provides an example that compares traditional manufacturing with AM. Under traditional manufacturing, material resource providers

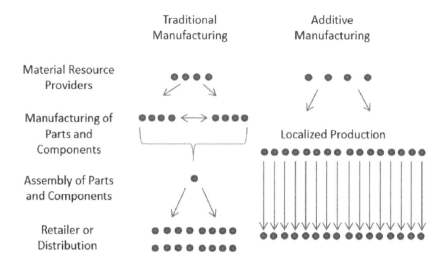

FIGURE 11.2 Example of traditional supply chain compared with the supply chain for additive manufacturing with localized production. (From Thomas, D.S. and Gilbert, S.W., NIST Special Publication 1176, 2014)

deliver to the manufacturers of parts and components, who might deliver parts and components to each other and then to an assembly plant. From there, the assembled product is delivered to a retailer or distributor. A disruption at any of the points in manufacturing or assembly may result in a disruption of deliveries to all the retailers or distributors if there is no redundancy in the system. AM with localized production does not have the same vulnerability. First, there may not be any assembly of parts or components. Second, a disruption to manufacturing does not impact all the retailers and distributors.

11.3.2 Well-Structured AM Cost

11.3.2.1 Factors Affecting Well-Structured AM Print Costs

These are the main factors that influence the printing cost of AM:

1. **Type of printing technologies:** 3D printing costs depend on the type of printing technology used. Fusion deposition modeling (FDM) is the cheapest technology among all existing AM technologies. We can find an affordable FDM printer from €100, while a selective laser sintering (SLS) machine will cost *at least* ten times more. FDM 3D printers are a cost-effective solution because of low machine price, affordable filaments, and low maintenance costs.
2. **Size of an object:** The size of the printed object influences the amount of filament it will require and how large the cost will be for an object.

3. **Infill:** In addition to the size, the infill density of the print should be determined. Usually, 3D prints are not solid parts with 100% plastic infill. The most common infill rate is 20%. Making your parts stronger and heavier will boost the cost accordingly. If parts do not need to be too strong and heavy, it is better to decrease the infill and hence, reduce the price. In addition, if you use break-away support, it will increase the price according to the amount of support you are using.

4. **Brand and type of a filament:** Different brands have different prices for their filaments, and sometimes, it contributes to the quality of the final product. It is possible to find a spool at approximately €10 per kilogram, but the final product might not have the smooth finish that the filament for €40 per spool could provide. In such cases, a compromise may be needed. In addition, some materials are more expensive than others. Acrylonitrile butadiene styrene (ABS) is one of the most common polymers, and the price per spool is one of the lowest. However, if we are looking into more complicated or sturdy materials, the situation could dramatically change. Modern metal-infilled filaments could cost up to several hundreds of euros per spool of 1–2 kilograms.

5. **Time:** It is important to consider time spent printing and electricity consumed. It might take a couple of days or even weeks to print a big object.

6. **Layer thickness:** The layer thickness influences the time spent on printing and consequently, the electricity costs involved.

7. **Price of a 3D printer:** To track the return on investment (ROI), it is important to consider the price of a 3D printer and how efficient it is. Cheaper machines with parts of a lower quality might require replacement more often, while expensive professional printers will last longer. Therefore, low upfront costs could result in bigger losses in the long run.

11.3.2.2 Well-Structured AM Cost Evaluation

11.3.2.2.1 General Considerations

An AM metal printer may cost a million dollars. The initial investment in materials can add up if more expensive alloys are used. Tens of thousands of dollars are paid for metal powder stock alone, depending on the selection. There are additional costs to consider, such as machine operator costs, material costs, consumables such as inert gas, utilities, depreciation, maintenance, and repair for the useful life of the machine. The cost of quality, documentation generation, and retention requirements all come into consideration.

While part size is easy to determine, the orientation of the part within the build chamber will affect both the volume and type of support structures required and the build time of the actual part. Laying a part down to minimize the z height and number of layers may speed up the build, but standing it up may allow more parts for other customers to be built at the same time. Angling the part with respect to the recoater blade may be required to ensure uniform powder spreading at the cost of both time and available build volume. Therefore, the customer specifies the build orientation.

Other factors to consider include single unit fabrication versus batch costs, part size, and the cost of a failed build when building either a single part or a batch. Is rework possible? If so, what are the costs? Material changeover costs will include additional chamber cleaning and material handling. In some cases, service providers dedicate specific machines to specific materials, therefore eliminating the need to fully clean the chamber when building the next part and eliminating the risk of contaminating the material of one build with that of the previous build. Finishing and post-processing costs will be a function of material support removal and a wide range of options selected by the customer.

If you plan to have parts built by a service provider, does the provider have multiple machines dedicated to specific materials? Do they have machines with build volumes that suit your component size? Will they choose the best orientation to achieve the best accuracy, or will they lay it down to reduce the build height and build time? Will your part be built at the same time as other customer parts with at-risk designs? Will they build a support structure on top of your part to stack fill the build volume? Will they be waiting for other customer orders to fill a build volume? If the process is interrupted and restarted, will you be informed and provided with the restart data? Assuming you are on a rapid prototyping schedule, what are the costs of those delays to you? How experienced is the provider with AM of those materials? If you order the same part a month later, will it be the same? These are all good questions, and certainly not a complete set, which may affect the cost of the part.

11.3.2.2.2 Cost Analysis of AM Steps

However, in AM, a comprehensive cost model has been proposed by Zohreh (2021) to consider all the cost incurred in all the steps of the AM process, namely, pre-processing, processing, and post-processing, for all types of AM technology. Figure 11.3 visualizes the three AM steps and their associated cost elements. Each of these steps consists of several cost elements. Pre-processing cost elements include the labor and overhead costs. Processing activity includes the material, machine, labor, energy, process control, and overhead costs. Material, machine, labor, energy, testing, and overhead costs are classified in the post-processing activity. Users have the option to remove any cost element from the comprehensive cost model according to their specific AM system.

1. **Pre-processing costs:**
   Pre-processing is the first step in AM. This step accounts for the labor cost and overhead cost. The labor cost includes the cost of converting a computer-aided design (CAD) file into an STL file, selecting part orientation and layer thickness, creating the support structures, and transferring the STL file to the AM system. After the file is prepared, it is transferred to the AM system to start printing.

   The next cost element in the pre-processing step is the overhead (Figure 11.3). Overhead cost is the expense that cannot be directly traced or identified in any particular unit, unlike manufacturing expenses such as raw material and labor. These are administrative and manufacturing overhead

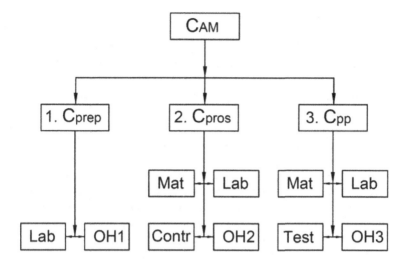

Total Cost : C_{AM} = 1. C_{prep} + 2. C_{pros} + 3. C_{pp}

Unit Cost : C_{AM}/N

Where, N = Sample size

FIGURE 11.3 The three AM steps and their associated cost elements. (Adapted from Zohreh, L., MSc Thesis, Department of Industrial, Systems and Manufacturing Engineering and the Faculty of the Graduate School of Wichita State University, 2021)

costs. Administrative overhead cost includes those costs that are not involved in the production of products, such as the building rent and maintenance, depreciation, utilities, external legal and audit fees, and licensing fees. Manufacturing overhead cost is all the costs that a factory incurs other than direct costs such as direct labor and direct materials. Everything that is called indirect cost, such as administration, light, rent, etc., should be assigned proportionally to the pre-processing time. The overall cost is calculated based on the overhead cost rate and the pre-processing time.

2. **Processing (printing) costs:**
Costs occurring during processing represent the majority of the costs in AM systems and include material, machine, labor, energy, process control, and overhead costs. Material cost includes direct material, support structures, and material depreciation costs. Machine cost includes processing costs, depreciation, and maintenance costs. The cost to monitor machine performance during the build process is also included. Overhead cost includes production and administrative costs (Figure 11.3).

Material cost in this step is not the same for different AM systems. Some AM systems utilize the same material to print the part and the support structures, and some use different materials. However, in some AM systems, such as powder bed fusion (PBF), no support structure is required. The only material used in this type of AM system is the build material. The material cost of the part and support structure is calculated as the product of the weight of the material used and its cost rate.

This cost accounts for the initial investment, maintenance cost, and scrap value. Also, both machine setup and the actual build times should be considered in calculating the cost. Typically, the AM machine is fully utilized during these two periods and cannot be used for any other purpose.

Tasks performed by the operator during pre-processing and processing are different and require different operator skills. While the building process is fully automated in most AM systems, operators may be required to start the system, add or change material, or observe build performance. In this case, labor cost can be calculated based on the labor cost rate and the time spent on these activities.

Another cost during the processing step is the cost of energy consumed by the AM system while building the part. Some systems consume high levels of energy, which cannot be ignored. On the other hand, some may have a very low level of energy consumption. By knowing the energy consumption of the system in kilowatt-hours, the energy cost can be estimated. It should be noted that this energy cost is only associated with the AM system, and any other energy consumed during processing is included in the overhead cost. For example, the cost of electricity consumed to light up the facility should not be considered as part of this cost.

The process control includes all activities undertaken to ensure the quality of the part as the AM system is building. Depending on the type of AM system and available resources, process control can be performed by either a process quality control technician or an automated process control device. If the control is performed automatically by an integrated device, the cost should have been considered as part of the machine cost. If a technician is monitoring the machine, this cost should be added to the labor cost.

The overhead cost is indirect costs like administration, light, and so on. It has to be assigned proportionally to set-up and build time (Tsb). This overhead cost rate may differ from the rate charged during the pre-processing because processing is usually carried out on the shop floor. Thus, they can have different rates.

3. **Post-processing costs:**

   These costs are typically incurred to bring the part to its final form. Depending on customer requirements, post-processing may require removing, cleaning, curing, assembling, and finishing operations. Some of these tasks are simple manual operations, while others may require special machines and operators with specific skill sets. Post-processing costs will include material, labor, machine, energy, testing, and overhead costs (Figure 11.3).

Material cost accounts for all extra materials used after constructing the part, such as chemical solutions used to clean the part or dissolve the support structure, or materials used to infiltrate the part to increase its density or strength.

Machine cost accounts for the machines and equipment utilized, such as curing apparatus, heat treatment, infiltration equipment, CNC, or wired electrical discharge machine (EDM). Multiple machines with different capabilities might be needed. Machine cost during post-processing should be determined based on the equipment needed and the time required to complete each operation. Most of these operations are performed at an assigned cost rate based on the initial investment and the useful life.

Labor cost accounts for all post-processing operations that should be done by an operator.

It depends on the time an operator spends to complete each operation, such as curing, infiltrating, matching, finishing, or painting the parts. Depending on how skilled an operator should be to successfully complete each task, different labor cost rates may need to be used.

Energy cost accounts for all energy consumed during infiltration, sintering, or machining.

All of the post-processing equipment and machines utilize electric power. For example, some AM parts can only be removed from the build plate by using a wired EDM, while others may require heat treatment in a furnace. Therefore, the cost of the energy consumed by these machines should be taken into consideration.

Testing cost is the cost of all inspection activities aimed at evaluating the quality of the final product. In this step, the manufacturer needs to verify characteristics such as dimensional accuracy, surface finish, and mechanical properties before shipping the part to the customer. Testing is required for all AM systems regardless of which manufacturing technique is utilized. Based on agreements between the manufacturer and the customer, testing can be done on all parts or just some of them as samples.

Lastly, overhead cost includes costs occurring due to administrative and production requirements. Again, the post-processing overhead rate may differ from the rates charged during the pre-processing because these tasks can be carried out in a different location.

Table 11.1 summarizes the mathematical formulae of the various pre-processing, processing, and post-processing costs of cost elements of AM steps, along with the cost parameters, as suggested by Zohreh (2021).

In the following, a numerical example from industry (Shuttle Aerospace Inc.) is given. For this example, a quantity of eight parts with identical weight, shape, size, and geometry were examined. The machine used for building these parts is an EOSINT M 280 (EOS 2014). It is a metal AM printer, which uses direct metal laser sintering technology. Figure 11.4 shows the printed shape for this example.

TABLE 11.1
AM Cost CAM

Cost of AM

Step	Cost Equations	Cost Parameters
Pre-processing: C_{prep}	$C_{prep} = CL_1 + C_{OH1}$ $CL_1 = O_1 * T_{prep}$ $C_{OH1} = H_1 * T_{prep}$	C_{L1} = Labor cost ($), C_{OH1} = Overhead cost ($), O_1 = Cost rate of pre-processing operator ($/h), T_{prep} = Time required to prepare the build (h), H_1 = Overhead cost rate ($/h)
Processing: C_{pro}	$C_{pro} = C_m + C_M + C_{L2} + C_E + C_{PC}$ $+ C_{OH2}$ $C_m = (W_1 * C_{m1}) + (W_2 * C_{m2})$ $C_M = [(C_1 + M - S)/T] *$ T_{sb} $C_{L2} = O_2 * T_{pro}$ $C_E = E * T_b *$ E_R $C_{OH2} = H_2 * T_{sb}$	C_m = Material cost required for the build ($), C_M = Machine cost ($), C_{L2} = Labor cost ($), C_E = Cost of energy consumed by the AM system ($), C_{PC} = Processing control cost ($), C_{OH2} = Overhead cost ($), W_1 = Weight of material used for the build (kg), C_{m1} = Build material cost rate ($/kg), W_2 = Weight of material used for the support structure (kg), C_{m2} = Support structure material cost rate ($/kg), C_T = Machine purchase value ($), M = Machine maintenance cost ($), S = Scrap value ($), T = Useful life of the AM machine (h), T_{sb} = Time required to set up and build (h), O_2 = Operator cost rate ($/h), T_{pro} = Time required for doing tasks (h), E = Energy consumed by AM system (kW), T_b = Built time (h), E_R = Energy cost rate ($/kWh), H_2 = Overhead cost rate ($/h), T_{sb} = Time required to set up and build (h)
Post-processing: C_{pp}	$C_{pp} = C_{m3} + C_{M3} + C_{L3} + C_{E3} +$ $C_T + C_{OH3}$ $C_{m3} = 1 - m \sum M_j * T_j^i C_{L3} =$ $C_{mi} C_{M3} = 1 - m \sum M_j * T_j^i C_{L3} =$ $1 - x \sum T_q * O_q C_{E3} = 1 - y \sum$ $E_k * T_k * E_{Rk} C_T = Q * A * n +$ $(1 - Q) [(N - n) * A + p * N *$ B]: Sample$C_T = A * N + [p *$ N * B]: $100\% C_{OH3} = H_3 * T_{pp}$	C_{m3} = Material cost ($), C_{M3} = Machine cost ($), C_{L3} = Labor cost ($), C_{E3} = Energy cost ($), C_T = Inspection cost ($), C_{OH3} = Overhead cost ($), C_{m3} = Cost of post-processing materials ($), W_i = Weight of the ith material i (kg), C_{mi} = Cost rate of the ith material ($/kg), M_j = Cost of utilizing machine j ($/h) = (Initial Cost + Maintenance - Scrap)/Useful life, T_j = Time machine j is used (h), T_q = Time required for operator q to complete the task (h), O_q = Operator q cost rate ($/h), E_k = Energy consumed by the machine k (kW), T_k = Time used by the machine k (h), E_{Rk} = Energy cost rate of the machine k ($/kWh), Q = Probability that the sample is accepted, A = Cost of testing one unit ($/unit), n = Number of units in the sample (unit), N = Total number of units, p = Probability of a nonconforming unit, B = Cost to repair or replace a single unit ($/unit), H_3 = Overhead cost rate ($/h), T_{pp} = post-processing time (h)
Total Cost: C_{AM}	$C_{AM} = C_{prep} + C_{pro} + C_{pp}$	

Source: Summarized from Zohreh, L., MSc Thesis, Department of Industrial, Systems and Manufacturing Engineering and the Faculty of the Graduate School of Wichita State University, 2021.

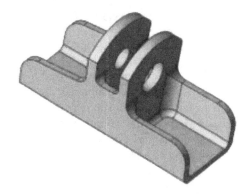

FIGURE 11.4 Isometric of part printed for illustrative example. (Courtesy of Shuttle Aerospace Inc.)

TABLE 11.2
Cost of AM Steps and Their Percentage Fraction of the Total Cost C_{AM}

Cost of step	Cost value ($)	Percentage of C_{AM}
C_{prep}	63	4%
C_{pro}	1365	86.7%
C_{pp}	147	9.3%
Total Cost C_{AM}	1575	100%

Source: Compiled from Shuttle Aerospace Inc. and National Institute for Aviation Research (NIAR).

To calculate the cost of manufacturing this part, two companies were contacted to work with.

1. Shuttle Aerospace Inc. in Wichita, Kansas provided the data to calculate per-processing and post-processing cost. This company found that:
   C_{prep} = $63
   C_{pp} = $147
2. National Institute for Aviation Research (NIAR), on the campus of Wichita State University (WSU), provided the data for processing step:
   C_{pro} = $1365

According to the results shown in Table 11.2, the percentage pre-processing cost, the percentage processing cost, and the percentage post-processing cost with respect to the total AM cost of the process amount to 4%, 87%, and 9%, respectively. The unit cost C_u is accordingly expressed by the equation

$$C_u = C_{AM}/n = 1575/8 = \$197$$

where n is the sample number.

11.4 GLANCE AT THE PRESENT AND THE FUTURE OF THE COST OF AM

Current research on AM costs reveals that this technology is cost-effective for manufacturing small batches with continued centralized manufacturing; however, with increased automation, distributed production may be cost-effective. Currently, research also reveals that material costs constitute a major proportion of the cost of a product produced using AM. Increasing adoption of AM may lead to a reduction in raw material cost through economies of scale. The reduced cost of raw material might then drive further adoption of AM. A number of factors complicate minimizing the cost of AM, including build orientation, envelope utilization, build time, energy consumption, product design, and labor. The simple orientation of the part in the build chamber can result in as much as a 160% increase in the energy consumed. Additionally, fully utilizing the build chamber reduces the per-unit cost significantly. Each of these issues must be considered in the cost of AM, making it difficult and complicated to minimize costs. These issues will likely slow the adoption of this technology, as it requires elaborate work.

As mentioned above, prices are expected to drop for both machines and materials as patents are expiring and the market is moving into a perfect competition. Machine functionality will improve in terms of speed, autonomy, repeatability, ease of use, and the ability to print with multiple materials simultaneously. Material quality is expected to improve as well. This all will lead to 3D printing becoming even more relevant in the general market, thanks to the new possibilities it provides. Statistics reveal that 3D printing will grow very quickly, most likely at a compounded annual growth rate of around 20% over the next decade.

11.5 REVIEW QUESTIONS

1. Draw a sketch to show the breakeven analysis of conventional manufacturing processes based on the Deloitte approach as compared with AM processes.
2. Categorize the production cost in AM into "well-structured" and "ill-structured" costs, and define each of them.
3. What are the factors affecting the well-structured AM print costs?
4. What are the three AM steps? Discuss their associated cost elements.
5. Explain the statement "The reduced cost of raw material might then drive further adoption of additive manufacturing".
6. Mention some of the issues that must be considered to minimize the cost of additive manufacturing.
7. What are the effective elements that affect the manufacturing cost of AM?
8. Explain why AM is not considered for mass production.

BIBLIOGRAPHY

Deloitte. (2014) 3D Opportunity: Additive Manufacturing Paths to Performance, Innovation, and Growth. Review issue.deloittereview.com

Holmström, Jan, Partanen, Jouni, Tuomi, Jukka, and Walter, Manfred. (2010) Rapid Manufacturing in the Spare Parts Supply Chain: Alternative Approaches to Capacity Deployment. *Journal of Manufacturing Technology*, 21(6), 687–697. https://doi.org/10.1108/17410381011063996

Khajavi, Siavash H., Partanen, Jouni, Holmström, Jan. (2014) Additive Manufacturing in the Spare Parts Supply Chain. *Computers in Industry*, 65, 50–63. https://doi.org/10.1016/j.compind.2013.07.008

Thomas, Douglas S. and Gilbert, Stanley W. (2014) Costs and Cost Effectiveness of Additive Manufacturing. A Literature Review and Discussion, Applied Economics Office Engineering Laboratory, National Institute of Standards and Technology, NIST Special Publication 1176. https://doi.org/10.6028/nist.sp.1176

Young, Son K. (1991) A Cost Estimation Model for Advanced Manufacturing Systems. *International Journal of Production Research*, 29(3), 441–452. https://doi.org/10.1080/00207549108930081

Zohreh, Lamei. (2021) A Comprehensive Cost Estimation for Additive Manufacturing. MSc Thesis, Department of Industrial, Systems and Manufacturing Engineering and the Faculty of the Graduate School of Wichita State University.

12 Environmental and Health Impacts of Additive Manufacturing

12.1 INTRODUCTION

In October, 2014, the Science and Technology Innovation Program sponsored a workshop to explore the environmental and human health implications of the growing field of additive manufacturing (AM). It has been identified that research is needed to examine these effects related to the key issues including energy use, occupational health, waste, and lifecycle (Rejeski and Huang 2014).

A variety of AM methods were examined to measure environmental impact, including lifecycle analysis (LCA), environmental impact scoring systems (EISS), and design for environment (DFE). It was found that the scarcity of research and the rapid evolution of this technology left a large number of unresolved issues. Hence, a joint effort is recommended from process control engineers, designers, and environmental specialist to assess this impact.

The AM processes may pose new health problems. Therefore, it is important to investigate the toxicological and environmental hazards that may occur during handling, using, and disposing of the materials used in various AM processes. These investigations can help achieve pollution prevention and reduction of occupational hazards and health risks. They may also prove to be a catalyst for greater acceptance of the AM industry (Huang et al. 2013).

Various studies on AM materials have concluded that severe skin reactions and eye irritation and allergies can occur when the operator comes in contact with these chemicals, either by inhaling the vapors or if the materials accidentally spill on the skin. Prolonged exposure to these chemicals may lead to chronic allergies, though nothing can be said about whether they can be fatal. Since the majority of the chemicals are long-chain molecules, their biodegradability is very poor, and the materials remain in the environment for extended periods of time. Poisonous gases like carbon dioxide (CO_2), carbon monoxide (CO), and nitrogen oxides are found to be emanated after the breakdown of these chemicals. It has also been predicted that noxious halocarbons (CFCs, HCFCs, CCl_4), trichloroethane (CH_3CCl_3), nickel, and lead compounds might emerge from the operations of AM machines. Therefore, the environmental impact of the AM industry is a subject of great concern. Even though some researchers have acknowledged the need for standardization of raw materials in the AM industry, the potential toxicity, environmental hazards, and chemical degradability of solvents used for their removal still remain a topic of considerable research potential (Huang et al. 2013).

DOI: 10.1201/9781003451440-12

Along with the few harmful after-effects of photopolymer liquid resin, not much is known about the effects of the solvents (propylene carbonate, tripropylene glycol monomethyl ether, and isopropanol) used to dissolve support structures left after making prototypes in stereolithography (SLA). Nonetheless, they are known to cause some symptoms, like skin burns and respiratory discomfort. AM machine operators need to be educated in the handling and disposal of these materials along with the handling of high-intensity laser beams. Safety equipment like masks, goggles, and working gloves must obviously be provided in the work area. Slowly and steadily, AM processes will surely become increasingly safe for the operators as new technological and safety features are developed and implemented in AM machines (Huang et al. 2013).

On the other hand, Wang et al. (2019) have reported that AM has reduced health and occupational hazards, such as fluid spills, wasted chip powders, extreme noise, and air pollution, that commonly result when using conventional methods such as casting and forging. However, AM may create new health problems. Environmental effects and human safety need to be addressed and regulated as the use of AM methods increases.

12.2 MATERIAL LIFECYCLE ASSESSMENT (LCA) OF AM

The AM material lifecycle starts with the extraction of raw materials from biological, inorganic, or petrochemical sources, which are processed through many chemical transformations and formulated into a feedstock for a three-dimensional (3D) printer. The printer operator uses this feedstock to produce a product that is ready for use as it is or following additional post-processing. The printed object can then be used for its intended application. Depending on the application, the 3D printed part may or may not be frequently handled by people. Ultimately, the final printed object must be disposed of or recycled along with excess materials from the printing process. This method of quantification has been standardized by two agencies: SETAC (Society of Environmental Toxicology and Chemistry) and UNEP (United Nations Environment Programme) under the ISO Standard 14 044. AM material lifecycle assessment may be grouped into six main stages, shown in Figure 12.1.

Therefore, a full materials assessment takes into account a lifecycle analysis (LCA) that covers the significant impacts associated with energy output and emissions in stages 1 to 6. LCAs are important for any organization to make internal improvements based on specific needs, but generalizing this process for vastly different AM technologies is a complex task. It is necessary to indicate important considerations and refer to current methodologies for evaluating the impact of how an AM material is manufactured (stages 1 and 2), and more importantly, to focus on comparative assessment of stages 3 to 6, because these are the stages where end users, operators of 3D printers, and the ecological sphere are most at risk of exposure to potential chemical hazards, where methodologies for assessment are still less developed.

STAGES 1 AND 2: RAW MATERIAL SOURCING, MANUFACTURING, AND FORMULATION

In summary, when scoring the material sourcing, manufacturing, and formulation stages in the AM material lifecycle, LCA metrics, green chemistry principles, and

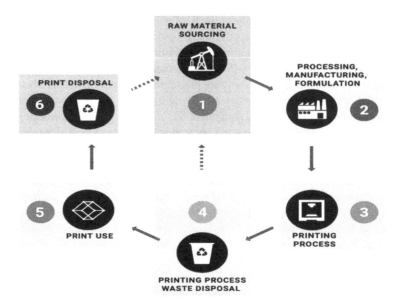

FIGURE 12.1   Six stages of the additive manufacturing (AM) lifecycle. (From Bours, J., Adzima, B., Gladwin, S., Cabral, J. and Mau, S.: Addressing Hazardous Implications of Additive Manufacturing: Complementing Life Cycle Assessment with a Framework for Evaluating Direct Human Health and Environmental Impacts. *Journal of Industrial Ecology*. 2017. 21(S1). S25–S36. Copyright Wiley-VCH Verlag GmbH & Co. KGaA. Reproduced with permission)

hazards should all be taken into account to give a comprehensive view of the sustainability of the material choice.

STAGE 3: PRINTING PROCESS

The most prominent AM material hazards are usually those affecting the operator and those in the surroundings of an additive manufacturing technology (AMT) during the printing process. Exposure and hazards vary somewhat depending on the type of process. All these technologies present some form of hazard to the operator of the printer. These hazards include dermal and environmental hazards, physical hazards (potential for explosion or fire), particulate emissions, including ultrafine particulates (UFP), and volatile organic compound (VOC) emissions. Postprocessing techniques also contribute, even within a single type of technology, to additional potential hazard exposure (Gibson et al. 2015). Although all these hazards may be present for each of the different AM technologies, the potential for exposure varies among these technologies. However, we will focus on three commonly used technologies, namely, extrusion-based systems, powder bed fusion (PDF) processes, and photopolymerization processes.

Extrusion-based systems, such as fused filament fabrication (FFF), and PDF processes, such as selective laser sintering (SLS), can generate chemicals from thermal decomposition as well as particulate emissions. Kim et al. (2015) found that FFF 3D printers, regardless of the material used, emitted nano-size particles at high concentrations, several aldehydes (including carcinogenic formaldehyde), phthalates, and VOCs such as toluene and ethylbenzene. Graff et al. (2016) recently determined that there was significant operator exposure to metal particles larger than 300 nanometers ($>10^8$ particles per cubic meter over 10-second cycles) when handling metal powders for SLS printing.

Lower-temperature photopolymerization processes, like SLA and digital light processing (DLP), typically utilize liquids that are not easily aerosolized and seldom contain components with significant vapor pressures (Carroy et al. 1997). Yet, because these materials are very reactive, they tend to be more toxic to humans and aquatic life, and their bioavailability tends to make them worse environmental toxins.

Post-processing procedures are often used after printing to make a printed object ready for use. SLA and DLP prints require washes with organic solvents, such as alcohols or propylene carbonate, to remove residual resin (monomers) from the printed parts. Solvent baths are sometimes used in FFF and SLS for removing supports or improving surface quality. Other post-processing processes include sand blasting, sealing parts, electrochemical polishing, bonding, and gluing (Gibson et al. 2015). A summary of the potential hazards for three common AM processes appears in Table 12.1.

In summary, for the printing process stage, materials are rated on their physical hazard, contact (dermal) and environmental hazard, the potential for exposure from UFP emissions, the potential for exposure from VOCs, and their post-processing score.

TABLE 12.1
Common AM Processes and Their Respective Exposures and Hazards

AM technology	Source	Main paths of exposure	Sources of hazard exposure
Extrusion based	Heated nozzle	Inhalation	Ultrafine particles, VOC emissions, explosion/fire burns
Powder bed fusion	CO_2 laser	Inhalation	Ultrafine particles, explosion, or fire
Photopolymerization	UV laser or projector	Dermal	Toxicity of compounds, solvents used to wash unreacted material

Source: Bours, J., Adzima, B., Gladwin, S., Cabral, J. and Mau, S.: Addressing Hazardous Implications of Additive Manufacturing: Complementing Life Cycle Assessment with a Framework for Evaluating Direct Human Health and Environmental Impacts. *Journal of Industrial Ecology.* 2017. 21(S1). S25–S36. Copyright Wiley-VCH Verlag GmbH & Co. KGaA. Reproduced with permission.

AM = additive manufacturing; CO_2 = carbon dioxide; UV = ultraviolet; VOC = volatile organic compound.

STAGES 4 AND 6: PRINTING PROCESS WASTE DISPOSAL AND PRINT DISPOSAL

The principal concerns related to disposal of AM outputs (prints, wasted feedstock, and other process by-products) are not immediate exposure to hazard but the long-term environmental impact. Even though hazardous waste must be treated according to local environmental regulations, these treatments are often costly, energy intensive, and do not completely eliminate the risk of environmental contamination (e.g. groundwater leakage or air pollution).

STAGE 5: PRINT USE

AM is increasingly used to produce marketable products. At the same time, the hazards of this stage of the AM lifecycle appear to be the least understood; risk depends on how the resulting 3D prints are used and in what setting (household, industrial, etc.). In the Print Use stage, the most appropriate categories to consider based on common uses of AM materials would address applications such as wearables, toys, household products, and industrial parts.

12.3 SUSTAINABILITY OF 3D PRINTED MATERIALS

12.3.1 POLYMERS

Polymers are by far the most popular and widely used materials in AM. They include thermoplastics, thermosets, elastomers, hydrogels, and polymer blends as feedstocks for lightweight engineering, architecture, food processing, optics, energy technology, dentistry, and personalized medicine.

The variants of feedstock used are powdered, liquid, and solid filament for the respective processes. One of the most promising applications is in the field of personalized medicine. The use of specific polymers for medical applications needs further investigation of health hazards. For example, additive manufactured acrylates are cytotoxic, and hence, their use could be hazardous to patients in the case of AM processing. Therefore, toxicological evaluations are needed to further investigate the use of specific polymers, such as methacrylates, for dental applications.

12.3.1.1 Polymer Disposal and Recycling

12.3.1.1.1 Polymer Disposal Methods

The total environmental impact of a product or process is decided by the end life scenario of polymers. The main approaches followed for disposing of polymers are landfill, incineration, degradation, recycling, and composting. Disposing of polymers in a landfill leads to land pollution and is toxic to humans and other life forms. Even though bio-based polymers decompose during landfill, the process degrades over time and emits methane gas, which mixes with the atmosphere. Petroleum-based polymers take hundreds of years to degrade by even 1% and pollute the environment. The incineration of polymer waste does not just produce harmful fumes and ash but also damages some biopolymers. The acceptance of incineration depends on a lot of factors. The problems caused by the process can be avoided with better incinerator design and separation of feedstock. Even though incineration with the energy recovery option reduces the environmental impact, composting and recycling have provided better impacts.

12.3.1.1.2 Polymer Recycling

Due to the increased usage of polymers and their low complete biodegradability, recycling of polymers has gained huge importance. Recycling will not only reduce the environmental impact but also decrease the resource depletion. Considering the waste management issues, recycling has demonstrated better performance ecologically as compared with other disposal processes. Recycled polymer waste can be used directly in the manufacture of new polymer products. However, recycling polymers has its own challenges. Low weight to volume ratio, heterogeneity of polymer wastes, and logistical operations related to the process are the main issues that outweigh the benefits of the recycling process. Only 6.5% of used plastics and 26% of post-consumer plastic wastes are recycled in the United States and Europe, respectively, and the overall polymer recycling on a global level is only 9% (Sanchez et al. 2017). Challenges in collection and transportation due to the low weight to volume ratio of polymers are the main reason, leading to low recycling rates and eventually, less economic benefit. A proof of concept for high-value recycling of waste polymers through distributed recycling revealed 69% to 82% embodied energy savings over centralizing recycling using an open source extruder (Baechler et al. 2013). Manufacturers found a successful procedure to fabricate new parts and demonstrated a means to convert scrap windshield wiper blades into useful, high-value, customized biomedical products: fingertip grips for hand prosthetics.

Polymer additive manufacturing is very promising for the current needs of the manufacturing sector coupled with the environmental impact, and many opportunities still exist for expanding the scope. Future work should focus not only on developing stronger and safer polymeric materials for AM, which fulfill the requirements of their role as highly functional structures, but also on carrying out recycling, to improve the quality of the recycled products.

12.3.2 COMPOSITES AND GREEN COMPOSITES

12.3.2.1 Composites

Composites built by fiber reinforced additive manufacturing (FRAM) are becoming popular in AM because of their light weight and superior mechanical properties. The composites are comprised of a matrix (continuous phase) and reinforcement (discontinuous phase or dispersed phase). Reinforcement materials such as carbon, glass, and Kevlar fibers are used to enhance the material properties of neat polymers. In terms of safety, the polymers and their micro- and nano-additives are potentially hazardous to human health. Reinforcement materials can cause dermal or inhalation toxicity during high-temperature melting of filament. Post-processing in the form of surface polishing and heat treatment to increase matrix–fiber adhesion also prompts exposure of skin, eyes, or lungs.

12.3.2.2 Green Composites

To reduce the accumulation of plastic waste and the harmful effects of the polymer composites and petroleum-based polymers, environment-friendly materials with cleaner manufacturing processes must be developed. Green composites represent the class of materials with good environmental qualifications. Green composites are a specific class of biocomposites in which biopolymers or bio-derived polymers are reinforced with natural fibers. An overview of biopolymers and natural fibers found in green composites is shown in Figure 12.2.

Despite all the positive attributes of green composites or natural fiber reinforced biopolymers, there are some disadvantages and limitations that need to be addressed. Additive manufacturing of green composites is currently very limited due to issues like difficulties in preparing composite feedstock filament because of nozzle clogging, void formation due to moisture absorption, fiber agglomeration and distribution, effect of fiber on curing and resolution, fiber orientation, fiber–matrix adhesion, light reflection from fiber, and uncured regions in SLA. Recent research has shown that despite being highly environment friendly, the green composites are not fully eco-compatible due to limitations in recyclability and biodegradability.

12.3.3 Metals and Alloys

Metal powders used for AM processes often pose toxicity, reactivity, combustibility, and instability hazards. Dust clouds, formed, for example, by accidental swirling of powders, have the potential to catch fire and explode. Other hazards include health-related risks resulting from inhalation, ingestion, and dermal exposure. The potentially explosive nature of dust formed from the metal powders has been amply demonstrated in a number of fire or explosion incidents in factories and workplaces. Additionally, metal dusts or powders, particularly from nickel and cobalt in different

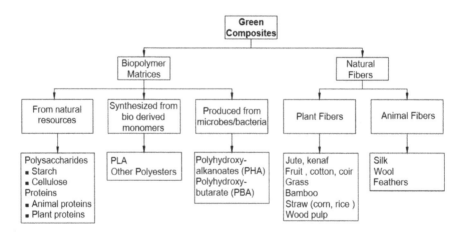

FIGURE 12.2 Green composites. (Adapted from Yaragatti, N. and Patnaik, A., *Materials Today: Proceedings*, 44(6), 4150–4157, 2021)

fractions, are known to be hazardous. It is noteworthy that standard guidelines have been developed because metal AM applications pose health and safety concerns.

12.3.4 Ceramics

Among the different types of ceramics used are Al_2O_3, ZrO_2, and $Ca_3(PO_4)_2$ (for bone tissue engineering applications). Additives such as lime, fly ash, and talc are used to enhance the mechanical properties and surface qualities of the final parts. Although ceramic powders may pose health hazards, dispersion of extra reinforcements (fibers and whiskers) may increase their propensity for dermal exposure or inhalation.

12.4 AM PROCESSES AND ASSOCIATED HEALTH AND SAFETY HAZARDS

AM methods fall into various process categories based on the form of feedstock material, processes (the mechanical forces and energies used to join the materials), and machine architecture. Overall consideration of these characteristics can assist in the development and conduction of effective hazard analysis. The following is an overview of the different AM processes and their potential hazards.

12.4.1 Types of Safety Hazards in AM Processes

The hazards of AM are listed below (Office of Environmental Health & Safety, 2022, Wayne State University):

1. Chemical Vapors: Many filaments and photopolymer resins have been shown to produce volatile organic compounds (VOCs) when heated in 3D printing processes. Exposure to VOCs may cause short-term health effects (headache, nausea, breathing problems, and eye/nose/throat irritation) and long-term health effects (cancer). Organic solvents used in post-processing pose an inhalation hazard.
2. Inhalable Particles: Plastic filaments produce inhalable nanoparticles (NPs) when heated during 3D printing. Additionally, the use of NP-containing media can emit inhalable NPs into the surrounding atmosphere. The health effects of NPs are not well understood, but preliminary research suggests that inhalation is associated with cardiovascular and pulmonary diseases. Acute and chronic inhalation exposure to metal powders can cause lung damage resulting in chronic respiratory diseases (such as asthma, chronic bronchitis, or emphysema), and some metals may be toxic or carcinogenic.
3. Dermal and Eye Exposure: Repeated skin exposures to photopolymer resins and some metals (such as nickel) can cause allergic dermatitis. Some dissolvable support materials are removed by placing prints in a heated corrosive bath containing sodium hydroxide or other corrosive chemicals. Exposure to these chemicals can cause serious chemical burns, scarring, and blindness.

4. Vapor Baths: Some filament printed objects can be smoothed or "polished" by placing them in a closed vessel filled with a small quantity of acetone or other organic solvent, which vaporizes and reacts with the plastic. The solvents are usually flammable and can cause symptoms when inhaled, such as headache, nausea, and respiratory tract irritation.
5. Biological Material: Printers using biological material can produce aerosols, which may be inhaled or deposited onto nearby surfaces.
6. Heat: Components such as UV lamps, motors, heat beds, and printheads become hot during operation and can cause burns when touched.
7. Flammability: Fine metal powders such as aluminum, aluminum alloys, steel, and titanium can spontaneously combust under normal atmospheric conditions (known as pyrophoricity). Thermoplastics can be flammable. Organic solvents (e.g. acetone) can combust when exposed to a heat source. Chemicals used in bed preparation, such as hairspray, are flammable. Printers can overheat and catch fire if thermal runaway protection is not activated in the firmware or is not working properly.
8. Inert Gas: Some 3D printers use inert gases, such as nitrogen or argon, to create an oxygen-deficient atmosphere in the printing chamber or use an inert gas as part of the aerosolization and deposition process. Inert gas leaks can displace oxygen in the room and present an asphyxiation hazard.
9. Electric Shock: 3D printers are high-voltage pieces of equipment, and interaction with unguarded electrical components (e.g. the UV lamp power supply or the printer power supply) may result in exposure to high voltage.
10. Mechanical Hazards: Hands and fingers may be pinched by moving printer components while in operation. Computer numerically controlled (CNC) post-processing of metal parts presents mechanical and noise hazards.
11. Sharps: To remove the support material, spatulas, razors, scalpels, and other sharps are commonly used. Printed metal parts may also have sharp edges. This can lead to cuts and abrasions.
12. Ultraviolet Light/Lasers: Eye exposure to the UV lights used in SLA printers can cause temporary or permanent vision loss. Directed energy deposition and powder bed fusion printers often use powerful Class 4 lasers, which can cause permanent eye injury from direct or reflected light.

12.4.2.1 Health and Safety Hazards in Material Extrusion

Materials used in the material extrusion process include acrylonitrile butadiene styrene (ABS), polylactic acid (PLA), polycarbonate (PC), and polyamide (nylon). Lately, fiber reinforced polymers and ceramic or metal have been developed as multiphase materials with improved functional properties. The emission of VOCs and particulate matter (PM) from thermoplastic filaments is reported in the literature. For ABS, the heating source may expose styrene or other hazardous chemical compounds to the environment. Additives such as fibers and powders could pose potential health hazards. There are safety concerns during processing and post-processing of the products. Hot build platforms and higher temperature of the nozzles can cause burns, while sanding, grinding, and polishing tools may cause physical injuries or inhalation exposure. The increasing use of acetone in vapor polishing to smooth out the surface also presents health hazards to the eyes and lungs. Among other important factors, the standards emphasize the need to evaluate material extrusion systems for environment safety.

12.4.2.2 Health and Safety Hazards in Vat Photopolymerization

In a vat photopolymerization (VP) process, the liquid resins (photopolymer) undergo a chemical reaction upon irradiation to become solid based on the cross sections of the CAD model. The process by which the photopolymers change their structural and chemical properties when exposed to light, usually within the ultraviolet (UV) (100–400 nm) and visible light (400–740 nm) region of the electromagnetic spectrum, is known as photopolymerization.

Depending on part complexity and other requirements, post-processing may include post UV curing, resurfacing, and dissolution or mechanical removal of support structures. Photopolymers present potential health hazards that may include the release of volatile or toxic elements and compounds such as acrylates and epoxies. These contaminants could be released before, during, and after the printing process. Another possibility of hazardous exposure is during the utilization of chemical materials to dissolve the support structures. UV light used for irradiation can cause skin cancer, skin aging, and eye damage, and can affect the immune system.

Apart from biomedical applications, the widespread use of photopolymers also raises questions pertinent to their toxicological properties and their short- and long-term health effects (e.g. allergenic reactions or chemical "burns") on persons exposed to them. For instance, methacrylate has high cytotoxicity due to Michael addition of the methacrylate group with amino or thiol groups of DNA or proteins in the human body. The hydrolysis of non-reacted groups forms unwanted methacrylic acid, decreasing pH locally, which may produce adverse biological effects on the surrounding tissues. It is reported that photopolymers produce different chemical compounds at various stages of printing. Caution is therefore required to limit exposure to toxic compounds. The use of personal protective equipment (PPE), i.e. gloves, gown, eye protection (e.g. goggles or face shield), and a disposable N95 filtering face piece respirator is strongly recommended to mitigate the harmful effects (e.g. irritation, sensitization, or toxicity) of liquid photopolymers and organic solvents for post-processing. For medical devices, a three-tiered approach to use only approved materials, apposite manufacturing parameters, and post-processing techniques may together guarantee optimal results for intended use. The current challenges with photopolymers also bring to the fore the need to consider recycling, which is central to sustainability, and safe disposal of used materials to reduce toxic waste and environmental pollution in the long term.

12.4.2.3 Health and Safety Hazards in Powder Bed Fusion

Depending on the type of energy source, the PBF process falls into two groups:

- Selective laser melting (SLM) or SLS
- Electron beam melting (EBM)

In SLM, metal powders are completely melted during the fabrication process, while EBM utilizes an electron beam to melt the metal powders. These processes require heated closed chambers either as internal (laser melting) or with high vacuum (EBM). Post-processing methods for these techniques include support removal, heat treatment, machining, and surface finishing (Chapter 7).

Since the feedstock is powdered, potential health hazards may arise from processing the various formulations. Feedstock powders, especially in a fine form, may be inhaled or cause skin sensitivity. Ionizing radiation from EBM can cause the release of volatile compounds to the atmosphere. Furthermore, during support removal and recycling of powdered materials, they may be exposed to environment during the post-processing times. The use of engineered nanomaterials and additives to enhance the sintering process also exposes users to these materials. Moreover, nitrogen or argon gas tubes may create mechanical hazards during the transportation, and asphyxiation hazards could be devastating in closed areas.

12.4.2.4 Health and Safety Hazards in Binder Jetting

The binder jetting (BJ) process dispenses liquid binding agent on a build platform covered with a layer of material in a powder form to selectively join and densify a two-dimensional (2D) pattern. The bed is then lowered vertically and covered with another powder material for a new 3D layer. Post-processing of BJ parts requires removal of excess powder from a build chamber, surface polishing, and sometimes post-annealing in a high-temperature furnace. The powdered materials may cause explosion and other hazards during inhalation and dermal exposure. The use of liquid binders like ethylene glycol may inflame or irritate the skin. Exposure to powder materials may lead to other related hazards during the post-processing, as discussed before.

12.4.2.5 Health and Safety Hazards in Material Jetting

Material jetting (MJ) creates 3D parts in a similar fashion as 2D inkjet printers. Liquid photosensitive material is jetted onto a build platform and solidified under UV light. Material availability is limited in MJ because it employs Drop-on-Demand printheads to deposit viscous liquids that form wax-like objects. Polymers and waxes are suitable due to their viscous nature and capability to be printed in droplets.

In MJ, the support material occupies the negative (empty) space in the final part and is removed during post-processing. Material curing occurs during printing or post-processing. The post-processing steps include removing remnant powder from printed parts using a closed cabin air pressure machine, surface modification methods (e.g. sanding, vibratory tumbling, or chemical polishing), as well as protocols for changing the color of the built parts (e.g., dying, electroplating, and painting). MJ processes present several potential hazards similar to photopolymerization processes due to the usage of similar materials and iterative processes.

12.4.2.6 Health and Safety Hazards in Sheet Lamination

Sheet lamination (SL) or laminated object manufacturing (LOM) is another AM process, which uses metallic sheets as feedstocks. SL utilizes a localized energy source such as ultrasonic or laser to join a stack of precisely cut metal sheets to build an object. With the application of ultrasonic waves and mechanical pressure, a sheet metal stacks at room temperature. The interfaces of stacked sheets are bonded layer by layer by diffusion rather than melting. Post-processing, such as traditional polishing, is used during or after the consolidation process to obtain a highly precise surface quality. Mechanical and thermal processes are frequently used in SL. These create additional health concerns, such as mechanical injury from the process and inhalation of diffused fumes or volatiles from binding agents. In the case of heavy additive manufacturing machines or machines that produce large-volume parts, mechanical injuries can occur due to improper loading and unloading. The method of preventing mechanical injuries is proper training for the working staff.

12.4.2.7   Health and Safety Hazards in Direct Energy Deposition

Direct energy deposition (DED) fabricates 3D objects utilizing a focused energy source (plasma arc, laser beam, or electron beam) to melt material deposited by a nozzle in the form of wire or in powder form. DED uses the characteristics of both material extrusion and powder bed fusion techniques. DED builds up the object layer by layer, which allows the manufacturing of complex parts without the need for support structures.

Potential hazards associated with DED process depend on the feedstock material used, the form of the material, and the welding mechanism. Potentially, the use of metals may create a toxic or sensitizing effect. In other words, dermal exposure or inhalation of powder material might occur during processing or post-processing steps. On the other hand, a joining mechanism such as an electron beam, plasma arc, or laser beam shows a potential burn hazard or may lead to vision damage due to exposure of the eyes to ionizing radiation. In addition, electric shock could occur in any step of the manufacturing. Electric shocks may occur due to loose wiring or loose electrical fittings in industrial machines. In the case of machines that use lasers or electron beams, electrical shocks are considered a life-threatening risk. Table 12.2 summarizes the potential hazards of each AM technology.

In addition to the hazards listed in the table, others could occur during maintenance and malfunction of AM hardware, such as electrical shock or mechanical injury. It is also important to note that noise and ergonomic hazards can be caused by 3D printers. Their use in offices and libraries may place constraints on the operating environment and generate heat, fumes, or airborne particulates. Similarly, workers may inadvertently transport materials beyond the workplace on their shoes, garments, and body, which could cause secondary health risks to others.

12.5   CONCLUSION

We should not forget the effects of the rapid development of additive manufacturing on the environment and safety. A critical technical review of the promises and potential issues of AM is beneficial for advancing its further development. It is concluded that:

TABLE 12.2
Potential Hazards of Each AM Technology

AM Technology	Material Binding Mechanism	Prominent Potential Hazards
ME (FDM)	Thermal heating to melt thermoplastic filament	Inhalation of VOCs, particulates, additives, burns
Vat photopolymerization	UV laser to cure photopolymer liquid resin	Inhalation of VOCs, dermal exposure to resins and solvents, UV exposure
PBF	High-powered laser to melt/sinter metal/ceramic/polymer powder material	Inhalation of VOCs, dermal exposure to powder, fumes, laser exposure, explosion
BJ	Adhesive to bind meta/ceramic/polymer powder material	Dermal exposure to powder and binders, inhalation of VOCs, explosion
MJ	UV light to cure photopolymer of liquid ink form	Inhalation of VOCs, dermal exposure to resins and solvents, UV exposure
SL (LOM)	Adhesives/US welding to bind metal/ceramic/polymer in rolled or sheet form	Inhalation of VOCs and fumes, shock, laser exposure
DED	Laser/electron beam heating	Inhalation/dermal exposure to powder, fume; explosion; laser/radiation exposure

Source: Adapted from Roth, G.A. et al., *Journal of Occupational and Environmental Hygiene*, 16(5), 321–328, 2019.

- The toxicological and environmental hazards as well as the safety issues of AM are not well known at present and should be the focus of further research.
- Potential health problems can be found in severe eye and skin irritation as well as allergic skin reactions and inhalation risks. Therefore, proper dust collection and air ventilation, as well as the use of protective gloves and safety glasses and masks, are highly recommended.
- AM feedstock production processes are not well documented in terms of their environmental performance, providing a highly relevant topic for future research.
- Materials developers and designers, printer operators, and print end users can create and choose the most appropriate and safe materials and AM processes based on their use cases.

Finally, the goal of the preceding discussion is to facilitate cooperation in the development and use of AM materials while minimizing their hazards and environmental impacts. Lifecycle analyses (LCAs) provide insight into the origin of the printing materials and their potential global impacts. It is hoped to create a path forward for safer, more sustainable materials and allow others to participate with additional data, feedback, and analyses.

12.6   REVIEW QUESTIONS

1. What are the most prominent potential hazards in ALS, BJ, PBF, and FDM?
2. Define green composites.
3. What are the variants of feedstock used in AM processing?
4. Mention ten of the safety hazards associated with AM.
5. Compared with metals and ceramics, polymers have their own challenges concerning their recycling. Explain this statement.
6. What are the disposal methods for polymers?
7. What is the importance of lifecycle analyses (LCAs) in industrial applications?

BIBLIOGRAPHY

Baechler, C., DeVuono, M., and Pearce, J. M. (2013) Distributed recycling of Waste Polymer into RepRap Feedstock. *Rapid Prototyping Journal*, 19(2), 118–125. https://doi.org/10.1108/13552541311302978

Bours, J., Adzima, B., Gladwin, S., Cabral, J., and Mau, S. (2017) Addressing Hazardous Implications of Additive Manufacturing: Complementing Life Cycle Assessment with a Framework for Evaluating Direct Human Health and Environmental Impacts. Journal of Industrial Ecology, 21(S1), S25–S36. https://doi.org/10.1111/jiec.12587.

Carroy, P., Decker, C., Dowling, J. P., Pappas, P., and Monroe, B. (1997) *Chemistry and Technology of UV and EB Formulation for Coatings, Inks and Paints, Prepolymers & Reactive Diluents (Volume II)*. Hoboken, New Jersey and UK: John Wiley.

Gibson, I., Rosen, D., and Stucker, B. (2015) *Additive Manufacturing Technologies: 3D Printing, Rapid Prototyping, and Direct Digital Manufacturing*. New York: Springer.

Graff, P., Stahlbom, B., Nordenberg, E., Graichen, A., Johansson, P., and Karlsson, H. (2016) Evaluating Measuring Techniques for Occupational Exposure during Additive Manufacturing of Metals: A Pilot Study. *Journal of Industrial Ecology*, 21(51), 5120–5129. https://doi.org/10.1111/jiec.12498

Huang, S. H., Liu, A., Mokasdar, P., and Hou, L. (2013) Additive Manufacturing and its Societal Impact: A Literature Review. *The International Journal of Advanced Manufacturing Technology*, 67(5–8), 1191–1203. https://doi.org/10.1007/s00170-012-4558-5

Kim, Y., Yoon, C., Ham, S., Park, J., Kim, S., Kwon, O., and Tsai, P. (2015) Emissions of Nanoparticles and Gaseous Material From 3D Printer Operation. *Environmental Science & Technology*, 49(20), 12044–12053. https://doi.org/10.1021/acs.est.5b02805

Rejeski, David and Huang, Yong. (2014) *Environmental and Health Impacts of Additive Manufacturing, Science and Technology*. Florida, USA: Innovation Program University of Florida Science and Technology, Wilson Center. http://nsfamenv.wilsoncenter.org/

Roth, G. A., Geraci, C. L., and Stefaniak, A. (2019) Potential Occupational Hazards of Additive Manufacturing. *Journal of Occupational and Environmental Hygiene*, 16(5), 321–328. https://doi.org/10.1080/15459624.2019.1591627

Sanchez, F. A. C., Boudaoud, H., Hoppe, S., and Camargo, M. (2017) Polymer Recycling in an Open-source Additive Manufacturing Context: Mechanical issues. *Additive Manufacturing*, 17, 87–105. https://doi.org/10.1016/j.addma.2017.05.013

Three-D Printing/Additive Manufacturing Safety. (2022) Office of Environmental Health & Safety at (313) 577 1200, 22-001F_3D Printing, Additive Manufacturing. Wayne State University.

Wang, J.-C., Dommati, H., & Hsieh, S.-J. (2019). Review of Additive Manufacturing Methods for High-performance Ceramic Materials. *The International Journal of Advanced Manufacturing Technology*, 103, 2627–2647. https://doi.org/10.1007/s00170-019-03669-3

Yaragatti, N., and Patnaik, A. (2021). A Review on Additive Manufacturing of Polymers Composites. *Materials Today: Proceedings*, 44(6), 4150–4157. https://doi.org/10.1016/j.matpr.2020.10.490

13 Fields of Application of Additive Manufacturing

13.1 CURRENT FIELDS OF APPLICATION

Additive manufacturing (AM) has been used across a diverse array of industries. New applications and benefits are expected to grow in time, and other applications will branch into significant subcategories. The applications of AM are expanding into numerous areas, such as aerospace, automotive, biomedical applications, food engineering, consumer products, energy, electronics, nanotechnologies, jewelry, architecture, entertainment, education, tooling, and repair. This chapter will briefly concentrate on some of these applications in major industrial fields.

13.1.1 AEROSPACE APPLICATIONS

Aerospace is an important industry that has traditionally applied AM since it was introduced. The primary advantage for production applications in aerospace is the ability to generate complex engineered geometries with a limited number of processing steps. Aerospace companies have access to budgets significantly larger than most industries. This is, however, often necessary because of the high-performance nature of the products being manufactured. Aerospace components are often usually made from advanced materials, such as Ti alloys, Ni-based superalloys, special steels, or ultra-high-temperature ceramics, which are difficult, costly, and time-consuming to manufacture using conventional processes. Additionally, aerospace production runs are usually small, limited to a maximum of several thousand parts. Therefore, AM technology is highly suitable for aerospace applications (Gibson et al. 2021).

Significant advantages can be realized if aerospace components are produced by additive manufacturing. These are:

- *Light Weight.* Anything that flies requires energy to get it off the ground. The lighter the component, the less energy is required. This can be achieved using lightweight materials with high strength/weight ratio, such as titanium and aluminum. More recently, carbon fiber–reinforced composites have gained popularity. However, it is also possible to address this issue by creating lightweight structures with hollow or honeycomb internal cores. This kind of topology optimization is quite easy to achieve using AM.
- *High temperature.* Both air- and spacecraft are subject to elevated temperature. Engine components are subject to very high temperatures, and innovative cooling solutions are often employed. Even internal components

are required to be made from flame-retardant materials. This means that AM generally requires its materials to be specially tailored to suit aerospace applications.

- *Complex geometry.* Aerospace applications can often require complex components. For example, a structural component may also act as a conduit, or an engine turbine blade may also have an internal hole for cooling. Furthermore, geometric forms of parts may be determined by complex mathematical formulae and should be produced accordingly.
- *Economics.* AM enables economical low production volumes, which are common in aerospace. Designers and manufacturing engineers need not design and fabricate molds, dies, or fixtures, or spend time on complex process planning for machining, as required by conventional manufacturing processes.
- *Digital spare parts.* Many aircraft have very long useful lives (20–50 years or longer), which places a burden on the manufacturer to provide spare parts. Instead of warehousing spares and store tooling over the aircraft's long life, the usage of AM enables companies to maintain digital models of parts.

The part that has received the most attention is a new fuel nozzle design for the CFM LEAP (Leading Edge Aviation Propulsion) turbofan engine, as shown in Figure 13.1. This new design is projected to have a useful life five times that of the original design, a 25% weight reduction, and additional cost savings realized through optimizing

FIGURE 13.1   GE Aviation fuel nozzle. (Courtesy of EOS GmbH)

the design and production process. Additionally, the fuel nozzle was engineered to reduce carbon build-up, making the nozzle more efficient. Each engine contains 19 fuel nozzles, and more than 4500 engines have been sold to date. This is claimed to save 454 kg of weight out of each engine. The nozzles are fabricated using the cobalt chrome material fabricated by electrical overstress (EOS) metal powder bed fusion (PBF) machines. Parts are to be stress relieved while still in the powder bed, followed by hot isostatic pressing (HIP) to ensure that the parts are fully dense (Gibson et al. 2021).

In 2002, a Boeing spin-off company, On Demand Manufacturing (ODM), was formed. Their first application was to manufacture environmental control system ducts to deliver cooling air to electronic instruments on F-18 military jets. They rebuilt several SLS Sinter-station machines in order to ensure that they could fabricate these parts reliably and in a repeatable manner. Shown in Figure 13.2 is an A320 nacelle hinge bracket that was originally designed as a cast steel part but was redesigned by Airbus to be fabricated in a lighter titanium alloy using PBF. Reportedly, they trimmed 10 kg off the mass of the bracket, saving approximately 40% in weight. Many more production applications of AM can be expected in the near future as materials improve and production methods become standardized, repeatable, and certified. Furthermore, Optomec recently used the laser engineered net shaping (LENS) process to fabricate complex metal components for satellites, helicopters, and jet engines.

13.1.2 AUTOMOTIVE APPLICATIONS

In the automotive industry, AM technology has been explored as a tool in the design and development of automotive components because it can shorten the development cycle and reduce manufacturing and product costs. AM processes also have been used to make small quantities of structural and functional components, such as engine exhausts, drive shafts, gear box components, and braking systems for luxury,

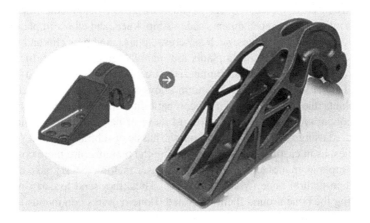

FIGURE 13.2 A320 hinge bracket redesigned for AM. (Courtesy of EOS GmbH)

low-volume vehicles. Unlike passenger cars, vehicles for motorsports usually use lightweight alloys (e.g. titanium) and have highly complex structures and low production volumes. Companies and research institutes also have successfully applied AM techniques to manufacture functional components for racing vehicles.

Since production volumes in the automotive industry are often high, AM has typically been evaluated as too expensive for production manufacturing, in contrast to the aerospace industry. Today, most manufacturers have not committed to AM parts on their mass-produced car models (Gibson et al. 2021).

In the metal PBF area, Concept Laser, a German company, introduced its X line 1000R machine recently, which has a build chamber large enough to accommodate a V6 automotive engine block. This machine was developed in collaboration with Daimler AG. According to Concept Laser, the 1000R is capable of building at a rate of 65 cm^3 per hour, which is fast compared with some other metal PBF machines. Additionally, the machine was designed with two build boxes (powder chambers) on a single turntable so that one build box could be used for part fabrication while the other could be undergoing cool-down, part removal, pre-heating, or other non-part-building activities.

For specialty cars of low-volume production, AM can be economical for some parts. Polymer PBF was used to fabricate some custom interior components, such as bezels, that were subsequently covered in leather and other materials. Typically, Bentley has production volumes of fewer than 10,000 cars for a given model, so this qualifies as low production volume.

13.1.3 BIOMEDICAL APPLICATIONS

Ample AM application opportunities exist in the biomedical field for the fabrication of custom shaped orthopedic prostheses and implants, medical devices, biological chips, tissue scaffolds, drug-screening models, and surgical planning and training apparatus. AM processes are well suited to biomedical industries, where there is no need for mass production but a particular necessity for customized implants that suit the anatomy of each specific patient (Avila et al. 2018). Orthopedic implants often refer to permanent joint replacements such as hip, knee, and elbow implants, or temporary fracture fixation devices, such as screws, plates, and pins (Jin and Chu 2019). About 50 % of the orthopedic implants are manufactured from metallic materials. The most commonly used metallic materials are titanium alloys, cobalt–chromium alloys, and some stainless steels. Titanium alloys are preferably employed in orthopedic implants due to their superior biocompatibility, corrosion, and wear resistance and mechanical properties that are close to those of natural bone. Ti alloys can be manufactured using AM processes without the hassle of subtractive processes and the need for dies, as in casting and forging, making the manufacture more cost-efficient. Titanium implants manufactured by selective laser melting (SLM) have even higher stiffness than natural bone (Mahmoud 2020). Thus, they tend to carry most of the load, leaving the bone around them unloaded. Bone requires continuous mechanical stimulation to remodel and regrow; otherwise, the bone will start reducing its mass by getting thinner or becoming more porous around the implant (Frost 1990).

13.1.4 APPLICATIONS IN THE FIELDS OF ELECTRONICS

In the electronics industry, AM processes have achieved new areas of accomplishment that paved the way for developing significant trends in the design aspects of the electronics industry. As the range of scalable AM processes expands, so will the market for 3D printed-electronics and new additive manufacturing systems. This section presents some important AM trends to anticipate in the near future and how they will affect electronics design and production.

- **Greater integration and customization:** The limitations of traditional printed circuit board (PCB) fabrication processes have constrained the creativity of layout engineers. Traditional PCB fabrication techniques require designers to work with an orthogonal interconnect architecture with components placed on surface layers of a board. While using multilayer boards with higher layer counts has given designers some freedom in routing and helps accommodate advanced devices with high I/O count, it has come at the cost of greater manufacturing costs and lead time. Using additive manufacturing systems for PCB production allows designers to break these traditional design rules, as any structure can be manufactured. This includes routing and laying out components with any configuration via or interconnect geometry, and designers can use any broad geometry. This aids the integration of additively manufactured PCBs into devices with any form factor to improve the performance of components, minimize size, optimize weight, and achieve complex and precise geometries while innovating. As more mobile and "Internet of Things" (IoT) devices take on increasingly complex form factors, including the addition of sensors, we can expect more electronics designers and layout engineers to see the value in creating their boards with an additive system.
- **On-demand Production:** Another aspect of customization is the ability to produce a single device on demand, such that it fits within a desired form factor. This is done by taking advantage of a unique quality of additive manufacturing systems, namely, that they do not require retooling. It also enables embedding of components such as sensors. This allows on-demand production of a single electronic device with fixed lead time and cost structure, significantly reducing the time to design and launch a product on the market. It also allows new devices to be easily produced with high mix and low volume directly from a manufacturer's digital inventory or a customer's 3D model. This capability will change the supply chain for electronic products and allow manufacturers to become instantly adaptable to customer demand.
- **Novel materials applications:** The research community has spent a significant amount of time and energy experimenting with advanced materials for use in a variety of additive manufacturing processes and for producing specialized devices. Electronics production relies on two types of material: an insulating dielectric substrate and conductive elements. Currently, the co-deposition of insulating and conductive nanoparticle suspensions is

the best option for producing a fully functional PCB in a layer-by-layer process, such as inkjet printing or aerosol jetting. Furthermore, newer polymer materials with low dielectric constant and semiconducting polymer materials, both with tunable electronic properties, are being adapted for use in nanoparticle suspensions. Electronics manufacturers should expect the range of new materials to continue expanding. This will enable precise production of more advanced devices from a variety of advanced materials at a higher scale and higher printing resolution, freeing designers from the constraints that traditional manufacturing imposes, primarily enabling any shape to be printed.

- **Printed antennas:** Another field of electronics is antennas, which can be printed directly into a helmet or onto a transmitting/receiving device, and connectors can be printed instead of mounted and connected. Figure 13.3 represents typical examples of applying AM to directly print (a) a circuit assembly designed to measure and report temperature onto a cylindrical component in a heating system (made by nScript) and (b) a faceted hemispherical fully electronic active antenna for aircraft, able to guarantee the Ku band link between an aircraft and a geostationary satellite in order to provide inflight entertainment services, complete with an radio frequency (RF) structure—a relatively complicated 3D printed device (Space Engineering S.p.A. & European Space Agency).

13.1.5 APPLICATIONS IN JEWELRY INDUSTRY

Additive manufacturing has become a well-known and widely used technology among designers and manufacturers of jewelry. Conventional fabrication methods such as investment casting are time-consuming and in some cases, lack precision compared with the quality demanded of the end product. Additive fabrication enables the fabrication of new products and geometries reducing manufacturing time, energy

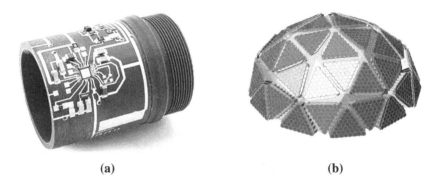

(a)   (b)

FIGURE 13.3 Typical examples of AM applications in electronics. (a) Circuit assembly on a cylindrical component, (b) faceted hemispherical antenna for aircraft. (Panel (a) from www .nscrypt.com/)

and labor costs. Today, thanks to additive manufacturing, an intricate piece of jewelry that used to take numerous days for a professional to make by hand can now be produced overnight on a laser melting (LM) machine. The machine builds up a net shape, layer by layer, in much the same way as a rapid prototyping system does, only instead of resin materials, the layers are made of metal powder. A finely woven mesh bracelet with thousands of tiny links could be grown in one piece on an AM machine using any accurately selected *Karat* (a measure of the fineness or purity) of gold powder. The resulting immaculate finish adds to the AM application, where the need for finishing is minimal, which considerably saves the cost of the precious metals used in this industry.

In the early stages of applying AM, the technology couldn't make that leap and produce high-quality "builds" with good surface finishes due to deficiencies in powders and the large sizes of the laser melting machines, which were not small enough to accommodate jewelry making. Now, thanks to smaller machines and most importantly, recent developments in precious metal powders, it can. Now, there are many suppliers that atomize various alloys of precious metal powder to achieve the particle sizes necessary to obtain the fine surface finish required in jewelry production. The most widely available precious metal powders are gold, silver, palladium, titanium, and platinum.

The majority of AM machines used for making jewelry with precious metals are laser LM types, especially direct metal laser melting (DMLM). Some of them, such as the Mlab Laser Melting machine developed by Concept Laser (since December 2016, Concept Laser has become part of GE Additive), are especially suited for link chain jewelry item processes. Instead of casting components separately and then soldering or laser welding them to arrive at the finished piece, the machine can eliminate countless labor hours by growing the pieces from powder. This was never possible before in the jewelry industry. Other additive manufacturing technologies are also applied in jewelry making, including selective laser sintering (SLS), stereolithography (SLA), and 3D printing. There are also many computational packages, or subroutines of the machine digital system, developed for jewelry design to help manufactures and customers to fabricate novel jewelry pieces. This tool is based on a customization concept, which has been of increasing interest during recent years.

While the potential for fully accepting AM technology as a tremendous advance for the jewelry industry is obvious, the response among industry experts so far has been mixed, with the machine and powder material cost as the biggest roadblock to implementation. The average investment in a laser melting machine suitable for jewelry production ranges anywhere between about $160,000 and $300,000 before accounting for the powder needed to operate it. The trend is to scale down the build chamber, control powder consumption, and develop computer-aided design (CAD) systems to make more than one design at a time to ensure that the technology will be more widely applicable.

Figure 13.4 demonstrates typical examples of additively manufactured precious metal jewel designs that are difficult, uneconomical, or even impossible to make using conventional manufacturing techniques. Figure 13.4a represents 18-Kt gold rings, part of a research project developed at Birmingham School of Jewellery.

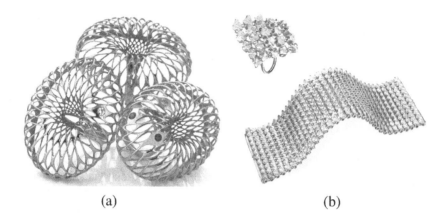

FIGURE 13.4   Typical examples of additively manufactured precious metal jewel designs. (a) 18-Kt gold rings, (b) additively manufactured 18-Kt yellow ring and bracelet. (Panel (a) courtesy of Frank Cooper, Birmingham School of Jewellery, UK/Cooksongold; panel (b) courtesy of Marie Boltenstern/Bolternstern GmbH)

Figure 13.4b demonstrates a ring and bracelet in Marie Boltenstern's 'Resonance' collection, consisting of 11 rows of additively manufactured 18-Kt yellow gold links joined together. Inspired by natural animal-scale structures, the bracelet moves and adapts to the wrist.

13.2   FUTURE TRENDS OF AM APPLICATIONS

Despite the significant developments in additive manufacturing technology, more insight is still required into the microscopic and macroscopic aspects of manufacturing processes as well as systems. Additionally, novel AM systems and standard processes need to be developed with a focus on the design of customized, complex, lightweight materials with high mechanical properties, multi-material structures, materials with multifunctional properties, electrically conductive, semiconducting, and insulating materials, biological tissues and biomaterials, and micro and nano-materials. The future trends of AM applications in different fields are based on these innovative material developments. In this section, only two key trends of the applications of these novel additively manufactured materials are presented. The first trend is based on the development of cellular lattice structure materials, and the second is based on the applications of novel smart materials, now known as 4D printing technology.

13.2.1   Cellular Lattice Structure Applications

Porous (cellular) materials are multiphase materials that consist of solid and gas/air phases simultaneously. These materials exist in many natural forms, such as bones, wood, cork, and honeycombs. The existence of the porous phase offers an exceptional combination of properties, such as low weight, high specific strength and stiffness, high energy absorption, large surface-to-volume ratio, and permeability. These characteristics encouraged engineers to mimic these materials' structures in polymeric, and later in metallic, products to exploit the advantages they offer to replace their solid counterparts in specific applications where high energy absorption and damping combined with light weight and other functional properties are required. Synthetic porous materials are commonly classified into two categories according to their structure. *Honeycombs* are non-stochastic structures based on 2D hexagonal prismatic cells and are used as sandwich structure airplane panels. The second category is based on stochastic structures, and their cells are three-dimensional, called *foams*. The shapes of unit cells in the foam structures are randomly generated, and the cell walls have random orientations in space, with the walls either separated (closed cell) or interconnected (open cell) according to the manufacturing technique. Foams are advantageous over honeycombs because of their 3D cells, which minimize the effect of anisotropy, and their manufacturing techniques are more diverse than honeycombs.

Due to the remarkable developments in AM technology in the last two decades, based on programmed dot-by-dot deposition in one layer and building the object in successive layers, a third category of *porous (cellular) materials* has been innovated, named *lattice structures*. The name is inspired by the lattice structure of engineering materials based on three-dimensionally repeated unit cells. This type of lattice structure can be regarded as the structure formed by an arrangement of lattice cells with a certain shape, topology, and size in three-dimensional Euclidean space. However, each unit cell, and even each strut in the lattice structure, can be set as the design variable and optimized to satisfy specific customized requirements functionally, which means that the mechanical properties of lattices can be more flexibly controlled than those of foams and honeycombs. Therefore, it can be concluded that lattice structures have better performance than foam structures and honeycomb structures, as they have many superior properties that foams and honeycombs do not have because of the unique property of tailoring.

Lattice structures are topologically ordered, three-dimensional, open-celled structures composed of one or more consecutively and repeatedly arranged interconnected cells. These cells are defined by the dimensions and connectivity of their constituent strut elements, which are connected at specific nodes. By tuning lattice structural parameters, such as cell topology (connectivity) or geometry (cell size and strut dimensions), the physical response of these structures can be significantly altered to exhibit properties unachievable by their parent materials, including acoustic, dielectric, and mechanical properties. AM lattice structures have been found to significantly outperform cellular structures produced by alternative manufacturing methods with equivalent porosity, particularly due to the greater geometric control and predictability provided by AM fabrication.

The commercial usefulness of these lattice structures has motivated much research, particularly for biomedical and aerospace applications. For biomedical applications, lattice structures can be used to reduce the stiffness of metallic medical implants to be closer to that of bone, thereby avoiding stress shielding while allowing fluid flow due to their porosity, with a large surface area-to-volume ratio to facilitate osseointegration. The high strength-to-weight ratio and thermal conduction properties of lattice structures make them attractive for aerospace applications. The reliable control of the collapse response of lattice structures due to their structural uniformity means that they may provide a technical advantage for energy absorption when compared with alternative metallic foams.

Lattice structures can generally be categorized based on their mechanical response as being either bending-dominated or stretch-dominated. Bending-dominated structures experience bending moments within their structure and so are compliant, whereas stretch-dominated structures experience axial loads, meaning that they are more stiff and strong than bending-dominated structures (Figure 13.5). A lattice structure's cell topology defines whether it will be bending- or stretch-dominated. A broad range of cell topologies have been investigated in the literature and are further categorized as being either strut-based or triply periodic minimal surfaces (TPMS).

13.2.1.1 Strut-Based Lattice Structures

The most common strut-based cell topologies that have been investigated are body-centered cubic (BCC) and face-centered cubic (FCC), or variations of these, such as the inclusion of z-struts (BCCZ and FCCZ), which are named after analogous crystalline structures of materials. Other strut-based topologies also exist, such as the cubic, octet-truss, and diamond, as shown in Figure 13.6. These strut-based topologies are often chosen for their simplicity of design, but strut-based topologies have also been generated from topological optimization to maximize the efficiency of material distribution within the lattice structure and fully embrace the opportunities presented by AM. Some of these structures are stretch-dominated and are stiff and strong, while others are bending-dominated and are compliant and deform more consistently, depending on the number of struts and nodes.

13.2.1.2 Triply Periodic Minimal Surface (TPMS) Lattice Structures

Lattice structures with unit cells based on TPMS include the Schoen gyroid, Schwartz diamond, and Neovius (Figure 13.7). The names associated with these cell types are those of the scientists who discovered them. The second examples of TPMS are the surfaces described by Schwarz in 1865, followed by a surface described by Neovius in 1883. In 1970, Schoen came up with 12 new TPMS based on skeleton graphs

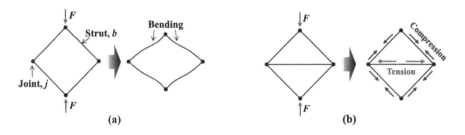

FIGURE 13.5   (a) 2D bending-dominated, (b) 2D stretch-dominated lattice structures.

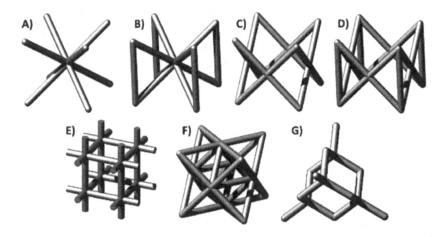

FIGURE 13.6 Strut-based lattice structures: BCC (A), BCCZ (B), FCC (C), FCCZ (D), cubic (E), Octet-truss (F), and diamond (G).

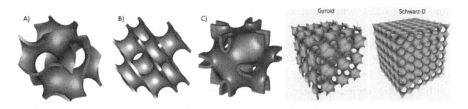

FIGURE 13.7 Triply periodic minimal surface unit cells: Schoen gyroid (A), Schwarz diamond (B), and Neovius (C). The two images on the right represent a larger volume of (A) and (B).

spanning crystallographic cells. This emphasizes that TPMS were known in the field of porous structures before being exploited by AM as innovative structures for novel applications.

TPMS lattice structures have potential advantages over strut-based topologies in terms of manufacturability and bone fixation. These structures potentially offer improved osseointegration over strut-based lattice structures. It has been proved by many researchers that gyroids have almost three times greater specific energy absorption than BCC structures with similar porosity, which is why this has become a commonly used lattice structure in AM of prosthetic implants. TPMS have also been of interest in architecture, design, and art.

13.2.1.3 Shell Lattice Structures

AM has enabled the design and manufacture of emerging cellular structures whose unit cells are composed of plates rather than struts. These lattice structures are referred to as "shell lattices". Closed-cell plate-based lattice materials have been shown to have superior elastic properties than strut-based open-cell structures with similar densities, though manufacture of these structures remains problematic for powder-based AM systems due to the requirement of powder removal. However, open-celled plate-based lattice structures have been designed, manufactured, and tested, and have been shown to exhibit superior strength and stiffness with very low densities. Figure 13.8 indicates a group of novel shell plate lattice structures that exhibit significantly higher elastic properties and good additive manufacturability compared with other competing topologies such as conventional truss-lattices, which opens a new channel for the design of lightweight mechanical metamaterials (Duan et al. 2020).

13.2.1.4 Functionally Graded Lattice Structure

A further development of the strut-based lattice structure has been implemented in AM. These were first named pseudo periodic lattice structures (also called conformal lattice structures). In this type of structure, each unit cell has the same topology, but its size is different (as shown in Figure 13.9). This gradual increase of the cell size calls for a gradient in the lattice structure, expressing gradual change of the density along the structure.

This approach has been further extended by building an object of gradient structure that can cope with the needs of practical applications, exhibiting either a discrete gradient or a regular gradient, as shown in Figure 13.10. For example, in the biomedical field, gradient lattice structures are similar to human bone tissue, which is more suitable for the growth of bone tissue than regular lattice structures. Gradation of density is also implemented in AM applications by using the same unit cell topology with a gradual increase in size, as presented in Figure 13.11. This is simply achieved by adding a linear term to the tessellating equation. For example, by appending a term to the equation, the lattice structure will exhibit a porosity gradient along the

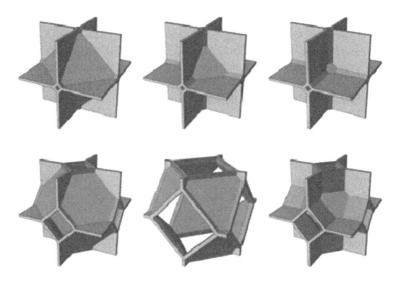

FIGURE 13.8 Novel shell plate lattice structures.

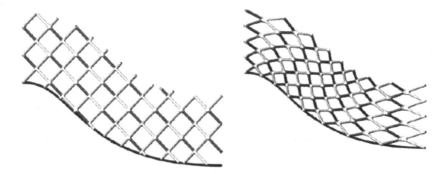

FIGURE 13.9   Periodic and pseudo periodic (conformal) structures.

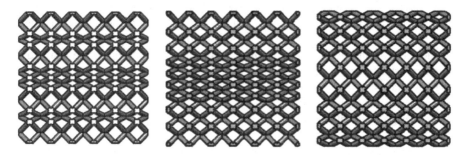

FIGURE 13.10   Functionally graded lattice structures with gradient topologies: (a) discrete gradient; (b) gradient increasing outwards; (c) gradient decreasing outwards.

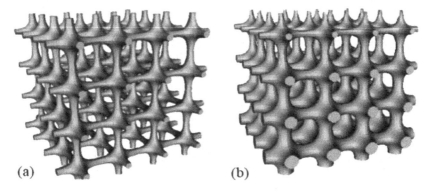

FIGURE 13.11   Lattice structure (a) without and (b) with porosity gradient. (From Tao, W. and Leu, M.C., In *2016 International Symposium on Flexible Automation (ISFA)*, 325–332, IEEE, Cleveland, OH.)

z direction. Because of the ability and flexibility of porosity control, more complex porosity distributions in implicit surface lattice structures can be achieved.

13.2.1.5 AM Processing of Lattice Structured Components

The excellent properties of lattice structures fill a gap in the manufacturing industry and provide unprecedented opportunities for structures with better manufacturing performance. The properties of lattice structures determine their wide application fields. Because of their light weight and high strength, lattice structures are often used in the structural design of aircraft, rocket, and other aerospace fields, as well as automotive fields. In addition, lattice structures have biocompatibility and high strength, which can be designed into the shape of human tissue and bone to replace diseased organs. Lattice structures have been extensively used in the medical field because of flexible mechanical properties and structural characteristics, which can meet specific requirements. Applying AM technology in the fabrication of lattice structured components provides promising opportunities to automatically and flexibly produce structures with complicated shapes and architectures that could not be fabricated by conventional manufacturing processes. In addition to the distinctive characteristics inherent in lattice structures, this technology contributes the following advantages:

- Design flexibility
- Wide range of sizes
- Using numerous kinds of materials (polymers, metallic alloys, ceramics, fibers, etc.)
- Setting up programs for automatic processing
- Saving energy and reducing cost

Different AM processes can be used, including electron beam melting (EBM) and SLS, but the most dominant is the SLM process. Typical specific applications of lattice structured products in different production areas are shown in Figure 13.12. Part (a) of the figure represents a one-shot, largely AM fabricated cylindrical fibre reinforced polymer (FRP)lattice structure for satellite application, developed by ATG Europe. Figure 13.12b demonstrates SLM manufactured, lightweight components with lattice structure for (i) a 316L stainless steel helicopter part and (ii) a control arm in the suspension system for a racing car, Institute for Laser Technology (ILT) in Aachen. Figure 13.12c shows an optimized cylinder head developed by FIT West Corp., with internal lattice structures fabricated using SLM. Because of the lattice design, 66% weight reduction was achieved. Meanwhile, the surface area was increased almost seven times due to the architecture, which contributes to a better cooling efficiency. In addition, the inherent porous feature of the lattice structure also makes it suitable for use as a filter. Figure 13.12d shows a successful orthopedic implant surgery of 3D printed sternum and rib cage fabricated by an EBM process and fitted inside a patient's body to replace and support damaged bones. The implant is able to regulate the mechanical properties to mimic those of human bones to avoid the "stress shield" issue and to enhance the strength of the interface. The internal

(a) ATG Europe developed lattice structure for satellite application. (ATG Innovation Ltd, Irland)

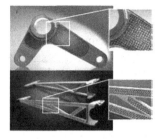

(b) Lightweight components with lattice structure for (i) helicopter part, (ii) racing car, Institute for Laser Technology (ILT) in Aachen.

(c) Lattice structure in a cylinder head, developed by FIT West Corp.

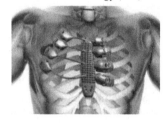

(d) biomedical implant

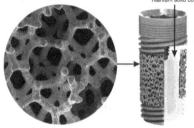

(e) Lattice structure tantalum surface made on titanium dental implant.

(f) Fine metal part designed by Bathsheba Grossman

FIGURE 13.12 Specific applications of lattice structured components made by AM.

pores are able to accommodate and guide the proliferation of living cells through the whole scaffold (Tao and Leu 2016). Figure 13.12e represents a lattice structure tantalum surface on a titanium dental implant, while Figure 13.12f demonstrates a fine metal part designed by Bathsheba Grossman.

13.2.2 4D Printing Technology Applications

Since the late 1980s, AM, commonly known as three-dimensional (3D) printing, has been gradually developed and extended worldwide. However, the microstructures fabricated using 3D printing are static. To overcome this challenge, four-dimensional (4D) printing is defined as fabricating a complex spontaneous structure that changes with time responding in an intended manner to external stimuli. Although 4D printing is mainly based on 3D printing, it is an entirely different development beyond the print, where the fabricated objects are no longer static and can be transformed into complex structures by changing their size, shape, property, and functionality under external stimuli, which makes 3D printing come alive. In this section, recent major progresses in 4D printing will be reviewed, including AM technologies for 4D printing, stimulation methods, materials, and applications, and the future prospects of 4D printing will be highlighted.

In 2013, Professor Tibbits (currently working for Massachusetts Institute of Technology [MIT]), first proposed the concept of 4D printing. He defined 4D printing as a new design of a complex spontaneous structure that changes with time due to interaction with the environment, marking the emergence of the concept of 4D printing. It is a purposeful evolution of 3D printing structure in shape, structure, and function, intended to effectively realize self-assembly, deformation, and self-repair. The process was later defined as the AM process that integrates smart materials into the initial form of printed materials for 3D printed structures/components. With the continuous development of research and technology, the definition of 4D printing will be more comprehensive. 3D printing is "pre-modeling + printing of the finished product", while the idea of 4D printing is to embed the design of the product into a flexible smart material based on 3D printing. Therefore, microstructures can be deformed according to the pre-designed track under specific time and activation conditions. Currently, 4D printing can create many objects that 3D printing cannot, and the color, volume, and shape of these objects can change with environmental conditions and stimuli, such as water and temperature. 4D printing has been successfully applied to some fields. For instance, bioprinting is an emerging technology whose greatest advantage lies in its ability to create 3D structures of living things, such as tissues, organs, nutrients, and cells. 4D bioprinting technology, which can be widely used in regenerative medicine, materials science, chemistry, and computer science, is emerging as the next-generation bioprinting technology. The main advantage of 4D bioprinting is that fabricated biostructures can alter their functionalities.

13.2.2.1 AM Technologies for 4D Printing

There are two commonly used AM categories for 4D printing:

- Extrusion-based methods, including fusion deposition modeling (FDM), direct ink writing (DIW), and inkjet.
- Vat photopolymerization methods, including SLA and digital light processing (DLP).

Stimulation Methods of (4D) Printing

The shape and function of fabricated structures can be changed according to one or more stimuli. There are two categories of stimuli, i.e. external stimuli and internal stimuli. External stimuli mainly include water/humidity, temperature, light, electric fields, and magnetic fields, while the main internal stimulus is the cell traction force.

1. **External Stimuli**

Water/Humidity Stimuli: Water and humidity were first used as stimuli in 4D printing. Materials sensitive to water or humidity are of great interest because of their ubiquitous presence and wide application. By using water as an external stimulus, the structure can be deformed underwater and restored to its original shape after drying. However, the degree of expansion/contraction of the humidity-sensitive material should be precisely controlled during the transition to maintain the integrity of the printed structure. Zhang et al. (2015) developed a material sensitive to water by modifying cellulose with stearoyl esters. A film was made using the material, and once the film was placed in an environment with a water gradient, bend deformation would occur due to the uneven absorption of water. Other researchers mixed cellulosic fibrils with acrylamide as a composite ink for printing the original flat structures, and the anisotropic swelling behavior could be regulated by the alignment of cellulose fibrils along printing pathways when immersed in water. The results indicated that the combination of materials and geometry can be controlled in space and time. Another application involves printing of droplets that can shrink under low osmotic pressure while swelling under a high osmotic pressure.

Temperature stimuli: Temperature is one of the most commonly used shape-shifting stimuli in polymer-based materials. Ge et al. (2016) printed a shape memory polymer (SMP) flower that could bloom when heated. The technology is also used to make smart grippers that do not require assembly. A late discovery by Bodaghi et al. (2017) showed that SMP structures can be preprogrammed by taking full advantage of the heating process in FDM printers. Researchers also developed a graphene-based bipiezoelectric structure that expands into a plate when heated and rolls back into a cylinder when cooled. Fiber-based glass-polymers exhibit a shape memory effect (SME) when heated above their glass transition temperature (Tg). The printing strip is initially flat in shape, and when heated above Tg, the shape memory material behaves as a rubber, and an external force is applied to the end of the band. The strip is then cooled to a lower temperature, where the shape memory material appears as a rigid solid. Due to the uneven thermal stress inside it, the strip is bent. If reheated, the shape memory material will become elastic again, and the strip will eventually return to its original flat state.

Light stimuli: Light is a common stimulus that regulates the polymer shape through remote induction. The polymer shape can be changed using a light trigger with different wavelengths. Since it doesn't cause any damage to the cells, such as increasing the temperature of the material, this stimulus can be used in biomedicine and drug delivery in vivo. For example, it was confirmed that the shape deformation of alginate/polydopamine (PDA)-based scaffold was induced by near infrared radiation (NIR). At room temperature, the alginic acid scaffold, which has been approved by the Food and Drug Administration (FDA) and has a good photothermal effect, folds slowly when dehydrated. It can quickly convert the absorbed light into heat, thereby accelerating the dehydration and deformation of the alginic acid scaffold. The bending process of the alginic acid/PDA bilayer could be controlled by the power and exposure time of light. What's more, it can be used to fabricate stents with well-controlled shape change. This method is also widely used in the field of 4D bioprinting, especially in the manufacture of self-folding 4D cell-laden structures. Unlike temperature and moisture, light is an indirect stimulus. It has been regarded as an effective activator for 4D printing technology thanks to its rich source of energy and its wireless and controllable properties. As an external stimulus intended to change the color of printed objects, light is advantageous because it can perform high-resolution control in space and time.

Electric field stimuli: Similarly to light, the electric field can also be used as a stimulus in remote control. When used as a stimulus, the electric field produces a resistive drive to fill a SMP with a conductive filler. A single carbon nanotube (CNT) has an electromagnetic induction shape memory effect. Under the action of an electric field, a CNT can be polarized by the electrons and aligned along the direction of the electric field. By applying an alternating current of 300 kHz, its original straight shape can be restored. Miriyev et al. (2017) demonstrated a soft, printed artificial muscle made from a mixture of silicone elastomer and ethanol. When an electric field is applied, heat is generated through resistance, causing the ethanol to evaporate. This phase shift from liquid to gas greatly increases the volume of the ethanol, thus expanding the entire matrix.

Magnetic field stimuli: Magnetically induced shape recovery can be achieved by doping SMP with magnetic nanoparticles (such as Fe_2O_3 and Fe_3O_4). AM is used to make magnetically induced thermoplastic SMP composites filled with Fe_2O_3 nanoparticles. The shape recovery of SMP composites can be induced by heating in an alternating magnetic field. Adding surface-modified superparamagnetic nanoparticles (Fe_3O_4, diameter: 11 nm) to the SMP matrix, using a polylactic acid (PLA) printing process, not only improves the mechanical properties and shape recovery of the material but also introduces a magnetic response to the 4D printed structure. Chen et al. 2019 fabricated a magnetic hydrogel octopus, using AAM-carbomer ink mixed with ferromagnetic nanoparticles, which can be driven remotely by a magnetic field and can move freely in a Petri dish.

2. *Internal stimuli or cell traction force stimuli*

Traction force is generated by the cells attached to the substrate. In biology, cell traction plays an important role in many processes, such as cell migration, proliferation, and differentiation. This has encouraged AM manufacture of a self-folding structure based on cell traction force. The material used for cell culture here should have sufficient flexibility and be able to maintain the state of cell adhesion under the action of traction. Due to the traction force of the cell, the flexible joint will be deformed, causing the flat panel to fold. The approximate folding angle can be determined by the number of cells on the microplate. The size of the folding angle is determined by the thickness and width of the flexible joint and has a certain relationship with the thickness of the microplate itself. One advantage of this method is that the folding behavior is induced by the cell without external force, which features high biological compatibility.

13.2.2.2 Materials Used in 4D Printing

Currently, materials widely used in 4D printing are SMPs and hydrogels. The main difference between these two types of materials is that changes in SMPs can be programmed after printing.

1. *Shape memory polymers (SMPs):* Since the discovery of SMPs in 1941, they have attracted the attention of many researchers. They show high stiffness and rapid response to stimulation, which can produce large, recoverable deformation after external stimuli. An SMP must consist of two segments, one highly elastic and the other capable of reducing its stiffness under certain stimuli. The latter can be a molecular switch or a stimulus-sensitive domain. After a particular stimulus, a switch/transform is triggered, and the strain energy stored in the temporary shape is released, resulting in shape recovery. Most materials printed in 4D have significant shape memory capabilities. Typical examples include polylactic acid (PLA), acrylonitrile butadiene styrene (ABS), and polyvinyl alcohol (PVA), which all have the ability to change shape when triggered by external conditions. Based on the shape memory effect, some new multifunctional SMP or nano-SMP composites have also been developed. The programmability of temporary shape distinguishes them from other deformed materials. They have potential applications in aerospace, biomedical/flexible electronics, and other fields. According to different shape memory mechanisms, SMP objects also have multiple shape memory effects and reversible shape memory effects, which can memorize multiple shapes and reversible shapes. 4D printed SMPs can not only carry out simple shape changes but also realize self-deformation, self-assembly, self-repair, and other functions by presetting a deformation scheme (including target shape, attribute, function, etc.). SMPs can be of thermal response type, light response type, or chemical response type.

2. *Chemical induction SMPs* are polymers that undergo deformation and recovery under the action of chemicals. The usual methods of chemical induction include pH change, equilibrium ion displacement, etc. The mechanism of polymer shape memory is different depending on the stimulus. For example, the stimulus method of pH value change is to soak the polymer in a hydrochloric acid solution, and the mutual exclusion between hydrogen ions will expand the molecular chain segment. When the equivalent NaOH solution is added to the system, the acid–base neutralization reaction will occur, and the molecular chain will shrink until the original length is restored.

3. *Hydrogels* are crosslinked hydrophilic polymers that do not dissolve in water. They are highly absorbent yet maintain well-defined structures. These properties underpin several applications, especially in the biomedical area. The term "hydrogel" was coined in 1894.

In many deformed materials, hydrogels not only have good biological affinity but also can be reversibly deformed in response to some stimuli. The swelling degree of hydrogels depends on internal properties, including crosslinking density, microstructural anisotropy, and hydrophilicity. In particular, an important factor affecting the choice of manufacturing process and the final product is the printability of the hydrogel. The advantage of using hydrogels is that they are biocompatible and easy to print using direct ink. Hydrogel is an easy-to-synthesize material with high biocompatibility, adjustable, high alignment, low cost, etc. It is a promising interface material for biomedical applications, including non-invasive diagnosis, implantation therapy, cell manipulation, and implants.

- *Thermally responsive hydrogels* are gels whose volume changes significantly when the ambient temperature changes, and the volume change is reversible. The most frequently used hydrogel material to date is poly (N-isopropyl acrylamide [PNIPAm]), which has been used in a variety of smart sensors and actuators. When the temperature of the aqueous solution is higher than its low critical solution temperature, the polymer network will fold, resulting in a smaller volume.

- *Light-responsive hydrogels* differ from other responsive hydrogels in that the light-triggered responsive behavior can be controlled remotely without the need for the gel to be in direct contact with the environment. The light-responsive hydrogels can provide the possibility of environmental stimulation based on the intensity of light or the directional illumination. Due to the high energy of short-wavelength light, many light-responsive hydrogels have ultraviolet activity.

- *pH-responsive hydrogels:* In this type of hydrogel, volume change mainly depends on the concentration of internal hydrogen ions in response to the change in pH. Since the pH value of the human body varies greatly, from the strong acidity of the stomach, through the approximate neutrality of the blood and colon, to the weak acidity of the vagina, pH-responsive hydrogels are widely used in biomedicine. When the pH value is high (alkaline), the hydrogel expands in volume, whereas at a relatively low pH (acidic), its volume will shrink. Through continuous adjustment of the polymer system, it can finally adapt to the physiological pH of the human body, offering the possibility of subsequent potential applications in the field of medical engineering.

13.2.2.3 Applications of 4D Printing
- *Drug Delivery:*
  The so-called ideal drug delivery system refers to the system used to release drugs at the location of the disease through the changes of the

environment and under controlled conditions. The ability of the oral drug delivery system to tailor the release of the drug depends mainly on the responsiveness of the printed object, and can use the change in the pH of the system to enable drug release at a specific location in the gastrointestinal tract. These studies indicate that 4D printing provides the ability to manufacture structures that can control the localization and release rate of drugs. With the development of smart materials that can respond to biological signals and pathological abnormalities in the body, the use of 4D bioprinting for drug delivery has become a reality.

- *Stents:*

    The most basic function of a stent is to support a hollow structure. For instance, stents open arteries that have become narrowed or blocked because of coronary artery disease. In the past, after manufacture, the stent has needed to be transplanted into the patient's body by surgery, which greatly increases the safety risk to the human body. Nowadays, with the advent of 4D bioprinting, stimuli-responsive materials smaller in size are being used to prepare scaffolds. After transplantation, with appropriate stimulation, the stent will automatically deform to a suitable size and shape, thus greatly lowering the risk of surgery. So far, a large number of materials and 4D bioprinting methods available for manufacturing stents have been developed. Kirillova et al. (2017) used 4D printing and hydrogel to make a self-folding stent with a hollow structure with a minimum diameter of 20 μm. When the Ca^{2+} ion concentration changes, the polymer can undergo a reversible shape change. For stents made from such biocompatible hydrogels, the cell survival rate is relatively high. An adaptive structure was prepared by Bodaghi et al. (2017), using 4D bioprinting. The structure is capable of self-expansion and self-shrinking under temperature changes. 4D bioprinting has opened up a new path for manufacturing smart stents. In the future, the demand for 4D bioprinted stents in the medical field will increase. Better biocompatibility and better adaptation to human biological characteristics need continuous future development.

- *Soft Robotics:*

    The 4D printing technology not only has the flexible processing performance of complex structural parts but also endows the material structure with unique intelligence and realizes the integration of structure and intelligence. In recent years, many studies on soft robots made using 4D printing technology have been reported. The development of soft programmable materials, engineering design, and other scientific fields has also promoted the rapid development of soft robots. Compared with the traditional robots composed of motors, pistons, joints, and hinges, soft robots are more portable and flexible. They can flexibly change in size and shape according to actual needs and can be added into more complex operations with higher safety and environmental compatibility. Therefore, soft robots have great application value and prospects in the medical and bionic fields.

- *Microfluidics:*

  Microfluidics have a wide range of applications in biomolecular and cell analysis, high-throughput screening, diagnosis, and treatment. The operation and control of microfluidic devices need to be achieved with low power consumption, low cost, and unconstrained miniaturization. These requirements can be satisfied by some micro and smart components. Therefore, the integration of soft and stimulus-responsive hydrogels in microfluidic devices is a trend of future development.

- *Tissue Engineering:*

  In order to create complex and dynamic tissues in vitro, 4D bioprinting needs to realize dynamic operations during the printing process. In recent research, the shape memory scaffolds have been proven to possess application potential in minimally invasive delivery of functional tissues. The shape memory scaffolds can be integrated with bioelectronics and biodegradation equipment, and control the transmission process in a wireless and precise manner. In addition, a major advance in the field of organ transplantation and tissue regeneration is the use of 4D bioprinting technology to implant stem cells directly into biological scaffolds. Miao et al. (2018) demonstrated a reprogrammable nerve guidance conduit designed and fabricated by stereolithographic 4D bioprinting for potentially repairing peripheral nerve injuries. The application of 4D printing for tissue engineering is still in the stage of proof-of-concept study, and this technique still has a long way to go for routine clinical application.

13.3 REVIEW QUESTIONS

1. List the main fields of application of AM.
2. Aerospace is an important industry that has traditionally applied AM since it was introduced. Explain this.
3. Compare aerospace and automotive concerning the production runs. State whether AM is equally adaptable for both applications.
4. What are the significant advantages that can be realized if aerospace components are manufactured by AM?
5. Justify this statement: Companies have successfully applied AM techniques to manufacture functional components for low-volume production and racing vehicles rather than mass-produced car modes.
6. Enumerate AM application opportunities existing in the biomedical field.
7. Why are titanium alloys widely employed in orthopedic implants?
8. Why are AM processes well suited to biomedical applications?
9. State the most important AM trends to develop design and production concepts in the electronics industry.
10. What is the role of returning to customized products by adopting AM techniques in the economic aspects of the electronics industry?
11. What are the differences between conventional cellular materials and cellular lattice structure materials?

12. Discuss some conventional manufacturing techniques for lattice structural materials, showing their limitations.
13. Comment on the differences between strut-based lattice structures and TPMS.
14. Describe different types of TPMS, and state why the gyroid type is the most widely used in orthopedic implants.
15. What are the distinguishing characteristics of shell plate lattice structures? Give examples of practical applications.
16. Classify functionally graded structures, and explain how they can meet design requirements of AM products.
17. Discuss how to fabricate a density gradient structure using SLM.
18. Search online for typical applications of lattice structures different from those presented in the text.
19. What are the basic features that distinguish 4D printing technology?
20. Present the stimulation methods of 4D printing, giving a typical example of each.
21. What are the materials used in 4D printing and their applications?
22. Present different fields of application of 4D printing, with examples of typical applications.
23. Enumerate three challenges that may threaten AM techniques in favor of conventional manufacturing techniques.
24. Identify two areas of research needed to enhance AM capabilities.

BIBLIOGRAPHY

Avila, J. D., Bose, S., and Bandyopadhyay, A. (2018) Titanium in Medical and Dental Applications ‖ Additive manufacturing of titanium and titanium alloys for biomedical applications. *Woodhead Publishing Series in Biomaterials*, 325–343. https://doi.org/10.1016/B978-0-12-812456-7.00015-9

Bodaghi, M., Damanpack, A. R., and Liao, W. H. (2017) Adaptive Metamaterials by Functionally Graded 4D Printing. *Materials and Design*, 135, 26–36. https://doi.org/10.1016/j.matdes.2017.08.069

Chen, Zhe, Zhao, Donghao, Liu, Binhong, Nian, Guodong, Li, Xiaokeng, Yin, Jun, Qu, Shaoxing, and Yang, Wei. (2019) 3D Printing of Multifunctional Hydrogels. *Advanced Functional Materials*, 29(20), 1900971. https://doi.org/1900971. 10.1002/adfm.201900971

Duan, Shengyu, Wen, Weibin, and Fang, Daining. (2020) Additively-manufactured Anisotropic and Isotropic 3D Plate-lattice Materials for Enhanced Mechanical Performance: Simulations & Experiments. *Acta Materialia*, 199, 397–412. https://doi.org/10.1016/j.actamat.2020.08.063

Frost, H. M. (1990) Skeletal Structural Adaptations to Mechanical Usage (SATMU): 1. Redefining Wolff's law: The Bone Modelling Problem. *Anatomical Record*, 226(4), 403–13. https://doi.org/10.1002/ar.1092260402

Ge, Qi, Sakhaei, Amir Hosein, Lee, Howon, Dunn, Conner K., Fang, Nicholas X. and Dunn, Martin L. (2016) Multimaterial 4D Printing with Tailorable Shape Memory Polymers. *Scientific Reports*, 6, 31110. https://doi.org/10.1038/srep31110

Gibson, Ian, Rosen, David, Stucker, Brent, and Khorasani, Mahyar. (2021) *Additive Manufacturing Technologies*, 3rd ed. Springer. https://doi.org/10.1007/978-3-030-56127-7

https://www.nscrypt.com/

Jin, W. and Chu, P. K. (2019) Orthopedic Implants. In *Encyclopedia of Biomedical Engineering*, edited by R. Narayan, 425–439. Amsterdam, The Netherlands: Elsevier Inc.

Kirillova, Alina, Maxson, Ridge, Stoychev, Georgi, Gomillion, Cheryl T., Ionov, Leonid. (2017) 4D Biofabrication using Shape-Morphing Hydrogels. *Advanced Materials*, 29(46), 1703443. https://doi.org/10.1002/adma.201703443

Mahmoud, Dalia. (2020) *Selective Laser Melting of Porosity Graded Gyroids for Bone Implant Applications*. PhD Thesis. Canada: McMaster University.

Miao, Shida, Cui, Haitao, Nowicki, Margaret, Xia, Lang, Zhou, Xuan, Lee, Se-Jun, Zhu, Wei, Sarkar, Kausik, Zhang, Zhiyong, and Zhang, Lijie Grace. (2018) Stereolithographic 4D Bioprinting of Multiresponsive Architectures for Neural Engineering. *Advanced Biosystems*, 2(9), 1800101. https://doi.org/10.1002/adbi.201800101

Miriyev, Aslan, Stack, Kenneth, and Lipson, Hod. (2017) Soft Material for Soft Actuators. *Nature Communications*, 8(1), 596. https://doi.org/10.1038/s41467-017-00685-3

Tao, Wenjin and Leu, Ming C. (2016) Design of Lattice Structures for Additive Manufacturing. In *2016 International Symposium on Flexible Automation (ISFA)*, 325–332. Cleveland, OH: IEEE. https://doi.org/10.1109/ISFA.2016.7790182

Zhang, Kai, Geissler, Andreas, Standhardt, Michaela, Mehlhase, Sabrina, Gallei, Markus, Chen, Longquan & Marie Thiele, Christina. (2015) Moisture-responsive Films of Cellulose Stearoyl Esters Showing Reversible Shape Transitions. *Scientific Reports*, 5, 11011. https://doi.org/10.1038/srep11011

14 Additive Manufacturing Characterization, Challenges and Needs, Future Trends, and Final Recommendations

14.1 CHARACTERIZATION OF AM

Together with the well-established "subtractive manufacturing", such as milling or turning, and "formative manufacturing", such as casting or forging, additive manufacturing (AM) provides the third supporting pillar of the entire manufacturing technology (Jandyal et al. 2022). As per the American Society for Testing and Materials (ASTM), AM has been divided into seven categories or techniques, which are vat photopolymerization, material jetting, binder jetting, material extrusion, powder bed fusion, sheet lamination, and directed energy deposition. The basics and principles of these seven categories of AM have been fully described and discussed with their related processes, along with their advantages and limitations. Then, the suitable technologies of additive manufacturing of engineering materials, such as polymers, composites, metallic materials, and ceramics, have been considered in detail. A comparison between these technologies has been provided in related chapters of the book.

AM, being a sustainable technology, has the potential to replace a number of conventional technologies. Based on the literature, it can be concluded that several AM technologies are compatible with most engineering materials. Each AM technology is associated with different advantages and disadvantages. As well as having the capability to handle complex and intricate shapes, AM of printed parts requires, in most cases, fewer post-processing operations. Among the 3D printing technologies, fused deposition modeling (FDM) is the most common technology, which is more suited to polymeric materials. The powder-based technologies, such as selective laser sintering (SLS) and binder jetting, face various issues, such as difficulty in transportation and storage of powders.

14.2 CHALLENGES AND NEEDS

At its 508th plenary session, held on 27 and 28 May, 2015, the European Economic and Social Committee in Brussels adopted the following opinions unanimously:

1. Additional research work is needed to expand the range of materials and the number of applications, and to improve the robustness, speed, productivity, and maturity of this technology. The steps toward a mature production process should be carried out in Europe in order to secure competitive position in the global markets and retain the economic benefits and high-quality jobs involved inside the European Union (EU).
2. The EU must facilitate investments in new AM equipment and should encourage the development of AM technology in open production systems that are flexible and easy to integrate with other production and finishing technologies, in order to enhance the number of applications and increase the turnover.
3. Strategic research is required to:
   * Transform AM into a serial production technology with next-generation machines
   * Integrate AM as a real production tool in the factory environment and systems
   * Extend the range of AM materials
   * Develop novel AM applications
4. Health and safety at work: There are very few studies about AM from the perspective of health and safety at work, and there is a real need for them due to:
   * Chemical risks, arising from volatile resins used in AM of polymeric parts, and volatile metallic or non-metallic additives in metallic powders
   * Chemo-physical risks arising from the use of powders, especially when those powders contain nanoparticles
   * Risk of explosion arising from the use of powders
   * Specific risks arising from the use of laser sources, electron beams, etc.

With the deployment of industrial AM applications, there is an urgent need for specific studies on risk assessment for workers in order to develop protection systems and standards. Safety training also needs to be developed for workers dealing with AM machines. This could be a part of the educational program to be improved or set up.

While AM processes have advanced greatly in recent years, many challenges and limitations remain to be addressed, such as limited materials available for use in AM processes, relatively poor part accuracy caused by the stair-stepping effect, insufficient repeatability and consistency of the produced parts, and lack of in-process qualification and certification methodologies. In order to realize AM's potential to usher in the "third industrial revolution", the products must be fabricated rapidly, efficiently, and inexpensively while meeting all stringent functional requirements. Many research efforts are needed to expedite the transformation from rapid prototyping to additive manufacture of advanced materials that boast material flexibility, the ability to generate fine features, and high throughput. In this context, the university–industry collaboration in AM must be considered.

Mass production could potentially be another business frontier. To compete with conventional mass production processes, AM technology needs to advance significantly in order to drastically reduce the cost of production, improve the performance of fabricated parts, and achieve consistency between parts. Despite misinformation about AM and its uncertain future, this technology is considered to be one of the most valued forms of manufacturing in history (Wohlers and Caffrey 2013).

Three areas that need further research attention are identified:

1. Materials play an important role in AM processes. At present, polymers, ceramics, composites, metals, alloys, and functionally graded, smart, and hybrid materials, etc. are widely utilized AM raw materials. These materials should have compatibility with particular AM machines. Poor mechanical performance, high cost, lack of availability of a suitable machine, health hazards associated with several materials, limitations on testing, as well as standardization and material characterization techniques, etc., are some aspects related to currently available raw materials that restrict the full exploitation of AM technology. Considerable research on materials for AM has been reported. However, the materials development for AM processes is facing some challenges. Anisotropy, mass customization, microstructural control, compositional control, variety, and other aspects remain as major restraints upon AM materials. To conclude, there is a long way to go before all the raw material aspects are fully addressed, and hence, several research avenues lie unexplored in this crucial aspect for the full-scale utilization of AM technologies.

2. More research is needed to accurately evaluate the energy consumption of various AM processes. A standardized procedure should be developed that takes into account various aspects of operating an AM machine. It is possible that when producing the same part, AM consumes more energy than conventional manufacturing processes. However, AM allows design optimization, which can lead to products with the same functionality but having less weight compared with those produced using conventional manufacturing processes. Taking into account supply chain simplification, the lifecycle energy consumption of such optimized and on-demand produced parts is likely to be comparable to, if not lower than, that of traditionally manufactured parts.

3. AM is perhaps the disruptive manufacturing technology being implemented by most industrial manufacturers today. Therefore, there is a need to better understand the potential occupational hazards of AM. The health effects of various AM materials and processes have not been well established. As the AM industry continues to evolve and expand, there is no doubt that government regulation is needed to safeguard the AM workforce. Research in this area will provide guidelines for such regulations. It will also enable the development of safe AM machines that can be used in a home or office environment. Lifecycle analyses (LCAs) provide insight into the origin of the printing materials and their potential global impacts. It is hoped to create a path forward for safer, more sustainable materials and allow others to participate with additional data, feedback, and analyses.

14.3 FUTURE TRENDS OF AM

14.3.1 GENERAL TRENDS OF AM

The new paradigm for solid freeform fabrication will evolve over the next decade as industrial engineering and applications develop and drive AM technology (AMT). Novel applications and needs, identified by independent designers and makers, will lead to greater adoption of AMT. The widespread application of plastic prototyping will lead the way for metal and ceramic AM processing. Additive/subtractive manufacturing (AM/SM) hybrid systems will proliferate in the next decade. AMTs have been in development for at least the past three decades, with the hard work of science and engineering being done at universities, national laboratories, and research laboratories. Advanced manufacturing has seen large investments, on the order of billions of dollars. The global impact of AM is gaining momentum, with high levels of funding seen at the levels of government, university, and major corporations (Milewiski 2017).

Industrial users and AM machine builders acknowledge the need to offer improvements to processing speed, accuracy, surface finish, material properties, quality, and repeatability. The development and improvement of new laser technology, electron guns, multiple beams, faster powder spreading, and fully integrated modeling and optimization and print software are seen each year at specialized exhibitions. Innovative solutions are rapidly evolving to reduce the cost and improve the quality of inks, binders, and build and support materials.

Because of recent developments in AM, there is no fundamental reason for products to be brought to markets through centralized development, production, and distribution. Instead, products can be brought to markets by individuals and communities in any geographical region. Many companies already use the internet to collect product ideas from ordinary people from diverse locations. However, these companies are feeding these ideas into the centralized physical locations of their existing business operations for detailed design and creation (Gibson et al. 2015).

In addition to the aforementioned, the following includes some hints that may be considered as future trends in the field of general industrial applications of AM techniques (Milewiski 2017):

- Databases relying on information gathered directly from production machines, in-process monitoring, embedded sensors, wireless connections, internet, and cloud resources will generate bigger databases and ultimately, data-driven solutions. Algorithms to analyze these data will provide insight and better understanding of all AM processes.
- Hybrid digital work cell systems will integrate AM with advanced computer numerically controlled (CNC) and non-destructive testing (NDT) inspection systems to provide work cells with unprecedented capacity and flexibility. Mobile hybrid systems incorporating AM, laser machining, CNC, and in-process inspection and real-time control will allow maintenance and repair on site for an ever-growing number of applications.
- Expandable or configurable build volumes for powder-based systems will be developed to resize the AM machines for producing large parts or high-aspect-ratio components.

- Cyber security of files and transfer of information will be protected by certification algorithms in software to ensure that digital product definition or the process data were not corrupted, hacked, or violated during storage or use.
- Validation of designs, processes, and products is needed to ensure safety and quality protection. Efforts are ongoing, within both corporations and consortiums, to share data along the value chain.
- Fabrication sequence translators, similar to those for translating 3D model file formats (e.g. STEP, IGES, and STL), will be developed as the subsequent generations of AM machines and software tools are retired or made obsolete.
- Automated systems for planning and control will optimize the selection of machine parameters (speed, laser power, powder flow, etc.) based upon data-driven algorithms and the growing databases generated in industry.
- Process optimization for design and machine function will more widely utilize finite element analysis (FEA), evolving toward the concept of finite element fabrication, where high-resolution thermal and mechanical models predict conditions such as full fusion, potential defect location, heat build-up, and microstructural evolution, and ultimately, predict quality and performance at every location in the part. Microstructural evolution and alloy design models will be employed for localized tailoring of microstructures and properties within the component.
- The establishment of localized 3D repair resources will enable on-site fabrication and repair, minimizing the need for skilled human intervention. Today, the repair of a bearing surface on a marine shaft or rebuilding a broken or worn ship propeller are already being performed on site in a less automated manner.
- Human augmented robots, capable of service and repair work using AM and other means, may be dispatched to work in remote or highly hazardous locations to perform difficult or dangerous tasks with a high level of precision and autonomy.
- AM process optimization algorithms will speed up AM deposition, reduce defects, improve accuracy, resolution, and surface finish, compensate for shrinkage, and produce functionally graded materials. The optimization of a huge number of parameters linked across a wide range of technologies (design, materials, process engineering, and fabrication) will result in significant savings in material, energy, and time.
- The development of designs to utilize functionally graded materials of metals, alloys, composites, and ceramics will continue.

14.3.2 FUTURE TRENDS FOR POLYMERS AND COMPOSITES

14.3.2.1 Future Trends for Polymers

The processing of polymeric materials in AM can be in any physical state (liquid/filament/powdered/sheet). With careful selection of processing technique and compatible polymer, these can be processed by almost all fusion-based AM processes. However, three AM techniques that are commonly utilized include photopolymerization, material extrusion, and material jetting. Thermoplastic polymers as well as UV-curable polymers both constitute the most prominently utilized polymeric AM materials. In addition, FDM, SLS, inkjet printing, etc. have proven capability to develop polymeric components and polymer composites.

14.3.2.2 Polymer Composites

The need for better materials in terms of strength, stiffness, density, and lower cost with improved sustainability is growing. Polymers have always attracted the attention of manufacturers due to their unique characteristics, like ease of production, availability, light weight, low cost, ductility, and long life. Parts printed from pure polymers have limited mechanical properties and functionalities and are not widely used as fully functional and load-bearing parts. Composites have emerged as one of the most sought-after solutions in the present and future to overcome the challenges presented by pure polymers. AM of polymer composites combines matrix and reinforcements to obtain useful structural or functional properties not attainable by any constituent alone, thus eradicating the problems generated by parts printed with pure polymers. Reinforcement with secondary metallic, ceramic, or polymeric inclusions in the form of fibers, whiskers, platelets, or particles in a host polymer matrix leads to the formation of a polymer matrix composite (PMC). PMCs offer enhanced material and mechanical properties and are serving to their full potential in various industries. Polymer composites have been broadly classified as particle-reinforced polymer composites and fiber-reinforced polymer composites (Yaragatti and Patnaik 2020).

14.3.2.3 Carbon Fiber–Reinforced Polymers

As a new technology, additive manufacturing of CFRPs is gaining significant importance in the manufacturing industry. AM of CFRPs combines the advantages of AM, such as customization, minimal wastage, low cost, fast prototyping, and rapid manufacturing, with the high specific strength of carbon fiber. Thus, it will create a new market for composite products. The promised applications of additively manufactured CFRPs in biomedical, electronics, and aerospace will highlight the importance of research and development in this field. Most research reviews provide a general insight into the AM of composite materials but are not focused on the AM of CFRPs (Adil and Lazoglu 2022).

Additively manufactured CFRPs are mainly categorized based on the reinforcement type, that is, short carbon fiber–reinforced polymers (short CFRPs) and continuous carbon fiber–reinforced polymers (continuous CFRPs). In the research, various techniques are proposed for the AM of short and continuous CFRPs, such as FDM, stereolithography (SLA), SLS, and laminated object manufacturing (LOM). Among these techniques, FDM is the most studied and applied technique, since its prints have shown superior mechanical properties, low cost, and high printing speed. SLA is one of the simplest methods and has demonstrated medical applications, such as spinal implants, knee implants, and bipolar components for invasive surgeries. SLS-CFRPs have shown promising results in ankle-foot orthosis designs due to higher bending strength and stiffness. As well as these, LOM has also shown great potential for the additive manufacturing of continuous CFRPs. So far, very few companies have commercialized the fabrication setups for continuous CFRPs (Hu et al. 2019).

Thermoplastic and thermosetting polymers are used as matrix materials in CFRPs. The most common thermoplastic matrixes are polylactic acid (PLA), acrylonitrile butadiene styrene (ABS), polycarbonate (PC), polypropylene (PP), polyamide (PA), polystyrene (PS), polyphenylsulfone (PPSU), polyether ether ketone (PEEK), polyaryl ether ketone (PAEK), and polyetherimide (PEI) (www.3dxtech.com/ and Ge et al. 2013). Thermoplastic matrix materials are commonly applied with the FDM technique. Only a few thermosetting resins have been reported for additive manufacturing of CFRPs; these include photo-curable resins, acrylic-based resins (Griffini et al. 2016), and cyanate ether (Chandrasekaran et al. 2017). The addition of carbon fibers to these polymers in the form of CFRPs leads to increased mechanical strength and stiffness, increased thermal conductivity, reduced thermal expansion, and reduced warpage compared with other reinforcements (Van de Werken et al. 2019). The particular combination of high-strength polymeric materials, that is, PEEK, with short and continuous fibers has enabled AM processes to produce lightweight composites for structural supports in aerospace applications, prototype demonstrations for the education and art sector, and organ and tissue repairs for the biomedical industry.

An important limitation of additive manufacture of CFRPs is that it offers a fiber volume fraction only up to 40–50% (Matsuzaki et al. 2016). This limitation yields low-strength structures as compared with conventional techniques, where fiber volume fractions up to 60–70% can be achieved (He and Gao 2015). To overcome the discrepancies, researchers are exploring new features that can revolutionize this technique in the future. Modification of current methods to incorporate similar fiber volume fractions may yield high-strength products. Similarly, the mechanical strength can be improved many-fold (Ueda et al. 2020). More research and development are required to develop new models to include the anisotropic nature, distribution, and fiber orientation of CFRPs to improve the quality and performance of prints (Goh et al. 2018).

14.3.3 FUTURE TRENDS FOR METALS AND ALLOYS

AM offers flexibility to fabricate simple as well as intricate metallic parts of virtually any complexity. It has numerous advantages over traditional metal processing techniques. The past two and a half decades have witnessed tremendous growth in metal AM as compared with the limited work accomplished in this field during the initial years of the advent of AM techniques. The modern industrial world has reached a stage where metal AM techniques have become the epi-center of interest for researchers as well as industry personnel. Powder bed fusion (PBF) and directed energy deposition (DED) are the two main commercial metal AM techniques. These technologies generally utilize powder as the feedstock material. However, wire-based feedstock materials are also utilized in DED techniques. These systems selectively melt metallic powders for part fabrication. Metallic materials such as titanium, alloys, steels, some grades of lightweight metal alloys (Al and Mg), Ni-based alloys, etc. are highly compatible with AM systems.

A. *Stainless steels* including austenitic, precipitation hardened, martensitic, duplex, etc. are processed via PBF-laser and DED-laser AM techniques. Grades of austenitic stainless steel including 304-, 316-, 304-, and 316L AISI types are most commonly used. AM produces fine-grained steel components as compared with conventional manufacturing techniques due to rapid solidification along with non-equilibrium conditions. Heat treatments are generally applied to AM produced steels to achieve desirable properties.

B. *Titanium alloys* are the most commonly researched AM materials. These alloys have excellent properties in terms of high strength-to-weight ratio, good fracture and fatigue resistance, good corrosion resistance and formability, etc. Therefore, they are widely utilized in the aerospace, automobile, and biomedical sectors. Various researchers have reported the fabrication of titanium components using different AM techniques such as PBF and DED. One of the most popular titanium alloys used for part fabrication via the AM route is Ti–6Al–4V. The main reason behind its extensive usage lies in its compatibility with numerous biomedical applications. Titanium has two phases in its pure form, commonly referred to as α and β, of which the former phase is strong with less ductility, while the latter is more ductile. Alloys that have both these ($\alpha + \beta$) phases possess high strength and formability. These two phases can be carefully adjusted to achieve the required properties in Ti alloys. To utilize them as bone mimics, the part density needs to be matched to the neighboring material, which is normally on the higher level, thereby requiring greater β phase content. As alloying elements, Al stabilizes the α phase and V stabilizes the β phase in Ti–6Al–4V alloy. The less stable α phase is formed when the β phase is quenched. Hence, during the fabrication of Ti–6Al–4V alloy prints, careful selection of the printing environment, specially ($\alpha + \beta$) phases, is needed to achieve the required properties, like strength, ductility, density, and corrosion resistance.

C. *Magnesium alloys* are promising materials for use as degradable biomaterials having similar stiffness to bone, which can minimize the stress shielding effects. The applications of Mg alloys are increasing at a rapid rate, including orthopedics, urology, cardiology, respirology, etc. AM of Mg alloys is attracting interest due to their ease of design as compared with traditional manufacturing techniques. AM has the capability to develop biodegradable implants. Several AM techniques, such as powder bed fusion, laser AM techniques, wire-arc AM, etc., being are utilized for the development of Mg alloy–based biodegradable materials.

D. *Aluminum alloys* are widely utilized in various engineering sectors due to their good strength-to-weight ratio and corrosion resistance. AM of Al alloys is still limited due to poor weldability and low laser absorption. Another possible reason is that Al alloys are melted during the fusion-based AM process, and there are more chances of hydrogen becoming solubilized and being entrapped during the subsequent solidification of the melt pool, which leads to the formation of pores. These solidification-related defects reduce the mechanical properties of the manufactured parts. To avoid these issues, the printing zone should be protected using additional shielding gas.

14.3.4 Future Trends for AMC

In recent decades, research challenges and industrial needs have massively promoted ceramics to become an exciting new area of application for 3D printing technologies. Research has mostly moved toward the use of slurry-based processes instead of dry powders to enable the fabrication of pore-free and crack-free components of ceramics with a high surface finish and a fine microstructure (Bandyopadhyay and Bose 2016).

While AM is generally capable of realizing relative complex geometries, this advantage is significantly compromised by the lack of microstructural quality control with the ceramic parts. Various issues such as porosity, purity, micro-defects, and interfacial defects commonly exist with AM ceramic structures, which still require extensive efforts to overcome. Due to the intrinsic staircase effect with the AM processes, the notch sensitivity issue of the ceramic parts is significant.

From traditional powder-based ceramic manufacturing, it is known that finer ceramic particles are preferred for achieving higher densities during the sintering densification. However, finer powder particles tend to exhibit lower flowability and become difficult to spread using the powder bed AM approach. On the other hand, with the powder suspension feedstock–based AM, due to the requirements for the rheological properties, the ceramic solid content is limited, which in turn, limits the achievable density of the densified parts and necessitates more complex sintering strategies such as liquid phase sintering or HIP.

In terms of ceramic material availability, the existing AM technologies demonstrate excellent material compatibility. New materials such as boron carbide (B_4C) and titanium boride (TiB_2) might also possess promising potential for AM adoption due to their high value-added ceramic armor applications. Additionally, in specific applications where exotic geometries such as cellular structures are desired, AM could also find immediate use. For example, it has been claimed that printed hydroxyapatite (HA) implants have the required porosity for osseointegration while achieving considerably higher mechanical strength compared with conventional implants with the same material and porosity. In the long term, ceramic AM will likely benefit from close collaborative research and development efforts between academia, system manufacturers, and industrial users.

Future trends in the field of ceramic printing techniques are listed as follows.

- One promising application is bone grafts. The fabrication of porous bone grafts using AMC is becoming a very popular research field.

- Bone tissue engineering requires bioactive and bioresorbable ceramics with a complex porous network mimicking the bone architecture. AMC can help in processing these structures with controlled pore size, pore–pore interconnectivity, and tailored volume fraction porosity. AMC can also be helpful to make patient-specific structures based on an individual's bone defect captured in a computed tomography (CT) or magnetic resonance imaging (MRI) scan (Darsell et al. 2003). A typical pore size of >300 μm and pore volume between 10% and 80% with extensive interconnectivity are considered ideal.
- Apart from bioceramics, porous ceramic structures can also be used for other applications such as filtration, sensors, and scaffolds for composites, which are all expected to benefit from the growth of AMC, where the micro- and macro-structures can both be controlled simultaneously.
- AMC is also expected to impact ceramic coatings for structural and functional applications, though these coatings may not be many layers thick but can be placed on demand in specific locations with tailored composition to change the surface properties. Such surface property changes may be needed to enhance resistance to high temperature, resistance to wear and corrosion, etc. In fact, some of these coatings can also be used to repair existing devices.
- AMC is still in its infancy. While novel techniques are becoming popular for thin micro-scale structures for flexible electronics and semiconductor devices, ceramic processing for large-scale bulk structures is still quite difficult.
- Further research and development is needed to make AMC a popular approach for direct low-volume manufacturing of structural, functional, or bioceramic prints (Bandyopadhyay and Bose 2016).
- To conclude with composite applications, approaches toward the fabrication of ceramic-loaded composites will probably find more real-world applications in the near future using a variety of AMC techniques.
- Finally, AM has successfully demonstrated a promising solution to addressing the limitations of conventional manufacturing approaches and unlocking new possibilities in designing and fabricating ceramic components with desired complex and intricate structures. The unitization of preceramic polymers as feedstock materials for AM processes has shown great potential for manufacturing high-performance components. There are a few hot topics that researchers have been working on or will continue to work on:
  a) With the rapid development of AM technologies and the urgent demands for advanced ceramic materials for various applications, such as high-temperature structure ceramics for hypersonic flight, electronic devices, thermal protection components, and healthcare devices, the development of preceramic feedstock materials for AM processes will continue to be one of the hot points for the fabrication of advanced ceramic components, especially the development of ceramic-based nanocomposites.

b) Enhancing versatile and multifunctional AM processes for manufacturing high-performance functional solid components. Integrating additional features, such as thermal energy, light, ultrasound waves, or other functions, during preceramic material printing may enable high-quality products with superior performance.

c) Developing highly dense near net shape advanced ceramic composites with low volumetric shrinkage and high performance is another hot topic. Volumetric shrinkage and porosity are still significant concerns, especially when preceramic polymer resins are adopted as feedstock materials. The addition of inert or active fillers or reinforcement materials has been demonstrated to effectively decrease the volumetric shrinkage of the printed ceramic components after pyrolysis and improve their properties. Exploring new composite feedstock materials for manufacturing near net shape high-quality ceramic components with superior performance properties using the AM process is desirable in the future.

14.3.5 FUTURE TRENDS IN POST-PROCESSING OF AM

Post-processing is a critical part of AM. It's the final step of the manufacturing process, where parts receive adjustments such as smoothing and strengthening. Despite all the advantages AM has over traditional manufacturing methods, it is still imperfect. Additively manufactured parts can still have poor surface finishes, and the production method can affect the mechanical properties of the component. The main objective of post-processing is to eliminate these potentially dangerous defects, and there are a few ways in which this can be achieved. These processes can include heat treatment, which is needed to reduce the stress on components before their removal from the build plate, separating the components from the support structure, as well as surface finishing procedures such as CNC machining, blasting, and polishing. The post-processing of AM parts can be just as vital as the fabrication itself.

In terms of metal AM, the post-processing technology of traditional manufacturing is still used. In order to further automate these technologies, some companies have also begun to implement robotic solutions that can install printing substrates, clean powder, unload parts, and perform post-processing. The goal is to replace all manual work in order to promote continuous and large-scale production. Although this development is encouraging, the pace of innovation in this field is still relatively slow. The number of advanced automatic post-processing solutions will certainly increase in the future so as to adapt to the growing development of the AM industry (Xing Peng et al. 2021).

14.4 FINAL RECOMMENDATIONS

Based on the identified technology gaps and research needs, recommendations for AM technology and research in terms of materials, design, modeling, sensing and control, process innovation, and system integration will be made. There should be a tight coupling among material development, process development, process sensing and control, and the qualification and certification of products fabricated by AM. As in the progression of many other emerging technologies, the greatest advancements will come at the boundaries of fundamental material science, physics, biology, lasers, electronics, optics, metrology, and control. Future studies can be undertaken to improve the 3D printing and to make the process more efficient and compatible with a wide variety of materials. The effect of various process parameters in different technologies can be studied by looking at the mechanical properties of the developed parts. The applications of these parts can be widened by making the 3D printing processes more user-friendly, efficient, and cost-effective (Jandyal et al. 2022).

For university–industry collaboration and technology transfer, recommendations may include incentivizing projects through funding, increasing research and development support, and increasing coordination efforts for public–private partnerships. For education and training, it is recommended to set up a university–community college partnership model with resource sharing and a teaching factory model to expose students directly to manufacturing enterprises. In addition, it is recommended to promote the public awareness of AM, using the internet to drastically increase outreach and resource sharing, and to establish publicly accessible AM facilities.

Finally, additive manufacturing companies leverage hardware, software, and materials innovations to improve the 3D printing efficiency. The report of Startus Insights, 2023 (www.startus-insights.com/) provides an overview of top additive manufacturing trends and innovations in 2023. They range from high-throughput 3D printing techniques and novel materials to additive manufacturing automation and high-volume production.

14.5 REVIEW QUESTIONS

1. Give examples for each of the subtractive, formative, and additive manufacturing techniques.
2. What are the main challenges and needs for AM?
3. List the main areas that need further research attention in AM.
4. What are the future trends in the field of general industrial applications of AM techniques?
5. Describe the future trends for polymers, polymer composites, and CFRP.
6. Describe the future trends for metals and alloys.
7. What are the future trends for AMC?
8. What are a few hot topics that researchers have been working on or will continue to work on?
9. Explain future trends in the post-processing of AM.

BIBLIOGRAPHY

Adil, Samia and Lazoglu, Ismail. (2022) A Review on Additive Manufacturing of Carbon Fiber-Reinforced Polymers: Current Methods, Materials, Mechanical Properties, Applications and Challenges. *Journal of Applied Polymer Science*, 140(7), e53476. https://doi.org/10.1002/app.53476

Bandyopadhyay, Amit and Bose, Susmita. (2016) *Additive Manufacturing.* Boca Raton, London, New York: CRC Press.

Chandrasekaran, Swetha, Worsley, Marcus, Duoss, Eric, and Lewicki, James. (2017). 3D Printing of High Performance Cyanate Ester Thermoset Polymers. *Journal of Materials Chemistry A*, 6, 853–858. https://doi.org/10.1039.C7TA09466C

Darsell, Jens, Bose, Susmita, Hosick, Howard L., and Bandyopadhyay, Amit. (2003) From CT Scan to Ceramic Bone Graft. *Journal of the American Ceramic Society*, 86(7), 1076–1080. https://doi.org/10.1111/j.1151-2916.2003.tb03427.x

European Economic and Social Committee. (2015) Additive Manufacturing, CCMI/131 – EESC. http://www.eesc.europa.eu

Ge, Qi, Qi, H. Jerry, and Dunn, Martin L. (2013) Active Materials by Four-dimension Printing. *Applied Physics Letters*, 103(13), 131901. https://doi.org/10.1063/1.4819837

Gibson, Ian, Rosen, David, and Stucker, Brent. (2015) *Additive Manufacturing Technologies 3D Printing, Rapid Prototyping, and Direct Digital Manufacturing.* Springer.

Goh, Guo Dong, Yap, Yee Ling, Agarwala, Shweta, and Yeong, Wai Yee. (2018). Recent Progress in Additive Manufacturing of Fiber Reinforced Polymer Composite. *Advanced Materials Technologies*, 4, 1800271. https://doi.org/10.1002/admt.201800271

Griffini, Gianmarco, Invernizzi, Marta, Levi, Marinella, Natale, Gabriele, Postiglione, Giovanni, Turri, Stefano. (2016) 3D-Printable CFR Polymer Composites with Dual-cure Sequential IPNs. *Polymer*, 91(174), S0032386116301884. https://doi.org/10.1016/j.polymer.2016.03.048

He, Hong-wei and Gao, Feng. (2015) Effect of Fiber Volume Fraction on the Flexural Properties of Unidirectional Carbon Fiber/Epoxy Composites. *International Journal of Polymer Analysis and Characterization*, 20(2), 180–189. https://doi.org/10.1080/1023666x.2015.989076

Hu, C., Sun, Z., Xiao, Y., and Qin, Q. (2019) Recent Patents in Additive Manufacturing of Continuous Fiber Reinforced Composites. *Recent Patents on Mechanical Engineering*, 12, 25–36. https://doi.org/10.2174/2212797612666190117131659

https://www.3dxtech.com/

https://www.startus-insights.com/

Jandyal, Anketa, Chaturvedi, Ikshita, Wazir, Ishika, Raina, Ankush, and Haq, Mir Irfan Ul. (2022) 3D Printing – A Review of Processes, Materials and Applications in Industry 4.0. *Sustainable Operations and Computers*, 3, 33–42. https://doi.org/10.1016/j.susoc.2021.09.004

Matsuzaki, Ryosuke, Ueda, Masahito, Namiki, Masaki, Jeong, Tae-Kun, Asahara, Hirosuke, Horiguchi, Keisuke, Nakamura, Taishi, Todoroki, Akira, and Hirano, Yoshiyasu. (2016) Three-Dimensional Printing of Continuous-Fiber Composites by In-Nozzle Impregnation. *Scientific Reports*, 6, 23058. https://doi.org/10.1038/srep23058

Milewiski, John O. (2017) *Additive Manufacturing of Metals: From Fundamental Technology to Rocket Nozzles, Medical Implants, and Custom Jewelry.* Springer Series in Materials Science, 258.

Peng, Xing, Kong, Lingbao, Fuh, Jerry Ying Hsi and Wang, Hao. (2021) A Review of Post-Processing Technologies in Additive Manufacturing. *Journal of Manufacturing and Materials Processing*, 5(38). https://doi.org/10.3390/jmmp5020038

Ueda, Masahito, Kishimoto, Shun, Yamawaki, Masao, Matsuzaki, Ryosuke, Todoroki, Akira, Hirano, Yoshiyasu, and Le Duigou, Antoine. (2020) 3D Compaction Printing of a Continuous Carbon Fiber Reinforced Thermoplastic. *Composites Part A: Applied Science and Manufacturing* 137, 105985. https://doi.org/10.1016/j.compositesa.2020.105985

Van de Werken, Nekoda, Tekinalp, Halil, Khanbolouki, Pouria, Ozcan, Soydan, Williams, Andrew, and Tehrani, Mehran. (2019) Additively Manufactured Carbon Fiber-Reinforced

Composites: State of the Art and Perspective. *Additive Manufacturing*, 31, 100962. https://doi.org/10.1016/j.addma.2019.100962

Wohlers, Terry and Caffrey, Tim. (2013) Additive Manufacturing: Going Mainstream, *Manufacturing Engineering*, SME, Dearborn, 150(6), 67–73.

Yaragatti, Neha and Patnaik, Amar. (2020) A Review on Additive Manufacturing of Polymers Composites. *Materials Today: Proceedings*, 44, 4150–4157.

15 Additive Manufacturing Case Studies

Additive manufacturing (AM) continues to see increased growth, with adoption in many industrial sectors and increased commercial market value. Market growth is attributed to continued technology and software improvements, growing availability of material types, and the ability to produce functional parts with complex geometries. The technology also provides the freedom to optimize the design of both the internal structure and the external contours of a component. Additive manufacturing has been driven by the biotechnical/medical, aerospace, and automotive industries due to a focus on research and high-value components.

Twelve case studies cover additively manufactured products extracted from different fields of industrial applications. The economic advantages and disadvantages in comparison to conventional technologies are described.

CASE STUDY 15.1: FUNCTIONAL 3D PRINTING MODEL OF A VALVE

In the first case study, a new design method to create moveable functional models is introduced. The required design changes are to be shown by the example of a throttle valve (Figure 15.1). The function of the lever and that of the shaft consist of closing or opening the valve by means of a rotational movement in order to control the flow rate of a fluid. In a standard 3D printing process, all components would stick together, making it impossible to move (rotate) the shaft and allowing only one valve setting.

As shown in Figure 15.2, the components that exercise a function are separated by a gap in the computer-aided design (CAD) system during pre-processing. For complex components, such as the valve with rotating shaft and lever, it may also be necessary to simplify the shape. Other components, such as the lever, can be printed separately in order to be subsequently joined to the shaft. The components are separated by these measures, and the loose powder applied to the gaps during the printing process can be removed. It should also be observed that the components are separated for the infiltration. At the end, the 3D model is available, which can be set variably according to the operation mode of the valve.

An additional work step is necessary for the creation of a functional 3D printing process, notably, the pre-processing of the CAD data. This reduces the amount of work required for changes, and the data remain consistent. The creation of a functional 3D printing process is more complex in comparison to a conventional 3D printing process, since the function has to be determined first, and the gap design also has to be implemented in the CAD system. In contrast, the saving due to the lower consumption of powder is marginal. However, the flexible functional model of variable setting is much more useful than the rigid conventional 3D printing model.

DOI: 10.1201/9781003451440-15

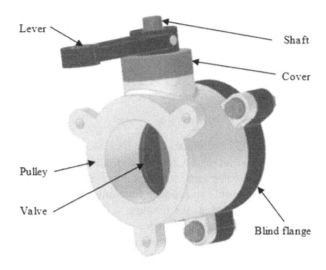

FIGURE 15.1 Case study: A metallic valve with rotating shaft and lever (Source: Junk Stefan and Tränkle Marco, 2011).

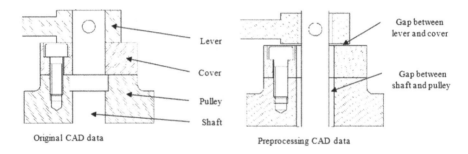

FIGURE 15.2 Changes in the CAD data (preprocessing) (Source: Junk Stefan and Tränkle Marco, 2011).

CASE STUDY 15.2: THERMOFORMING OF PLASTIC SHEETS

A thermoforming tool was constructed for an automobile model as an example. In doing so, the external shape was taken from CAD before the additional design steps were performed (Figure 15.3). First, a cavity was created in the interior of the tool in order to access the vacuum channels. The cavity also allowed the amount of powder required to be reduced considerably. A total of ten channels were then designed for the vacuum in the tool. They initially had a diameter of about 12 mm, which can be reduced to a diameter of 1.5 mm. It should be observed here that it is very difficult to remove the loose powder in the event of long channels with small diameters. The spacer under the tool (height of 1.5 mm) and the blind holes for the fastening screws were also designed.

A tool designed in this way and manufactured by means of 3D printing technology was used for thermoforming (Figure 15.4). It was established that this tool was

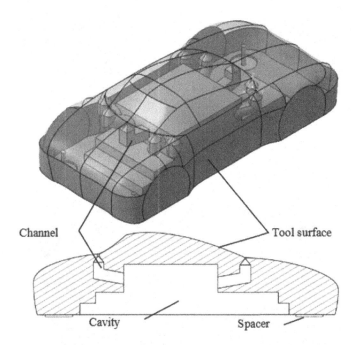

FIGURE 15.3 CAD model of a forming tool: 3D view and cross-section (Adapted from European powder metallurgy association, EPMA 2013).

able to withstand the thermal loads due to the heated plastic sheet (approx. 180°C). A small series of deep-drawn sheets was manufactured.

Besides the new design options provided by the additive manufacturing technologies, their economic advantages can also benefit the manufacture of tools. This includes, in particular, the quick implementation of the tools and hence, short delivery times and development cycles. However, costs can also be reduced, since the low procurement and operating costs of 3D printers make the hourly machine rates considerably lower than for similar technologies based on laser technology or conventional technologies. These advantages compensate abundantly for the relatively high costs for powder and infiltration.

CASE STUDY 15.3: VACUUM PERMEATOR

The geometry is impossible to produce by a conventional manufacturing process. The material is stainless steel (AISI 316L), Part weight and dimensions: 2 kg, $10 \times 10 \times 20$ cm. The vacuum permeator is needed in the industrial energy sector. The part (Figure 15.5) is printed using laser beam melting (LBM) technology.

CASE STUDY 15.4: AERO ENGINES BORESCOPE BOSSES FOR A320NEO GEARED TURBOFAN TM ENGINE

This part (Figure 15.6) is needed for aerospace applications.

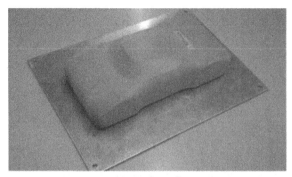

a. Mounted mould

b. Thermoformed plastic sheet

FIGURE 15.4   Use of rapid tooling (Adapted from European powder metallurgy association, EPMA 2013).

FIGURE 15.5   Prototype of 316L vacuum permeator for ITER made by LBM, this part is impossible to produce by conventional processes (Adapted from European powder metallurgy association, EPMA 2013).

FIGURE 15.6 MTU Aero Engines borescope bosses for A320neo geared Turbofan TM engine (Adapted from European powder metallurgy association, EPMA 2013).

MATERIAL, DIMENSIONS, TECHNOLOGY, AND VENDOR

- Industry user sector: Aerospace industry
- Material: Nickel alloy 718
- Part dimensions: Volume: 15,600 mm³, L × B × H Boundary Box: 42 × 72 × 36 mm
- Additive process used: Laser beam melting (SLM) on an EOS machine
- They form part of MTU's low-pressure turbine case and allow the blading to be inspected at specified intervals for wear and damage using a borescope

BENEFITS OF AM TECHNOLOGY

- Series production of up to 2000 parts per year
- Lower development production lead times and lower production costs
- Suitable for producing parts in materials that are difficult to machine, as, for example, nickel alloys
- For complex components that are extremely difficult, if not impossible, to manufacture using conventional methods
- Tool-free manufacturing and lower material consumption

CASE STUDY 15.5: SUPPORT TO SATELLITE ANTENNA

Figure 15.7 shows a titanium alloy support to a satellite antenna, manufactured using Poly-Shape electron beam melting (EBM) technology.

MATERIAL, DIMENSIONS, TECHNOLOGY, AND VENDOR

- Industry user sector: Aerospace
- Material: Ti6Al4V
- Part height and weight: 380 mm, 3.3 kg
- Additive process used: EBM, Poly-Shape

FIGURE 15.7 Ti6Al4V support to satellite antenna made by EBM with a lightweight design made by topology optimization (Adapted from European powder metallurgy association, EPMA 2013).

BENEFITS OF AM TECHNOLOGY

- Weight reduction: 55%
- Tool-free manufacturing and lower material consumption

CASE STUDY 15.6: MEDICAL HEARING AID

Figure 15.8 illustrates a medical hearing aid made of a titanium alloy and printed using SLM technology.

HEARING AID, 3D SYSTEMS:

Industry user sector: Biomedical
    Material: Ti6Al4V (grade 23)
    Part height: 15 mm
    Additive process used: Laser beam melting

BENEFITS OF AM TECHNOLOGY

- High accuracy
- High biocompatibility
- High mechanical resistance
- High flexibility of manufacturing

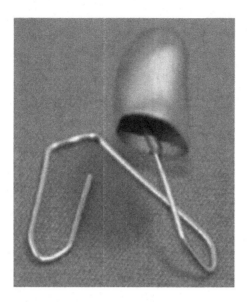

FIGURE 15.8   Medical hearing aid (Adapted from European powder metallurgy association).

CASE STUDY 15.7: CLEANABLE FILTER DISC

Figure 15.9 shows a stainless steel 316L cleanable filter disc.

CLEANABLE FILTER DISC

- Industry user sector: Medical instruments
- Material: Stainless steel 316L
- Part dimensions: Ø55 mm
- Additive process used: Laser beam melting
- Traditional methods of creating filters often result in gaps between the securing steel ring and the mesh, as well as the weft and warp strands of the woven wire. Known as "bugtraps", these can quickly gather bacteria and dirt
- By using additive manufacturing, CAM removed these bugtraps from the design, meaning that the filters can be cleaned much more easily, decreasing downtime for the customer as well as the requirement for replacement parts

BENEFITS OF AM TECHNOLOGY

- No recesses in part compared with conventional woven wire mesh equivalent
- Less contamination through particulate build-up
- Easier to clean to a high standard, decreasing customer downtime and costs
- Design size can be easily altered to suit customer requirements, including a change in aperture size, without the creation of new tooling

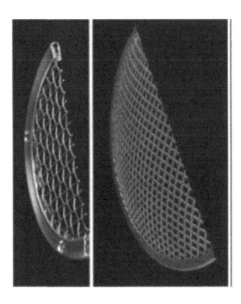

FIGURE 15.9   Cleanable filter Disc (Adapted from European powder metallurgy association, EPMA 2013).

FIGURE 15.10   Heat exchanger (Adapted from European powder metallurgy association, EPMA 2013).

CASE STUDY 15.8: HEAT EXCHANGER

Figure 15.10 shows a prototype heat exchanger for a motor sport using an LBM technology.

AUTOMOTIVE AND CAR RACING

- Prototype of heat exchanger
- Industry user sector: Motor sport

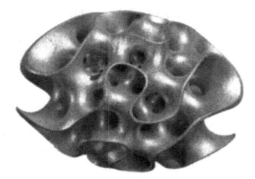

FIGURE 15.11   Fine metal part designed by Bathsheba Grossman (Courtesy of Höganäs AB - Digital Metal®).

- Material: Al Si 10Mg
- Additive process used: Laser beam melting
- New design with self-supporting integrated cooling fins on outside surfaces and turbulators inside cooling tubes to disrupt the flow of the cooled fluid
- Produced on an EOS M290 machine

BENEFITS OF AM TECHNOLOGY

- Maximum heat transfer
- Compact and scalable design

CASE STUDY 15.9: RYGO SCULPTURE

Figure 15.11 illustrates a fine metal part designed by Bathsheba Grossman.

RYGO SCULPTURE

- Industry user sector: Fashion and design
- Material: Stainless steel 316L
- Part dimensions: 25 × 25 × 30 mm
- Additive process used: Precision inkjet on powder bed
- Bathsheba Grossman is an artist recognized for her 3D printed art and sculptures. Not many of her complex designs can be produced in any other way than by additive manufacturing

BENEFITS OF AM TECHNOLOGY

- High level of resolution and surface quality
- Effective mass customization of designs
- Possible to achieve very thin walls and sections

CASE STUDY 15.10: JEWELRY PLATINUM HOLLOW CHARMS

Figure 15.12 shows platinum hollow charms for the jewelry industry, made of 950‰ platinum powder alloy.

Consumer Goods, Platinum Hollow Charms

- Industry user sector: Jewelry
- Material: 950‰ platinum powder alloy
- Part dimensions: 31 parts of 2.8 g each (2.4 g after polishing)
- Additive process used: Laser beam melting
- Platinum has always been difficult to use with casting. With the SLM technique, it's possible to match its fashion effect with the maximum freedom of shape, also preserving light weight to make it affordable

Benefits of AM Technology

- Hollow parts costs less than full parts made of a cheaper material
- Maximum customization, making it exclusive
- Eco-friendly production process

CASE STUDY 15.11: CONFORMAL COOLING CHANNELS

One of the most widely studied applications is conformal cooling. Conformal cooling channels follow the external geometry to provide more effective and consistent heat transfer. Figure 15.13 illustrates a schematic of a conventional cooling channel (left) and a conformal cooling channel (right) in an aluminum product produced by SLM. Early research showed that conformal cooling improves process efficiency and quality. Industrial injection molding case studies have confirmed these benefits with

FIGURE 15.12 Jewellery platinum hollow charms (Courtesy of Progold S.p.A).

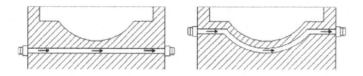

FIGURE 15.13 Schematic of conventional cooling channel (left) and conformal cooling channel (right) (Adapted from Altaf et al. 2013).

reports of reduced lead time, more uniform temperature distributions, reduced cycle times, improved quality, reduced reject rates, reduced corrosion, longer maintenance intervals, and overall cost savings.

CASE STUDY 15.12: PRODUCTS PRODUCED FROM FUNCTIONAL GRADED MATERIALS (FGMS) BY AM

These case studies illustrate the diverse applications and benefits of utilizing FGMs through AM processes. The ability to customize material properties within a single part offers significant advantages across various industries, including improved performance, enhanced functionality, and better adaptability to specific application needs.

1. **Aerospace Components: Turbine Blades as Illustrative Product**

Aerospace companies utilize FGMs with AM to produce turbine blades with optimized material composition. These blades have tailored material properties, such as enhanced strength at high temperatures, achieved by strategic placement of different composite materials in a single print.

2. **Biomedical Implants: Customized Bone Implants as Illustrative Product**

FGMs combined with AM techniques are used to create patient-specific bone implants. These implants are designed with varying material densities to mimic the mechanical properties of natural bone, promoting better integration and reduced risk of rejection.

3. **Automotive Parts: Lightweight Structural Components as Illustrative Product**

FGMs allow the production of lightweight yet durable automotive parts, like chassis components. AM processes are employed to precisely control material composition and distribution, leading to improved fuel efficiency and overall vehicle performance.

4. **Sporting Equipment: High-Performance Bicycle Frames as Illustrative Product**

FGMs are utilized in manufacturing bicycle frames through AM processes. These frames have areas of reinforced fibers strategically placed to optimize strength, stiffness, and weight, catering to the specific demands of professional athletes.

In these cases, FGMs offer designers and engineers the ability to customize material properties within a single part, leading to products with improved performance, reduced weight, and enhanced functionality. These examples demonstrate the versatility and potential of combining FGMs with AM processes to create innovative and high-quality products across various industries.

BIBLIOGRAPHY

Altaf, K., Rani, A. M. A., and Raghavan, V. R. (2013) Prototype Production and Experimental Analysis for Circular and Profiled Conformal Cooling Channels in Aluminum Filled Epoxy Injection Mold Tools. *Rapid Prototyping Journal*, 19(4), 220–229.

European Powder Metallurgy Association (EPMA). (2013) European Additive Manufacturing Group, EAMG, Introduction to Additive Manufacturing Technology, A guide for Designers and Engineers. www.epma.com/am

Junk, Stefan and Tränkle, Marco. (2011) Design for Additive Manufacturing Technologies: New Applications of 3D-Printing for Rapid Prototyping and Rapid Tooling, International Conference on Engineering Design, ICED11, August 2011. Technical University of Denmark.

Index

Printed in the United States
by Baker & Taylor Publisher Services